Species	*Range*	[illegible]	[illegible]
Australian night parrot *(Geopsittacus occidentalis)*	West and South Aus[illegible]	O[illegible] co[illegible]	[illegible]
Paradise parrot *(Psephotus pulcherrimus)*	Queensland and New South Wales, Australia	Total population *c.* 150	Full
Kakapo *(Strigops habroptilus)*	South Island, New Zealand	Less than 100 specimens left	Full
Strigiformes			
Seychelles Island owl *(Otus insularis)*	Mahe Island, Seychelles	Last collected 1940	None
Soumagne's owl *(Tyto soumagnei)*	Malagasy Republic	May now be extinct	None
Caprimulgiformes			
Puerto Rico whippoorwill *(Caprimulgus noctitherus)*	Puerto Rico	Rediscovered in 1961	Partial
Piciformes			
Imperial woodpecker *(Campephilus imperialis)*	Sierra Madre, Mexico	Facing extinction	None
Passeriformes			
Noisy scrub-bird *(Atrichornis clamosus)*	Mt Gardner, Western Australia	Total population *c.* 80 and increasing	Full
Hawaiian crow *(Corvus tropicus)*	Confined to Mt Hulalalai, Hawaii Island	Less than 20 specimens left	Full
Reunion cuckoo shrike *(Coracina newtoni)*	Reunion, Mascarene Islands, Indian Ocean	Verging on extinction	Full
Seychelles magpie-robin *(Copsychus seychellarum)*	Frigate Island, Seychelles	Probably extinct	None
Rufous-headed robin *(Erithacus ruficeps)*	Cameron Highlands, Malaya Tsinling Mts, China	Known from only 4 specimens	None
Puaiohi *(Phaeornis palmeri)*	Kauai, Hawaiian Islands	Very rare and localised	Partial
Grand Cayman thrush *(Turdus ravidus)*	Grand Cayman Island, Caribbean Sea	Not seen since 1938 and probably extinct	None
White-necked rock fowl *(Picathartes gymnocephalus)*	Sierra Leone, West Africa	Extremely rare and numbers decreasing	None
Eyrean grass-wren *(Amytornis goyderi)*	South Australia	Rediscovered in 1962 but no records since	
Rodriguez warbler *(Berrornis rodericana)*	Rodriguez, Mascarene Islands, Indian Ocean	Total population *c.* 10–20	None
Seychelles warbler *(Bebrornis sechellensis)*	Cousin Island, Seychelles	Less than 50 specimens left	None
Semper's warbler *(Leucopeza semperi)*	St Lucia Island, Caribbean Sea	Last sighting 1947	None
Watut leaf warbler *(Sericornis nigroviridis)*	Morobe, eastern New Guinea	Known only from the type specimen collected 1964	None
Chatham Island robin *(Petroica traversi)*	Little Mangare, Chatham Islands	Facing extinction	Full
Kauai O-o *(Moho braccatus)*	Kauai, Hawaiian Islands	Total population reduced to 1 pair	Full [3]
Kauai akialoa *(Hemignathus procerus)*	Kauai, Hawaiian Islands	Critical	None
Palila *(Pstittirostra bailleui)*	Confined to Mauna Kea, Hawaii	Last sighted in 1940	Partial
Ou *(Psittirostra psittacea)*	Confined to Kauai and Hawaii Islands	Very near extinction	None
Slender-billed grackle *(Cassidix palustris)*	Confined to marshes near Mexico City, Mexico	Probably extinct	None

Sources: *The Red Data Book*, vol. 2. *Aves*. Curry-Lindahl (1972).

Notes:
[1] Confined to a small *partly* artificial reserve
[2] Seven British records before 1887
[3] The world's rarest living bird

THE **GUINNESS** BOOK OF ANIMAL FACTS AND FEATS

Second Edition

by Gerald L Wood

'There can be nothing in literature more unreliable than accounts of the size and weights of animals, gathered at random. The first estimate is often a guess, which immediately acquires an air of accuracy by being expressed in figures. It does not usually get into print until it has been rolled over many tongues, and during the process it increases in size like the snowball which is rolled along the ground.' – Dr Francis Hobart Herrick (1895)

Guinness Superlatives Limited
2 Cecil Court, London Road, Enfield, Middlesex

OTHER GUINNESS SUPERLATIVES TITLES

Facts and Feats Series:

Air Facts and Feats, *2nd ed.*
John W R Taylor, Michael J H Taylor and David Mondey

Rail Facts and Feats, *2nd ed.*
John Marshall

Tank Facts and Feats, *2nd ed.*
Kenneth Macksey

Yachting Facts and Feats
Peter Johnson

Plant Facts and Feats
William G Duncalf

Structures – Bridges, Towers, Tunnels, Dams . . .
John H Stephens

Car Facts and Feats, *2nd ed.*
edited by Anthony Harding

Business World
Henry Button and Andrew Lampert

Music Facts and Feats
Robert and Celia Dearling with Brian Rush

Guide Series:

Guide to Freshwater Angling
Brain Harris and Paul Boyer

Guide to Mountain Animals
R P Bille

Guide to Underwater Life
C Petron and J B Lozet

Guide to Formula 1 Motor Racing
José Rosinski

Guide to Motorcycling, *2nd ed.*
Christian Lacombe

Guide to French Country Cooking
Christian Roland Délu

Guide to Bicycling
J Durry

Other titles:

English Furniture 1550–1760
Geoffrey Wills

The Guinness Guide to Feminine Achievements
Joan and Kenneth Macksey

The Guinness Book of Names
Leslie Dunkling

Battle Dress
Frederick Wilkinson

Universal Soldier
Martin Windrow and Frederick Wilkinson

History of Land Warfare
Kenneth Macksey

History of Sea Warfare
Lt.-Cmdr. Gervis Frere-Cook and Kenneth Macksey

History of Air Warfare
David Brown, Christopher Shores and Kenneth Macksey

The Guinness Book of Answers
edited by Norris D McWhirter

The Guinness Book of Records, *23rd ed.*
edited by Norris D McWhirter

The Guinness Book of 1952
Kenneth Macksey

Published in Great Britain by
Guinness Superlatives Ltd, 2 Cecil Court,
London Road, Enfield, Middlesex

ISBN 0 900424 60 5

Set in Monophoto Bembo Series 270,
printed and bound in Great Britain by
Jarrold and Sons Ltd, Norwich

CONTENTS

ACKNOWLEDGEMENTS

The superlatives of the Animal Kingdom are so diverse and prolific that no single person could possibly compile a book of this nature without leaning very heavily on the works and researches of other people. I am therefore immensely indebted to the many men and women, some of them anonymous, who have contributed directly or indirectly to the text through their own zoological writings.

Similarly, this book could not have been compiled without the friendly co-operation of the Zoological Society of London, and I would like to thank the Librarian, Mr Reginald Fish, and his assistant, Mrs Susan Bevis, for their invaluable assistance and the generous loan of a mountain of books, journals and scientific papers.

Apart from published sources, a great deal of material has also been derived from conversations and correspondence with many animal specialists, and I would like to express my special appreciation to the following people who offered valuable suggestions and corrections: Miss Aisla M Clark (Echinoderms) and Mr John E Hill (Chiroptera) of the British Museum (Natural History); Dr Ray Gambell and Mr Sidney Brown of the Whale Research Unit, London; Dr W Rydzewski, Editor of *The Ring* (Warsaw); Mr Forrest G Wood of the Naval Undersea Center at San Diego, California; Dr Brian Yalden and Mr Edmund Seeyd of Manchester University; Dr Peter Ward of the Quelea Investigations Project in Maiduguri, Nigeria; Prof. Frederick A Aldrich of the Marine Sciences Research Laboratory, Memorial University of Newfoundland, St John's; and Mr John D MacDonald of the Union Whaling Company Ltd in Durban, South Africa.

Sincere thanks are also due to Mrs Barbara E Wolman, President, Northwest Newfoundland Club, Bothell, Washington, for keeping me fully up to date on dog weight-hauling competitions in North America; Mr Ron Pawlowski of Mount Isa, Queensland, Australia for unpublished data on the Estuarine crocodile; Mr J Hasinger of Philadelphia, Pennsylvania, USA for material on the largest tiger ever recorded; Mr Robert C Goodden of Worldwide Butterflies Ltd, Sherborne, Dorset for the latest information on British butterfly populations; and David J Moon, Editor of the Exotic Entomology Group, for valuable advice on superlative insects.

Hundreds of other people also very kindly supplied information and material for this new edition, but as a list of their names would be much too long to reproduce here, I hope I will be forgiven if I express my gratitude to them in one collective 'thank you'.

Finally, I wish to thank Miss Beatrice Frei and Mr David Roberts of Guinness Superlatives Ltd for their valuable assistance during the preparation of this book and, of course, my wife/secretary Susan for her enthusiastic support at all times.

INTRODUCTION

Since the first edition of this book was published in 1972 an astonishing number of new superlative items have materialised, and much of this fresh information has been incorporated into the revised text.

Who would have believed, for instance, that the Etruscan pygmy shrew would lose its title of 'smallest living mammal' to a new species of bat discovered in Thailand; that a woman would shoot an 8·63 m *28 ft 4 in* long Estuarine crocodile in Australia; that a captive pair of Orang-utans would live together for 54+ years; that measurements up to 6·70 m *22 ft* would be recorded for a giant earthworm found in the Transvaal, South Africa; or that the remains of a huge pterosaur (winged lizard) with an estimated wing span of more than 15·24 m *50 ft* would be discovered in Texas, USA?

I have also devoted some space to endangered wildlife, because many of the creatures featured in this book are now on the verge of extinction.

The mass destruction of animal life by man began with the age of discovery. Since 1600 at least 300 of our higher vertebrates have disappeared from the face of the earth, and today at least another 1000 species are threatened with the same fate despite the splendid efforts of the conservationists.

The baleen whales are a classic example of this shameful war of attrition. These gentle marine giants are extremely vulnerable to human predation because they have a long gestation period and normally give birth to a single calf – by 1975 the total stocks of the Blue whale, Fin whale, Sei whale and Minke whale had been reduced to 25000, 95000, 116000 and 150000 respectively – and if a 10-year ban on all commercial killing is not brought in very soon by the International Whaling Commission then nothing can save the world's largest creatures from extinction.

The status of some land animals is almost equally precarious, and this is mainly due to the rapid expansion of urban civilisation since the beginning of this century.

The outlook is particularly gloomy in Brazil, where at least 77 species of mammal, bird and reptile are known to be in great danger and dozens of others are threatened. And yet, despite the gravity of the situation, the hunting of exotic animals is still permitted in 13 of the country's 21 states.

In Australia nearly 2000000 kangaroos and wallabies are killed each year for use as pet food, and in the state of Victoria alone 7 out of 15 species have already been wiped out. In Queensland, which has been dubbed the 'slaughter state', at least 600000 of these helpless victims of commercial exploitation are butchered annually, and at the beginning of this year the federal government *approved* the culling of a further 400000 because Queensland farmers claimed there had been a population explosion and the marsupials were threatening their crops.

The time has now come for civilised man – and I use this term in its broadest sense – to turn away from senseless destruction and concentrate instead on painstaking, long-range ecological planning on a worldwide basis. If he treated the living environment with the same reverence that he does the cultural treasures of the past there would be no problems; unfortunately, however, man's future survival is not dependent on his ability to preserve valuable artifacts, but on his willingness to live with nature in peaceful co-existence.

Finally, if this book can settle arguments on the extremes of the animal world then its primary purpose will have been achieved. If there are any factual errors I would appreciate having them drawn to my attention.

July 1976

LIST OF CORRESPONDENTS

Keith Adams
Prof Frederick A Aldrich
William B Allen, Jnr
Dr Dean Amadon
Mrs Virgie Arden

Sam Barnes
Anthony Best
Gordon Blower
James Bond
Prof Osmond P Breland
Stanley Brock
Sidney Brown
Dr A N Brunner

Dr Guy Chauvier
Keith Chivers
Dr Malcom Clarke
Prof J Cloudsley-Thompson
Dr Doris M Cochran
Dr Harold Coggan
Dr Edwin H Colbert
Dr Roger Conant
L R Conisbee
Dr W P Crowcroft
Dr K Curry-Lindahl

H E Mr A E Davidson
Dr R K Dell
H J de S Disney
Herndon G Dowling
Prof William A Dunson
Dr Arne S Dyhrberg

Dr F C Fraser

Dr Ray Gambell
Prof V G Geptner
Dr Willis J Gertsch
Prof Perry Gilbert
Dr Coleman J Goin
Robert C Goodden

Miss A G Grandison
Dr Bernard Grzimek

Carl-Heinrich Hagenbeck
Sir Edward Hallstrom
Miss Sybil Hamlet
David J Hasinger
Ralph Helfer
John E Hill
Randall Hincliffe
William A Hooper
Dr I Howell-Rivero
Gordon Hull

R W Ingle
Mrs Nan Ingleton

Dr James Jensen
J Strand Jones

Dr F Wayne King
Miss Judith E King
Dr Heinz-Georg Klos

Frank Lane
G H Locket
Mrs Lyn Logue
Victor Louis
Arthur Loveridge

J G MacDonald
Dr A Mackay
Dr Harrison Matthews
R M McKenzie
Dr G R McLachlan
Mrs Herbert H Miller, Jnr
James R Montgomery
D J Moon
G S Mottershead
Terence Murphy
Prof George S Myers

Dr Murray A Newman
Dr Masaharu Nishiwaki
Dr H Nott
Adrian Nyoka

Ron Pawlowski
R Marlin Perkins
Dr Tory Peterson
Dr Jorge Sabater Pi
Prof George Pilleri
Dr Giovanni Pinna
Dr J C Poynton

Dr William C Schroeder
Dr Albert Schwartz
Miss Elspeth Sellar
Dr Henry W Setzer
Edmund Seyd
Dr Sylvia Sikes
Ken Sims
Gary Smart
Dr Stewart Springer
Dr Margaret M Stewart

Prof Gilbert L Voss

Dr Peter Ward
Dr H Wackernagel
Dr J W Watson
Dr Bernard Whitkop
C Williams
Mrs Barbara E Wolman
Forrest G Wood

Dr D W Yalden
I H Yarrow

Section I

MAMMALS

(class Mammalia)

A mammal is a warm-blooded, air-breathing vertebrate or backboned animal. It has a four-chambered heart and the body temperature is maintained by an insulating covering of hair or blubber. The brain is large and well developed. Young are usually born alive and are nourished on milk secreted from their mother's mammary glands.

The earliest known mammals were shrew-like creatures which lived about 190 million years ago, but practically nothing is known about their history. Their remains were first discovered at Thaba-ea-Litau, Lesotho, in 1966.

Living mammals are divided into (1) the primitive monotremes or egg-laying mammals; (2) the marsupials or mammals with pouches in which the embryonic young are nurtured soon after birth; and (3) the placentals whose young remain in the womb nourished by means of a placenta until they have reached a comparatively advanced stage of development.

These subclasses, in turn, are split up into 19 orders comprising 122 families and 4230 species. Rodents account for 42 per cent (1790 species) of this total, and bats for 23 per cent (981 species).

The largest and heaviest mammal in the world, and the largest animal ever recorded, is the Blue whale or Sibbald's rorqual (*Balaenoptera musculus*), which is also called the 'Sulphur-bottom whale' because of the yellow film of microscopic plants (diatoms) often found on its underside.

Three races are recognised: the Northern blue whale (*B. m. musculus*) of the North Atlantic and North Pacific Oceans; the larger Southern blue whale (*B. m. intermedia*) of the Southern Hemisphere; and the Pygmy blue whale (*B. m. brevicauda*) of the southern Indian Ocean and Antarctic waters (Scheffer & Rice, 1963).

The average sizes of the three sub-species (both sexes) at physical maturity are given in the following table:

Race	*Male*				*Female*			
	m	*ft*	*tonne*	*ton*	*m*	*ft*	*tonne*	*ton*
B. m. musculus	22	*73*	64	*63*	23	*77*	78	*77*
B. m. intermedia	24	*79*	85	*83.5*	26	*86*	109	*107*
B. m. brevicauda	20	*67*	56	*55*	22.5	*72*	67	*66*

The largest accurately measured blue whale on record (length taken in a straight line from the point of the upper jaw to the notch in the tail) was a female brought into the Cia Argentina de Pesca shore station, South Georgia, some time between 1904 and 1920 which measured 33·58 m *110 ft $2\frac{1}{2}$ in* (107 Norwegian fot) (Risting, 1922).

Another female caught near the South Shetlands, Falkland Island Dependencies, in March 1926 was 33·27 m *109 ft 4¼ in* long and an enormous male 32·64 m *107 ft 1 in* (Risting, 1928).

Seven other blue whales (all females) measuring in excess of 30·48 m *100 ft* were also taken in the Southern Ocean between 1922 and 1925. The largest specimen, measured at South Georgia by the Norwegian Tonsberg Whaling Company in 1924, was 31·74 m *104 ft 2 in* long.

Obviously measurements taken at fixed stations are much more reliable than those taken on floating factories. The measurement from the point of the snout to the notch in the tail is known as the 'total' length. The 'over-all' length (projection of the lower jaw to the notch in the tail) is somewhat longer, but as the lower jaw is hardly ever in its natural position when the whale is lying on the flensing platform, this measurement is rarely taken.

According to the *International Whaling Statistics* (Oslo) 13 blue whales measuring between 30·48 and 31·08 m *100* and *102 ft* in length have been caught in the Southern Hemisphere since 1930.

Shortly after the First World War a female blue whale measuring 31·18 m *102 ft 4 in* was brought into the shore station at Donkergat, Saldanha Bay, Cape Province, South Africa, and another one processed there at about the same time was longer than the 30.48 m *100 ft* flensing platform. Both whales were killed in Cape waters (Green, 1958).

The largest accurately measured Northern blue whale on record was a 29·87 m *98 ft* female which accidentally entered the Panama Canal from the Caribbean on 23 January 1922 and was killed by machine-gun fire after

A 28·04 m 92 ft *long model of a Blue whale (American Museum of Natural History)*

The 7·46 m 23 ft 5 in *long unborn foetus of a Blue Whale, the largest animal the world has ever seen (National Institute of Oceanography)*

threatening shipping. Parts of the body subsequently came ashore at Santa Isabel between Nombre de Dios and Cape San Blas, where the second and third cervical vertebrae were found by Mr Mitchell Hedges, the English explorer and big-game fisherman, who later presented them to the British Museum (Natural History), London (Harmer, 1927).

Another female measuring 29·26 m *96 ft* was found floating dead in the North Sea in 1868 and towed into Vardo, Norway, but this measurement may have been in Norwegian fot, in which case the length would have been 30·14 m *98 ft 11 in* (Collett, 1886).

In the early days of whaling lengths up to 36·57 m *120 ft* were reported for blue whales taken in northern waters, but these measurements are now considered unreliable.

Four blue whales have been stranded on British coasts since 1913, at least two of them after being harpooned by whalers. The last occurrence (one about 18·28 m *60 ft*) was at Wick, Caithness, Scotland, on 15 October 1923 (Harmer, 1927). According to Mr Sidney Brown of the Whale Research Unit of the National Institute of Oceanography (cited by Fraser, 1974), small numbers of blue whales were captured by whalers operating from the Shetland Islands and the Outer Hebrides between 1903 and 1914 and between 1920 and 1929. Most of the kills (57 in 1920) occurred west of St Kilda. Later, in 1950–51, a station was reopened at West Loch Tarbert in Harris, Outer Hebrides, and during this period six blue whales were captured.

The Pygmy blue whale was not described until 1961. This race has a proportionately longer trunk than the Northern and Southern blue whales and is correspondingly heavier (most of this extra weight is derived from its bulkier digestive organs), but the tail region is noticeably smaller (Ichihara, 1961, 1966).

The largest accurately measured Pygmy blue whale on record was a 24·1 m *79 ft* female captured by a Japanese whaling expedition in the Antarctic during the 1960/61 season. The male does not appear to exceed 21·6 m *70 ft 10 in*.

The first piecemeal weighing of a Blue whale took place at Balaena shore station, Hermitage Bay, Newfoundland, in May 1903 under the supervision of Dr F A Lucas, Director of the American Museum of Natural History, New York. The rather slender 23.46 m *77 ft* specimen scaled 64 tonne/*63 ton* including blood, which accounted for 8 per cent of the total weight. The flesh weighed 40 tonne/*ton*, the blubber 8 tonne/*ton*, the blood, viscera and baleen 7 tonne/*ton* and the bones 8 tonne/*ton* (Andrews, 1916). A cast of this whale is in the US National Museum, Washington, DC.

On 27 January 1948 Lt-Col Winston C Waldon (1950) of the US Army, General MacArthur's personal representative in charge of the First Fleet of the 1947/8 Japanese Antarctic Whaling Expedition, had a 27·12 m *89 ft* female weighed piecemeal aboard the 10200 tonne *10000 ton* factory ship *Hashidate Maru* in the Ross Sea.

The whale tipped the scales at 136·4 tonne *300707 lb* or *134·2 ton* including an estimated 8.7 tonne/*ton 19500 lb* for the blood and stomach contents. Its measurements were as follows: maximum bodily girth 13·25 m *43 ft 6 in*; jawbone length 6·96 m *22 ft 10 in*; flukes 6·04 m *19 ft 10 in*; fins 2·89 m *9 ft 6 in*. Weight: blubber 19970 kg 19·6 tonne/*ton*; meat 61430 kg 61·4 tonne *60·3 ton*; bones 18590 kg 18·4 tonne/*ton*; tongue 3090 kg 3·03 tonne/*ton*; skull 4610 kg 4·08 tonne/*ton*; jawbone 1990 kg 1·9 tonne/*ton*; vertebrae 7320 kg 7·1 tonne/*ton*; ribs 1470 kg 1·44 tonne/*ton*; and fins 1170 kg 1·14 tonne/*ton*.

It took 80 men 3 h 45 min to flense, dismember and weigh this specimen, and the blubber, bones and meat were weighed on four platform scales, each capable of registering 91 kg *200 lb*. Twenty-five other men were working below in the processing section.

According to Lt-Col Waldon this whale was a 'lean' individual, and this is borne out by the small yield of blubber (19·6 tonne/*ton*). He added: 'Larger and heavier whales were caught by the expedition, but they were not officially weighed and recorded.'

A very corpulent 29·49 m *96 ft 9 in* female brought into the shore station at Prince Olaf, South Georgia, in 1931 and processed by the Southern Whaling and Sealing Company was computed to have weighed 166·6 tonne *163.7 ton*, inclusive of blood, judging by the number of cookers that were filled by the creature's blubber, meat and bones. The total weight of the whale was believed to have been 177 tonne *174 ton* (Laurie, 1933).

On the principle that the weight should vary as the cube of the linear dimensions, a blue whale 15·24 m *50 ft* long should weigh about 20 tonne/*ton*, and a 30·48 m *100 ft* example about 163 tonne *160 ton*. Because of seasonal fluctuations in the thickness of the muscles and blubber, however, there are wide variations in the weights of individual blue whales at any given length. Also, a large female can lose as much as 51 tonne *50 ton* while nursing a calf.

Blue whales inhabit the colder seas and migrate to lower latitudes in winter for breeding. New-born calves measure 6·5–8·6 m *21 ft 3½ in–28 ft 6 in* long and weigh up to 3000 kg 2·95 tonne/*ton*. A calf's growth from a barely visible ovum weighing a fraction of a milligramme *0·000035 oz* to a weight of about 26 tonne/*ton* in 22¾ months (made up of 10¾ months' gestation and the first 12 months of life) **is the most rapid growth in either the animal or the plant kingdom**. This is equivalent to an increase of 30000 millionfold.

Since the turn of the century it has been calculated that at least 340000 blue whales have been killed, 97 per cent of them in the Antarctic. In 1930 there were still an estimated 100 000 blue whales roaming the world's oceans, and in the 1930/31 whaling season alone 29410 were caught in the Antarctic (Tomilin, 1967). By 1953, however, the stocks had been reduced to an estimated 15000 through unrestricted slaughter. In 1960 the species was given partial protection, and seven years later members of the International Whaling Commission, including Japan and the USSR, were banned from catching the blue whale. The Japanese promptly got round this by starting up whaling companies in Chile and Peru – both non-member countries – the following year and, up to 1970, they were still killing blue whales under flags of convenience (Small, 1971).

The USSR's record of compliance with whaling regulations also leaves a lot to be desired, and as they have never allowed international inspectors on their ships, it is more than likely that they have been catching blue whales in small numbers since the world-wide ban was introduced.

In 1968 the world population of the blue whale was estimated by cetologists to be no more than 600–700 (500–600 in the Antarctic and about 50–100 elsewhere). Fortunately these figures have since been found to be much too low.

According to Dr Ray Gambell (1972), head of the Whale Research Unit of the National Institute of Oceanography, more recent scientific assessments of existing stocks of blue whales indicate that the *exploitable* stock in the Antarctic is about 7000, with another 1000–2500 in the North Pacific and *c.* 500 in the North Atlantic. Later the same authority (pers. comm. 12 January 1973) said he had calculated from recent age composition data that the total stocks of blue whales were about 50 per cent bigger than exploited stocks, which means the world population (including the so-called 'Pygmy' blues, which have an estimated stock of 9000) is now somewhere between 17 500 and 19 000 animals.

The only other whale which has been credited with measurements in excess of 30·48 m *100 ft* is the now very rare Greenland right whale (*Balaena mysticetus*). Two enormous specimens killed in the 16th–17th centuries reportedly measured 30·3 m *99 ft 5 in* and 30·6 m *100 ft 3½ in* respectively, but Masaharu Nishiwaki (1972), the Japanese cetologist, says these measurements were either exaggerated or taken along the curve of the body. Of 322 Greenland right whales killed by Scoresby (1820), the largest measured 17·7 m *58 ft 8½ in*, but he did mention one caught near Spitzbergen which was a full 21·33 m *70 ft*. Another example taken off the Kamchatka coast, NE USSR in 1846 measured 21 m *68 ft 11 in* (Tomilin, 1967). This species, however, is proportionately much bulkier than the blue whale, and it has been calculated that a 21·33 m *70 ft* individual in good condition would weigh about 111 tonne *109 ton*.

Although absolute age determination in whales has not yet been resolved, it is probable that the longest-lived mammals after man (up to 114 years) are the Blue and Fin whales. Studies of the annual growth layers of laminations found in the wax-like plug deposited in the outer ear indicate a maximum life-span under natural conditions of 90–100 years (Nishiwaki, 1972).

Until quite recently the greatest recorded depth to which a whale had dived was 1133 m *620 fathoms 3720 ft* by a 14·32 m *47 ft* bull Sperm whale (*Physeter catodon*) found entangled with a submarine telegraph cable running between Santa Elana, Ecuador, and Chorillos, Peru, on 14 October 1955. At this depth the whale withstood a pressure of 762 kg/cm² *1680 lb/in²* (Heezen, 1957).

It is now known, however, that large bulls can dive much deeper than this in pursuit of food. The Russian cetologist, A A Berzin (1972), for instance, mentions a telegraph cable linking up Lisbon, Portugal, and Malaga, Spain, which was found damaged at a depth of 2200 m *7217 ft*. 'In this case,' he writes, 'there was no intact corpse in the cable, but from the character of the damage and the remains of the body it was assumed that . . . the culprit was a sperm whale.'

In 1970 American scientists, by triangulating the location clicks of sperm whales, calculated that the maximum depth reached by this species was 2500 m *8202 ft*, but even this nadir has been beaten!

On 25 August 1969 Dr Malcolm Clarke (pers. comm. 14 September 1971) of the National Institute of Oceanography, Wormley, Surrey, accompanied a pilot on spotter patrol for the Union Whaling Company at Durban. During the flight he carried out some observations on two big sperm whales in an effort to determine the duration of their dives. One of them remained below for 53 min and the other for a record-breaking 1 h 52 min. Soon afterwards both whales were caught by the whaling fleet, and inside the stomach of the bull which had remained submerged for nearly two hours were found two small sharks, which must have been swallowed about an hour earlier. These were later identified as *Scymnodon* sp., a small family of selachians which are restricted to the sea-floor. At this point from land – 149 km *93 miles* – the depth of water is in excess of 3193 m *1746 fathom 10476 ft*, which now suggests that the Sperm whale sometimes descends to a depth of over 3048 m *10 000 ft* when seeking food.

The only other marine mammals known to dive deeper than 914 m *3000 ft* are the Killer whale (*Orcinus orca*), the Bottle-nosed whale (*Hyperoodon rostratus*) of the North Atlantic, and the Berardius (*Berardius bairdii*) of the North Pacific. In 1965 a killer whale was found entangled in a telegraph cable off Vancouver Island, British Columbia, Canada at a depth of 1029 m *3378 ft* (Scheffer, 1970), and bottle-nosed whales have been known to take out 200–700 fathoms 366–1280 m *1200–4200 ft* of line in a few minutes when harpooned (this species usually dives vertically). The blue whale and the fin whale rarely descend to depths greater than 100 m *328 ft* because the krill on which they feed is not found in abundance below this level (Kooy-

A Pilot whale clamping a special 'lift device' to a spent torpedo in deep water (US Navy Undersea Center, San Diego)

man & Andersen, 1969).

In 1968 the US Navy started training whales to locate and retrieve valuable cylindrical objects like spent torpedoes or air-to-surface missiles lost in water deeper than 914 m *3000 ft*. The experiment, called 'Project Deep-Ops', involved the conditioning of two Killer whales (*Orca orcinus*) and two Pilot whales (*Globicephala scammoni*) at the Naval Undersea Center's laboratory in Hawaii.

The mammals were directed to the sunken hardware by an acoustic homing device and trained to clamp a 'hydrazing gas generator lift device' to its shell. This would inflate a balloon and raise the object to the surface. The whale would then swim back to the recovery ship and claim its reward – a tasty fish meal.

Unfortunately, although the bull killer whales were trained in all system behaviours and worked in the open ocean they did not take too kindly to training pressure and there were 'behavioral control problems'. 'Ahab', for instance, the larger of the two, (5·79 m *19 ft* and 2494 kg *5500 lb*), often refused to make more than one dive, even at the shallowest depths; other times he would spit out the lift device or grabber and start lob-tailing. On 4 June 1971, however, he surprised everyone by diving to a depth of 259 m *850 ft* and remaining below for 7 min 40 s, but after that performance he refused to go below 167 m *550 ft* and was dropped from the programme. 'Ishmael' (5·18 m *17 ft* and 2041 kg *4500 lb*) was even more inconsistent and refused to dive deeper than 152 m *500 ft*. Finally, on 19 February 1971 he decided to call it a day and took off for home (i.e. in the vicinity of Seattle, Washington). A big air-sea search was mounted, but the killer whale was never seen again!

One of the pilot whales 'Morgan' (3·65 m *12 ft* and 544 kg *1200 lb*) was much more co-operative. On 9 July 1971 he dived to a depth of 503 m *1654 ft*, where he deployed a practice grabber, and 19 days later he made a volunteered dive (without grabber) to a depth of 609 m *2000 ft*.

At the end of 'Project Deep-Ops' the Navy concluded that it was possible to train whales (particularly pilot whales) to retrieve 272 kg *600 lb* objects from 304 m *1000 ft* or 136 kg *300 lb* objects from 608 m *2000 ft* of water (Bowers & Henderson, 1972).

The fastest-swimming whale over short distances is probably the Sei whale (*Balaenoptera borealis*) which, when harpooned, 'dashes off at a tremendous pace for perhaps a third of a mile or less' (Andrews, 1916). The same writer says this species can attain a speed of 57·2 km/h *30 knots 34·5 miles/h* on the surface during its initial rush.

Recent observations of fin whales in the Antarctic frightened by the echo-sounding Asdic apparatus of whale-catchers suggest that their top speed is also in the region of 57·2 km/h *30 knots 34·5 miles/h*, and Nishiwaki reports that this species can maintain a speed of 33·12 km/h *18 knots 20·71 miles/h* for 30 min before tiring.

During the 1947/8 whaling season Dr R W Gawn (1948), a British scientist, carried out some speed tests on the blue whale in Antarctic waters. He discovered that it could maintain a velocity of 36·8 km/h *20 knots 23 miles/h* for 10 min when frightened, 26·6 km/h *14·5 knots 16·68 miles/h* for 2 h and could keep ahead of a whaling ship travelling at 18·3 km/h *10 knots 11·5 miles/h* all day. He also calculated that a 27·43 m *90 ft* blue whale would need to develop 520 hp in order to propel its enormous bulk along, or about 4 hp per tonne/*ton* of body-weight.

The fastest marine mammal over short distances is the Killer whale (*Orcinus orca*). In 1960 Johannesen & Harder timed a large bull at 57·2 km/h *30 knots 34·5 miles/h* as it approached their ship. It then circled the vessel, which was travelling at 37·9 km/h *20·6 knots 23·7 miles/h*, for 20 min. Other observers (Maxwell, 1952; Blond, 1953) believe that this powerful marine predator can reach 64 km/h *40 miles/h* for short bursts when chasing prey. (Despite its name, the killer whale is classified as a dolphin.)

Speeds of 55·2–58·8 km/h *30–32 knots 34·5–36·8 miles/h* have been reported for Common dolphins (*Delphinus delphis*) riding the bow waves of destroyers (in bow-wave riding the dolphin attains its propulsive power from the moving vessel and is squeezed along by the water pressure), but this is not true swimming. Nevertheless, this highly streamlined creature is still one of the swiftest of all the cetacea.

The head of the 8·23 m 27 ft *long female Grey whale 'gigi', the largest marine mammal ever held in captivity (*Sea World*)*

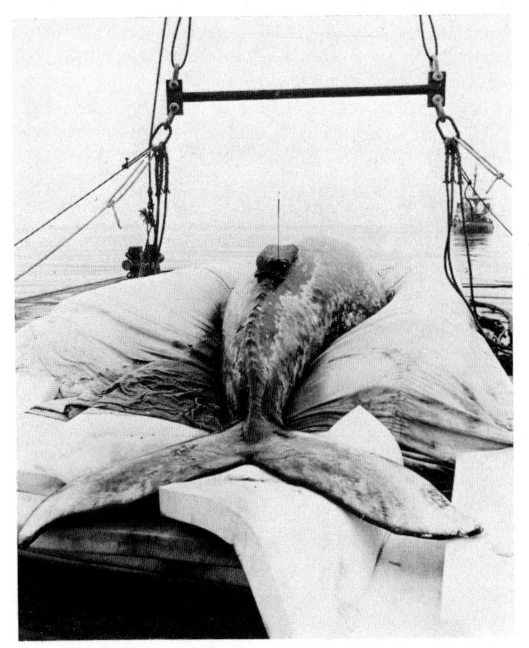

*The female Grey whale 'Gigi' off the San Diego coast prior to being released (*Sea World*)*

In June 1938 Tomilin (1967) observed a school of some 200–300 individuals moving at a tremendous rate near Novorossisk, a city on the Black Sea coast, European Russia. In order to determine the speed of one of these dolphins, he attached a 100 m *328 ft* line to the tail of a just-captured 1·6 m *5 ft 3 in* long female and then released her. 'When she reached the end of the line', he said, 'her speed was 7 m/sec 44·08 km/h *27·55 miles/h*. The line, capable of withstanding the stress of 33 kg *73 lb* snapped immediately.'

The Spotted dolphin (*Stenella attenuata*) is another rapid swimmer and has been clocked at 39·36 km/h *21·4 knots 24·62 miles/h* (Nishiwaki, 1972), and the Pacific white-sided dolphin (*Lagenorhynchus obliquidens*) has been credited with speeds up to 40·49 km/h *22 knots 25·31 miles/h* (Bruyns, 1971).

The largest marine mammal ever held in captivity was a female Pacific gray whale (*Eschichtius gibbosus*) named 'Gigi' at Sea World, Mission Bay, San Diego, California, USA. She was captured in Scammon's Lagoon, Baja California, Mexico, on 13 March 1971 by a colleting expedition from Sea World and arrived (by boat) in San Diego four days later. She was then approximately 6–10 weeks old and measured 5·53 m *18 ft 2 in* in length (weight 1950 kg *4300 lb*). By 30 March she had increased to 6·04 m *19 ft 10 in* and 2268 kg *5000 lb*, and ten weeks later she was 6·70 m *22 ft* and 3401 kg *7500 lb*.

On 13 March 1972, when she was one year old, and rapidly outgrowing her surroundings, Gigi was returned to the sea about 6·4 km *4 miles* north of Port Loma. She was then 8·22 m *27 ft* long and weighed an estimated 6350 kg *14 000 lb*. Her release was timed to coincide with the annual 11 200 km *7000 mile* northward migration of the gray whale herds from the calving grounds in the lagoons of Baja California to the Arctic. The whale was fitted with a compact radio and instrumentation package developed by the Naval Undersea Research and Development Center in San Diego, and for five days her movements were tracked by a research vessel and aircraft as she joined up with the herd and moved northward (Janet Rogers, pers. comm. 17 September 1973).

In February 1976 Gigi was spotted off the Mexican coast, and appeared to be quite happy.

The largest marine mammal held in captivity today is a bull killer whale named 'Orky' at Marineland of the Pacific, Palos Verdes Estates, California. He was captured in Pender Harbour, British Columbia, Canada, on 12 April 1968 and was moved to Marineland by air on 10 May. At that time he was 5·18 m *17 ft* long and weighed 1814–2041 kg *4000–4500 lb*. By September 1972 he had increased to 6·70 m *22 ft* and 3743 kg *8252 lb*, and in January 1976 he measured 7·21 m *23 ft 8 in* and weighed an *estimated* 4989 kg *11 000 lb* (Otten, pers. comm. 1 Sept. 1976).

Ishmael the bull killer whale pictured before his escape (US Navy Undersea Centre, San Diego)

The largest living terrestrial animal is the African bush or savannah elephant (*Loxodonta africana africana*) which was formerly considered a full species. The average adult bull stands 3·20 m *10 ft 6 in* at the shoulder and weighs 5·6 tonne/*ton*. The average adult cow is much smaller, standing 2·59 m *8 ft 6 in* at the shoulder and weighing about 2·5 tonne/*ton*. (Most authorities give the weight of an adult cow as about 4 tonne/*ton*, but this figure is incorrect.)

At one time the bush elephant ranged over the entire African continent, but commercial exploitation, the expansion of cultivated land and the degeneration of grassy plains to desert, have greatly reduced its range over the past hundred years. Today its numbers (about 300000) are largely concentrated in wildlife parks and game reserves in central, eastern and southern Africa.

The largest accurately measured African bush elephant on record, and the largest recorded land animal of modern times, was an enormous bull shot by J J Fenykoevi, the Hungarian big-game hunter, 76·8 km *48 miles* NNW of Macusso, Angola, on 13 November 1955. This giant measured 4·01 m *13 ft 2 in* in a projected line from the highest point of the shoulder to the base of the forefoot while lying on its side, which means that its standing height must have been *c.* 3·81 m *12 ft 6 in* at the shoulder. (Height measurements of wild elephants taken after death are about 5 per cent greater than the living height because the great weight tends to spread the body out laterally in all directions.) Other measurements included an over-all length of 10·10 m *33 ft 2 in* (tip of extended trunk to tip of extended tail) and a maximum bodily girth of 5·99 m *19 ft 8 in*. The weight was estimated at 10886 kg 10·7 tonne/*ton 24000 lb*, which seems reasonable for an elephant of this size.

On 6 March 1959 the mounted specimen was put on display in the rotunda of the US National Museum in Washington, DC.

Another huge African bush elephant known as 'Dhlulamithi' ('Taller than the Trees') may have equalled the 'Fenykoevi Elephant' in size. Experienced game-rangers in Rhodesia familiar with this animal said he stood at least 3·65 m *12 ft* at the shoulder, and one of his giant tusks was so long that it trailed along the ground.

According to the Shangaan tribesmen of SE Rhodesia a notorious ivory-poacher and hunter named Bvekenya Barnard, with 300 elephants to his credit (or rather discredit) tried to track down this massive bull in 1929. Eventually he spotted the huge pachyderm and inflicted a slight shoulder wound, but the great beast then disappeared into the bush. A few days later Barnard had the elephant in his sights again, but he was reportedly so overwhelmed by the animal's extraordinary size that he hung up his guns, vowing never to hunt again. Some hunters, however, assert that this just does not fit the Barnard character.

From that day onwards, the story goes, hunters and game-rangers alike swore never to shoot this lord of the bush, even if he wandered into a tsetse-fly control corridor where animals are automatically shot on sight. Over the years there were unconfirmed sightings of him in Portuguese East Africa and South Africa, as well as Rhodesia, and he appeared to be domiciled in the Nyamatongwe area where all three countries are close to each other (J King, pers. comm. 2 December 1966).

In 1959 Mr B J Olivier, a roads overseer, who had worked in the lowveld for many years, spotted Dhlulamithi near Chipinga Pools in SE Rhodesia, and there were several other sightings up to January 1965. After that the enormous bull dropped out of sight until August 1967 when he was killed *just outside* the Gona-Re-Zhou Game Reserve at Nuanetsi, SE Rhodesia, by a high-ranking South African policeman named V R Verster.

According to Allen Wright (1972), who was then District Commissioner at Nuanetsi, the tragedy occurred when Warden Timothy Braybrooke, who was in charge of the game reserve and responsible for the protection of the animals, including Dhlulamithi, was instructed to take the South African visitor on a hunting trip in the VIP area just east of the Lundi River. As they were in sight of the towering Clarendon Cliffs the two men suddenly saw a huge bull elephant with enormous tusks directly in front of them.

'Braybrooke recognised the bull at once', said Wright. 'It was obviously the mammoth from the Nyamatongwe area that we were all trying to protect. He knew that this was Rhodesia's greatest tusker strayed temporarily from his unfenced haven just a few thousand yards away across the dry Lundi, but what reply was he to make to his guest's excited demands to be allowed to shoot this animal?'

'This was the classical game-warden's dilemma – should he take it on himself to declare that although the elephant was in the hunting area, it could not be shot because it was a protected animal, or should he comply with the letter of the Ministerial permit the South African held authorising him to shoot any elephant in that particular area? If he refused to allow the elephant to be killed, would his superiors be critical of his actions? He had a few seconds to make up his mind and his decision was quickly followed by the crack of an express rifle and some 70 years of history thrashed and struggled in the Lowveld dust.'

When Wright heard about the shooting a few days later he was 'angrier than I have ever been in my life', and for weeks he saw red every time he spotted a member of the Wild Life Department. Today, however, the bitterness has been dulled by the passing years, although he admits that 'deep down inside me the resentment remains'.

Because of the embarrassment caused by the killing of this legendary animal, the whole episode was hushed up by the Rhodesian Government, and no field measurements were taken. The tusks, however, were secured and later presented by Col Verster to the South African Embassy in Washington, DC, USA.

Mr Anthony Best, current editor of the famous *Rowland Ward's Records of Big Game* told the compiler (pers. comm. 13 February 1973) that they measured 2·54 m *8 ft 6¼ in* (60 kg *132 lb*) and 2·32 m *7 ft 7½ in* (48 kg *107 lb*) respectively. Although these figures are somewhat disappointing compared to some of the top tusks recorded from East Africa some 70 years ago, Dhlulamithi is still the largest tusker ever recorded south of the Zambesi River.

In August 1960 P K van der Byl shot another huge tusker in the Cuando River region which measured 3·83 m *12 ft 6½ in* in the prone position (standing height about 3·63 m *11 ft 11 in*) and had an over-all length of 9·13 m *29 ft 11½ in* (Rowland Ward, 1962).

Other African bush elephants for which shoulder heights of 3·65 m *12 ft* or more between pegs have been claimed include: a 3·65 m *12 ft* bull shot by William Cornwall Harris (1839) in South Africa in 1836, and another one killed there in *c.* 1849 by William Cotton Oswell which measured 3·71 m *12 ft 2 in*; an enormous bull shot by the Duke of Edinburgh (Prince Alfred, second son of Queen Victoria) in the Knysna Forest, Cape Province, South Africa, in 1875 which reputedly measured 3·96 m *13 ft* at the shoulder and 9·75 m *32 ft* over all. (The girth of this animal was given as 8·52 m *28 ft*, but this must have been post-mortem distension) (Bisset, 1875); and a bull shot by A Haig on the Blue Nile (Sudan) which was estimated by him – on the basis of the circumference of the elephant's forefoot – to have measured over 3·96 m *13 ft*. (The ears of this specimen were enormous, measuring 1·97 m *6 ft 5½ in* in vertical diameter and 1·24 m *4 ft 1½ in* in transverse diameter, and probably constitute a *record* in themselves) (Lydekker, 1907).

In 1929 a mighty bull said to have measured 3·65 m *12 ft* at the shoulder was found dead on the Nile bank just below the Murchison Falls, Uganda, apparently from old age (Pitman, 1953), and the famous 'Mohammed' of Marsabit Mountain Reserve, Kenya, was credited with the same shoulder height between pegs at the time of his death in 1960. The latter measurement, however, should be taken with a pinch of salt, because it is well known that the elephants of northern Kenya are small. Ample proof of this is provided by the even more celebrated 'Ahmed' of the same reserve, who was described by the world's newspapers, television and certain commercial tour operators as 'the world's largest living elephant'. The only thing that made this pachyderm unique, however, was the fact that President Kenyatta of Kenya issued a protective decree in 1970 which virtually made this elephant a living national monument. Ahmed died on 17 January 1974 through a combination of old age and deterioration of feed, and Dr A Mackay (pers. comm. 21 April 1975) Director of the National Museums of Kenya, where this elephant is now on display, said that the creature measured a very modest 2·99 m *9 ft 10 in* at the shoulder (between pegs?).

In *c.* November 1969 Dr Sylvia Sikes (1972), the world's leading authority on the African elephant, shot a huge bull ('the biggest elephant I have ever set eyes on') near Portofino, Lake Chad, Central Africa, which measured 3·60 m *11 ft 9¾ in* at the shoulder. The pachyderm, a member of an eight-bull herd which lived in the swamps along the Nigerian shores, had a forefoot circumference of 1·80 m *5 ft 11 in* (equivalent to a shoulder height of 3·96 m *13 ft*), thus indicating that swamp- and lagoon-dwelling elephants have disproportionately large feet owing to the nature of their habitat. The weight of this giant was calculated to be 6200 kg *13 668 lb* 6·10 tonne/*ton* based on a 0·5 g *0·107 oz* heart-weight/100 g *3·54 oz* body-weight index.

The shoulder height of an adult African bush elephant corresponds roughly to the width across the expanded ears or six times the forefoot diameter.

The largest African bush elephant ever held in captivity was probably the famous 'Jumbo', who was acquired by London Zoo on 26 June

1865 from the Menagerie du Jardin des Plantes, Paris, in exchange for an Indian rhinoceros. (The elephant had been caught originally south of Lake Chad in the French Sudan.) In 1879 he measured 3·28 m *10 ft 9 in* at the shoulder, and in January 1882, when he was purchased by Phineas T Barnum, the famous American showman, he allegedly stood 3·35 m *11 ft* at the shoulder (weight 6·5 tonne/*ton*) and could reach (forefeet off the ground) an object 7·92 m *26 ft* above him with his trunk (Bartlett, 1898).

The true standing height of Jumbo will never be known for certain because Barnum refused to allow anyone to measure his prize exhibit. He claims the elephant measured 3·55 m *11 ft 7 in* in March 1883 and 3·65 m *12 ft* shortly before his death on 15 September 1885 (hit by a train at St Thomas, Ontario, Canada), but both these figures were exaggerated for commercial reasons.

According to 'Professor' Henry Ward of Ward's Natural Science Establishment in Rochester, NY, who was asked by Barnum to mount Jumbo for exhibition purposes, the elephant measured 3·45 m *11 ft 4 in* at the shoulder between pegs (maximum bodily girth 4·97 m *16 ft 4 in*), which suggests that the animal must have stood about 3·28 m *10 ft 9 in* in life as reported earlier; but even this measurement is on the low side. In 1883 Robert Gillfort, a pole-jumper with the Barnum and Bailey Circus, attempted to find out the true height of Jumbo by casually standing his pole alongside the elephant and carefully noting the mark on the wood that corresponded with the creature's highest point at the shoulder (Hornaday, 1911). The measurement came out at 3·28 m *10 ft 9 in*, which means that Jumbo, who was still growing at the rate of 25 mm *1 in* or so a year, must have stood a full 3·35 m *11 ft* at the shoulder at the time of his death. (The skeleton of Jumbo in the American Museum of Natural History, New York – presented by Barnum in 1889 – has a mounted shoulder height of 3·17 m *10 ft 5¼ in*, which is equivalent to a measurement of 3·40 m *11 ft 1¾ in* in the flesh.)

After touring North America for two years the model of Jumbo was dismantled and the 697 kg *1538 lb* hide reconstructed, stuffed and sent to the Barnum Museum at Tufts University, Massachusetts, where it was destroyed by fire in April 1975.

Another African bull elephant named 'Tembo' at Copenhagen Zoo, Denmark (received in 1936 as a three year old) was also of comparable size. At the time of his death in August 1970 – he was destroyed after turning violent – he reportedly measured 3·70 m *12 ft 2 in* at the shoulder and weighed 6250 kg *13 778 lb* 6·15 tonne/*ton* (*International Zoo News*, No. 94). Unfortunately this shoulder measurement was exaggerated, although the poundage was accurate.

Dr Arne S Dyhrberg (pers. comm. 4 May 1972) Managing Director of Copenhagen Zoo, said this elephant had to 'slightly' lower his head when passing through the entrance (height 3·34 m *10 ft 11½ in*) to his stall, which means Tembo must have had a height (normal position) of *c.* 3·40 m *11 ft 2 in*. This is equivalent to a shoulder height of 3·30–3·32 m *10 ft 10 in–10 ft 11 in* because, in large bulls, there is only some 76–101 mm *3–4 in* difference between the two measurements.

In January 1975 a height of 3·78 m *12 ft 5 in* was reported for a bull African bush elephant named 'Varaux' at the Georg von Opel Zoo, Kronberg im Taunus, West Germany, but this measurement is so extreme it must have been taken to the crown of the *raised* head. Photographic evidence suggests that the shoulder height of this rangy pachyderm which, incidentally, is still growing, is somewhere in the region of *c.* 3·35 m *11 ft*, and this measurement is also supported by the weight, which has been estimated at 6 tonne/*ton*. Varaux was captured in the Big Game Range, Arusha, Tanzania, and arrived at the German zoo in 1955.

The largest adult cow African bush elephant to be accurately measured between pegs was probably a specimen shot on 25 March 1964 in Murchison Falls National Park, Uganda, during 'cropping' operations. It measured 2·93 m *9 ft 7¼ in* at the shoulder and scaled 2942 kg *6485 lb* 2·89 tonne/*ton*. Another cow measuring 2·83 m *9 ft 3½ in* and weighing 2375 kg *5236 lb* 2·33 tonne/*ton* was shot in the same park three weeks later (Laws *et al*, 1967).

The largest cow African bush elephant ever held in captivity was probably a specimen named 'Sudana'. She was captured on the south-eastern slopes of Mount Kilimanjaro, Tanzania, on 6 May 1929 when aged two, and arrived at New York Zoological Gardens (Bronx Zoo) on 9 November 1931. In 1947 she measured 2·61 m *8 ft 7 in* at the shoulder, and recorded the same height on 15 September 1958. She was destroyed on 11 August 1962 after being rendered immobile by arthritis (Crandall, 1964).

This size was matched by the famous 'Dicksie' of London Zoo (received October 1945 from Kenya) who measured 2·61 m *8 ft 7 in* in July

1965 when aged 25. She died tragically on 6 September 1967 from heart failure after falling (or was she pushed by a jealous comrade?) into the moat surrounding her enclosure.

Another cow called 'Alice' purchased by London Zoo as a mate for Jumbo on 9 September 1865 and later shipped to the USA, reputedly measured 'a few inches over 9 ft in 1886', but this claim has never been substantiated. She died in the fire at P T Barnum's winter quarters at Bridgeport, Connecticut, on 20 November 1887.

The African forest elephant (*Loxodonta africana cyclotis*), which is found in the rain forests of Guinea, French Equatorial Africa and the Congo, is a shorter animal than the bush elephant, although it is stockier and proportionately heavier. The average adult bull measures 2·35 m *7 ft 8½ in* at the shoulder and weighs about 2·5 tonne/*ton*, and the average cow 2·10 m *6 ft 10½ in* and 1·75 tonne/*ton*.

The greatest standing height at the shoulder recorded for an African forest elephant is 3 m *9 ft 10 in* for a bull measured at Api by Commandant Pierre Offermann (1953) while Director of the Belgian Congo army-run elephant training centre. This centre was at Api from 1900 until 1926, and then at Gongala-na-Bodia.

A 2·85 m *9 ft 4¼ in* bull shot at the Api station in 1906 and immediately cut up and weighed was found to total a surprising 6000 kg *13 227 lb* 5·89 tonne/*ton* (the weight of the blood and stomach contents was estimated at 340 kg *748 lb*), which is slightly more than the average weight of a bull bush elephant measuring 3·20 m *10 ft 6 in* at the shoulder.

The largest African forest elephant ever held in captivity was a cow named 'Doruma' received at the New York Zoological Gardens (Bronx Zoo) in October 1946 from the training centre at Gongala-na-Bodia as a gift from the Belgian Congo Government (Bridges, 1946). On 27 June 1963 her height was recorded as 2·34 m *7 ft 8½ in* (Crandall, 1964), and in October 1969 she measured 2·36 m *7 ft 9 in*. She died in the spring of 1970 when her weight was estimated to be 3175 kg *7000 lb* (3.12 tonne/*ton*) (Conway, pers. comm. 8 December 1970).

The status of the African pygmy elephant (*Loxodonta africana pumilio*), which is found in the swampy forests of Gabon and the Congo, is still uncertain, but although this animal has been credited with sub-specific rank (Noack, 1906; Hornaday, 1923), the modern consensus of opinion is that pygmy elephants (maximum shoulder height 1·98 m *6 ft 8 in*) are a race of undersized *forest elephants* living in an unfavourable environment.

The greatest true weight recorded for an elephant is 6640 kg *14641 lb* 6·53 tonne/*ton* for a huge bull shot by Captain Hewlett (with Dr G Crile) near Ngaruka, Tanzania, on 26 December 1935 (Benedict & Lee, 1938; Crile, 1941). The body was cut up into 176 parcels and weighed piecemeal on scales capable of registering 272 kg *600 lb*. The shoulder height of this beast was not recorded, but as the front foot measured 1·52 m *60 in* in circumference, it must have stood about 3·35 m *11 ft* in life.

In 1965 a relationship between dressed hind-leg weight and live body-weight was demonstrated for 26 bush elephants (13 of each sex) killed in the Murchison Falls National Park during cropping operations by Uganda National Parks rangers; and another 12 elephants (six of each sex) were killed in Kenya's Tsavo National Park the following year to check the accuracy of the earlier findings. According to Laws *et al* (1967) the hind-leg weights of the 38 elephants accounted for between 5·3 and 6·3 per cent of the total live weight and could be used to predict live weight more accurately than by carefully weighing the animal in pieces with considerable blood and fluid losses.

In the Murchison Falls experiment the greatest weight recorded for a dressed hind-leg was 308·4 kg *678 lb 8 oz* for a bull which weighed 5002 kg *11 025 lb* 4·92 tonne/*ton*, while in Tsavo National Park the maximum weight was 294 kg *649 lb* for a 5000 kg *11 023 lb* 4·92 tonne/*ton* specimen.

The greatest true weight recorded for a cow African bush elephant in the field is 3232 kg *7126 lb* 3·18 tonne/*ton* for a specimen killed in Murchison Falls National Park on 16 January 1964. The shoulder height of this animal was 2·59 m *8 ft 6 in* (Laws *et al*, 1967). Another cow (shoulder height 2·69 m *8 ft 10 in*) killed in the same park 15 days previously weighed 3133 kg *6906 lb* 3·08 tonne/*ton*.

Several races of the Asiatic elephant (*Elephas maximus*) have been described (Deraniyagala, 1955 lists eight), but only four are generally recognised. They are: *E. m. indicus* of India, Burma, Thailand, Vietnam and Borneo, where it is thought to have been introduced a few hundred years ago; *E. m. maximus* and *E. m. ceylanicus* of Sri Lanka; and *E. m. sumatranus* of

Sumatra (Chasen, 1940; Morrison-Scott, 1951).

The average adult bull Indian elephant (*E. m. indicus*) stands 2·74 m *9 ft* at the shoulder and weighs 4·5 tonne/*ton*, and the average adult cow 2·28 m *7 ft 6 in* and 2·5 tonne/*ton*.

The two races of Ceylon elephant (*E. m. maximus* and *E. m. ceylanicus*) average rather larger and heavier than the typical race from the Indian mainland, while *E. m. sumatranus* is slightly smaller.

The shoulder heights attained by the Asiatic elephant have been greatly exaggerated in the past. This is mainly due to the fact that measurements were often taken by throwing a tape over the shoulders of the animal. The ends were then taken down to the outside of each front foot and half the subsequent length taken as the elephant's height.

Francis Benedict (1936) says an 2·43 m *8 ft* cow elephant measured by this method recorded a height of 2·69 m *8 ft 10 in*, while a 3·04 m *10 ft* bull came out at 3·45 m *11 ft 4 in*! It is also known that some mahouts (elephant-drivers) would measure their charges from the ground to the crown of the head and then claim the result as the elephant's shoulder height.

Twice the circumference of an Asiatic elephant's forefoot when resting on the ground and enlarged by pressure gives approximately its height at the shoulder (up to six per cent error), but this rule does not always apply in the case of young growing animals. The formula for the African bush elephant is twice the circumference of the forefoot plus ten per cent.

The largest recorded Asiatic elephant to be accurately measured between pegs was a bull (*E. m. maximus*) shot by Mr W H Varian at Chalampia Madua, in the North Coast Province of Sri Lanka in 1882. The following measurements were taken immediately after death: height at arch of back 3·58 m *11 ft 9 in*; at the shoulder 3·37 m *11 ft 1 in*; length from tip of extended trunk to tip of extended tail 7·92 m *26 ft*; girth of body at thickest part 6·80 m *22 ft 4 in*; estimated about 8 tonne/*ton* (Osborn, 1942).

In June 1950 Mr H Mant shot an elephant (*E. m. maximus*) in a Sri Lanka swamp which had a forefoot circumference of 1·87 m *6 ft 1½ in*. Normally this would be equal to a shoulder height of 3·73 m *12 ft 3 in*, but Deraniyagala (1955) also says swamp-dwelling elephants often have disproportionately large feet because of the nature of their habitat.

Turning to more recent records, Patrick Stracey (1963) a former Chief Conservator of Forests in Assam, gives details of four bulls measuring in excess of 3·28 m *10 ft 9 in* between pegs. One of them, shot by L Mackrell at Mismuri, in the Goalpara Division of Assam, measured 3·33 m *10 ft 11 in*. Another huge tusker shot by the Maharajah of Mysore in the Mysore jungle during the Second World War was exactly 3·35 m *11 ft*, and the same measurement was reported for a bull shot by Lalji, Kumar of Gauripur, in the Garo Hills in 1945. (The height of this elephant could not be taken because it fell in an awkward position, but its forefoot measured 1·67 m *66 in* in circumference.) The fourth bull, a rogue tusker, also measuring 3·35 m *11 ft*, was shot by the Maharajah of Talcher in Khenkanal, Orissa, in 1953.

On 21 May 1965 Duncan Hay shot a very large tusker near the Reserve Forest, Lakhimpur, Assam, which measured 3·34 m *10 ft 11½ in* at the shoulder (circumference of forefoot 1·66 m *5 ft 5½ in*).

The largest Asiatic elephant ever held in captivity was the Maharajah of Nepal's famous tusker 'Hari Prasad', which measured 3·28 m *10 ft 9 in* at the shoulder in 1957 (Stracey, 1963). Another huge tusker owned by the Rajah of Nahan-Sirmout in the Punjab measured 3·24 m *10 ft 7½ in* in *c.* 1870.

J Corse (1799), a former Superintendent of the East India Company at Tiperah, East Bengal, who probably saw more elephants than any other European of his time, only encountered one bull measuring over 3·04 m *10ft* at the shoulder, and this was a magnificent tusker owned by Asaph-ul-Daulon, Vizier of Loudh, which was 3·21 m *10 ft 6½ in*. He did hear of one owned by the Nabob of Dacca which allegedly stood 4·26 m *14 ft* (*sic*) at the shoulder, but when he investigated this 'superlative' he found that the animal measured exactly 3·04 m *10 ft*.

G P Sanderson (1882) offered a reward of a brand-new rifle to anyone who could produce an Asiatic bull elephant measuring 3·04 m *10 ft* or more at the shoulder, but the largest he saw was 2·99 m *9 ft 10 in*, and Osborn (1936) says none of the 150 adult bull elephants measured in Bengal in 1923 reached 3·04 m *10 ft*, although there were a few over 2·89 m *9 ft 6 in*.

The famous tusker 'Chandrasekharan', owned by the Maharajah of Travancore, measured 3·04 m *10 ft* at the shoulder in 1913, and shortly after his death on 10 August 1940 the Trivandrum Museum taxidermist reported that his height between pegs was 3·23 m *10 ft 7 in* (obviously some shrinkage due to old age).

In the film *Elephant Boy* (1937), starring Sabu, adapted from Rudyard Kipling's famous story 'Toomai of the Elephants', the part of 'Kala Nag' was played by a large tusker owned by the Maharajah of Mysore. This animal stood 3·01 m *9 ft 11 in* at the shoulder and weighed 5 tonne/*ton*.

The largest Asiatic elephant ever held in a zoo was the celebrated 'Bolivar', who was presented to the King of Italy by the Maharajah of Oudh in 1871. The beast was then three years old and stood 1·52 m *5 ft* high. When King Victor Emmanuel died on 9 January 1878 the elephant was sold to Carl Hagenbeck, who in turn transferred it to Adam Forepaugh, the American circus-owner, in 1881 (Flower, 1931). In January 1889 Forepaugh presented the impressive bull to Philadelphia Zoological Garden, Pennsylvania, at which time he measured 2·74 m *9 ft* at the shoulder. On 3 June 1890 he recorded a weight of 3991 kg *8800 lb* 3·92 tonne/*ton* (Conklin, 1890). Shortly before his death on 31 July 1908 aged *c.* 40 years Bolivar measured 3·04 m *10 ft* at the shoulder and weighed a light 5443 kg *12 000 lb* 5·35 tonne/*ton* (Benedict, 1936). His skeleton was later mounted and put on display in the entrance hall of the Academy of Natural Sciences in Philadelphia.

Another Asiatic bull named 'Ziggy' acquired by Chicago Zoological Park (Brookfield Zoo), Illinois, USA, in the summer of 1936, is also of comparable size. On 12 July 1966 this dangerous bull – on 26 April 1941 he tried (unsuccessfully) to gore his keeper and spent the next 29 years doing penance in his stall – was *cautiously* measured by a local engineering firm who established that the elephant stood 2·94 m *9 ft 8 in* at the shoulder and weighed an estimated 5443 kg *12 000 lb* (5·35 tonne/*ton*) (Dr W Crowcroft, pers. comm. 4 December 1970). In March 1975 this crusty old bachelor was trapped for almost two days in a 3·04 m *10 ft* deep moat round his new paddock. In the end he extricated his *c.* 5896 kg *13 000 lb* 5·80 tonne/*ton*? by walking up a ramp which rescue workers had built with 42 tonne/*ton* of gravel.

'Big Charlie', an Asiatic bull owned by Butlin's Ltd, reputedly stood 3·20 m *10 ft 6 in* at the shoulder and weighed 7·5 tonne/*ton* in 1957 when he was 25 years old, but these figures were commercial exaggerations. In reality he measured 2·92 m *9 ft 7 in* and weighed *c.* 5 tonne/*ton*. This pachyderm was so devoted to his trainer Ibrahim that when he died suddenly at Butlin's Holiday Camp at Filey, Yorkshire, from pneumonia in 1959 the grey giant refused to eat and

The famous American circus elephant 'Tusko', the heaviest Asiatic bull ever held in captivity (Karl K Knecht)

became completely unmanageable. In the end the RSPCA were called in to end his suffering.

In 1932 a shoulder height of 3·15 m *10 ft 4 in* and a weight of 7500 kg *16 534 lb* 7·38 tonne/*ton* were reported for a huge tusker named 'Harry' (b *1880*) at the Berlin Zoo, West Germany, but these figures will never be substantiated because the zoo records were destroyed by Allied bombing during the Second World War (Prof Heinz-Georg Klos, pers. comm. 23 November 1970).

The greatest true weight ever recorded for an Asiatic elephant is 6492 kg *14 313 lb* 6·38 tonne/*ton* for the famous American circus elephant 'Tusko'. This magnificent bull, who stood 3·09 m *10 ft 2 in* at the shoulder, died in Woodland Park Zoological Garden, Seattle,

'Zebi', the largest female Asiatic elephant ever held in captivity (Bristol, Clifton and West of England Zoological Society)

USA, on 10 June 1934 aged *c.* 40 years (Lewis, 1955), but spent practically all of his life in circuses. Another 3·09 m *10 ft 2 in* bull living in the elephant stables at Mysore in 1953 scaled 6198 kg *13656 lb* 6·09 tonne/*ton* on a weighbridge (Rensch & Harde 1955).

The largest Asiatic cow elephant ever held in captivity was a specimen named 'Zebi' presented to Bristol Zoo by the Maharajah of Mysore in 1868. At the time of her death in January 1910 she reportedly stood 3·04 m *10 ft* at the shoulder and weighed *c.* 5 tonne/*ton*, and these figures are borne out by photographic evidence (Flower, 1931; Greed, pers. comm. 15 December 1970).

The greatest true weight recorded for an Asiatic cow elephant is 4163 kg *9180 lb* 4·09 tonne/*ton* for a 2·38 m *7 ft 10 in* example named 'Big Modoc' owned by the Ringling Brothers-Barnum & Bailey Circus (Benedict, 1936). Three other cows in the same circus weighed 4126 kg *9098 lb*, 4106 kg *9053 lb* and 4082 kg *9000 lb*.

James Clarke (1969) estimates that between 200 and 500 people are killed by elephants in Africa annually, and Carrington (1955) puts the yearly death-toll in India at *c.* 50. Circus and zoo elephants, 90 per cent of them Asiatic, probably account for another ten deaths annually.

In 1949 two rogue elephants killed five people near Kasungu, Malawi, in the space of 24 hours (Debenham, 1955). Three years later a one-tusked elephant killed 27 people in Bangladesh before it was shot, and Clarke says one highly trained work elephant in India killed 18 men and *still managed to avoid execution* because it was considered too valuable an animal to be destroyed. There is also a record of a rogue elephant in northern Zululand killing 12 natives in quick succession during a temporary bout of insanity.

In September 1973 a rogue elephant bolted from the jungles of Zaire across the border into the Sudan and killed eleven people, and ten months later a herd of 150 drunken elephants stampeded through a village in West Bengal killing five villagers and injuring twelve others after drinking liquor stored in an illicit still.

Although the elephant is the strongest terrestrial animal, it is interesting to note that pound for pound it is weaker than man. This was proved beyond any shadow of doubt in 1964 when, as a tourist attraction, a tug-of-war was arranged between an Asiatic elephant weighing 3492 kg *7700 lb* 3·43 tonne/*ton* and a platoon of 50 soldiers weighing a total of 2893 kg *6380 lb* 2·84 tonne/*ton* in the NE province of Surin, Burma. Most of the bets were on the elephant, but Gale (1974) says the soldiers won hands down.

The tallest living animal is the Giraffe (*Giraffa camelopardalis*), which is now found only in the dry savannah and semi-desert areas of Africa south of the Sahara.

Nine races are generally recognised, although up to 13 have been described, and the tallest are the Masai giraffe (*Giraffa c. tippelskirchi*)of Kenya and Uganda; the Baringo giraffe (*Giraffa c. rothschildi*) of NW Kenya and SE Uganda; the Southern giraffe (*Giraffa c. capensis*) of SW Africa; the Angolan giraffe (*Giraffa c. angolensis*) of Angola and SW Africa; and Thornicroft's giraffe (*Giraffa c. thornicrofti*) of northern Rhodesia.

The average adult bull stands 4·87–5·18 m *16–17 ft* in height (tip of forehoof to tip of 'false' horn with neck erect) and weighs 1078–1270 kg *2376–2800 lb* 1·06–1·25 tonne/*ton*. The cows are smaller than the bulls, and are always lighter in build. They average about 4·41 m *14 ft 6 in* in height and weigh about 562 kg *1240 lb*.

The greatest realiable height recorded for a giraffe between pegs is 5·87 m *19 ft 3 in* (standing height about 5·79 m *19 ft*) for a Masai bull shot by Caswell in Kenya (Shortridge, 1934). It was thus 1·21 m *4 ft* taller than a London double-decker bus. This animal was not weighed piecemeal, but it must have scaled nearly 2 tonne/*ton*. Less credible heights of up to 7·01 m *23 ft* have been claimed.

The greatest reliable height recorded for a cow giraffe between pegs is 5·13 m *16 ft 10 in* for an example of the southern race shot by Henry R Bryden in the desert country near the Botletli River, Ngamiland, in 1889.

The tallest giraffe ever held in captivity was a massive Masai bull named 'George' who arrived at Chester Zoological Gardens on 8 January 1959 from Kenya when he was an estimated 18 months old. At the age of six years he measured 5·48 m *18 ft*, and when he stopped growing a year later his head almost touched the roof of the 6·09 m *20 ft* high Giraffe House.

Incidentally, George was something of a character and he certainly gave GPO engineers a lot of headaches. Apparently he was very fond of licking the telephone wires which ran past his enclosure because he liked the salty flavour . . . and the 50 volt tingle . . . and for six months he caused the telephone system at Chester Zoo to go

'George', the tallest giraffe on record, browsing with his family – note the telephone wires (Kenneth W Green)

completely haywire. Every day phones rang for no reason and calls got crossed. Eventually the trouble was traced to the huge animal and the wires were raised another 914 mm *3 ft* – much to George's annoyance.

The bull died on 22 July 1969 (G S Mottershead, pers. comm. 24 November 1970).

The tallest cow giraffe ever held in captivity was probably a specimen named 'Rosie', a member of the Baringo race, who arrived at Whipsnade Zoological Park in June 1934 from Kenya. She stood approximately 5·18 m *17 ft* (Manton, pers. comm. 30 October 1970).

The smallest living mammal excluding bats is Savi's white-toothed pygmy shrew (*Suncus etruscus*), also called the 'Etruscan shrew', which is found along the coasts of the northern Mediterranean and southwards to Cape Province, South Africa (Rode, 1938; Morrison-Scott, 1947). Not much is known about this tiny species, and most of the published information has come from studies of owl pellets in which this animal's teeth, fur and bones have been found.

Mature specimens have a head and body length of 36–52 mm *1·32–2·04 in*, a tail length of 24–29 mm *0·94–1·14 in* and weigh 1·5–2·5 g

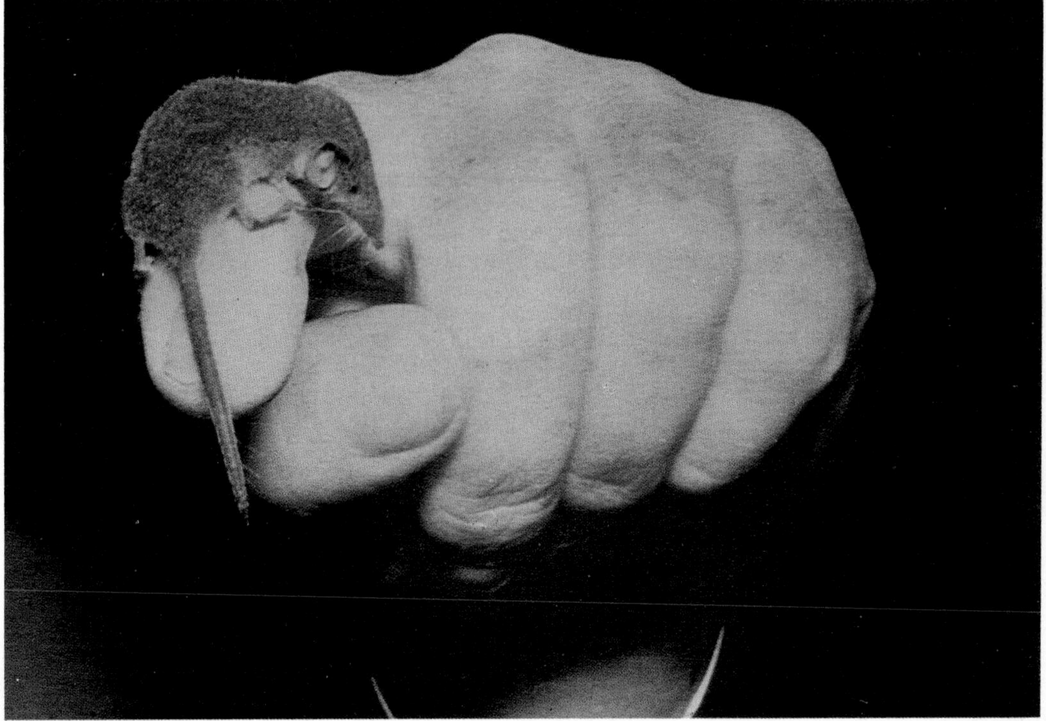

Savi's white-toothed pygmy shrew, the world's smallest non-flying mammal (Claus Konig)

0·052–0·09 oz (Van der Brink, 1955). Some idea of this creature's minute size can be gauged by the fact that it can *travel through tunnels left by large earth-worms*!

Its nearest rival for diminutiveness is the very rare Least shrew (*Sorex minutissimus*) which has been recorded twice in Finland and also in the Valdai Hills and the Province of Moscow, USSR (Siivonen, 1956). Mature specimens have a head and body length of 35–55 mm *1·37–2·08 in*, a tail length of 21–32 mm *0·82–1·2 in*, and weigh 1·5–4 g *0·052–0·141 oz*.

Another strong contender is the North American pygmy shrew (*Microsorex hoyi*). Mature specimens have a head and body length of 58–78 mm *2·28–3·07 in*, a tail length of 27–31 mm *1·06–1·22 in* and weigh 2·2–3·8 g *0·078–0·134 oz* (Walker *et al*, 1964).

The smallest land mammal found in the British Isles is the European pygmy shrew (*Sorex minutus*). Mature specimens have a head and body length of 43–64 mm *1·69–2·51 in*, a tail length of 31–46 mm *1·22–1·81 in* and weigh 2·4–6·1 g *0·084–0·213 oz* (Crowcroft, 1954; Shillito, 1960). This species has been recorded from the top of Ben Nevis (1343 m *4406 ft*), Britain's highest mountain.

Because of their high rate of activity shrews are prodigious eaters and will consume their own body-weight in food a day. Blossom (1932) claims the Masked shrew (*Sorex cinereus*) eats 3·3 times its own weight daily, but Crowcroft (1957) says 'this figure is so high that I suspect some food hoarding may have escaped notice'.

The smallest totally marine mammal is the protected Sea otter (*Enhydra lutris*), which is found in the coastal waters off California, western Alaska and the Komandorski and Kurile Islands. Two races are recognised: the Northern sea otter (*Enhydra l. lutris*) and the Southern sea otter (*Enhydra l. nereis*), but the distinction between them rests entirely on minor cranial differences (Merriam, 1904). Adult specimens – males are a bit larger and heavier than females – measure 1·20–1·56 m *3 ft 11¼ in–5 ft 1½ in* in total length and weigh 25–38·5 kg *55–85 lb*.

The rare Marine otter (*Lutra felina*), also called the 'Chingungo', which ranges along the entire coast of Chile south to Tierra del Fuego, and northwards to the coast of northern Peru, is much smaller, but it is not strictly marine because it ascends freshwater rivers in search of the prawns on which it feeds. An adult male taken at the southern end of Chiloe Island, 48 km *30 mile* west of the Chilean mainland, by a Field Museum of Natural History (Chicago) expedition in 1923 measured 910 mm *35·82 in* total length and weighed only 4·08 kg *9 lb* (Osgood, 1943).

The smallest marine mammal found in British waters is the Common porpoise (*Phocaena phocaena*). Adult males measure 1·67–2·13 m *5 ft 6 in–7 ft* in length and weigh 77–136 kg *170–300 lb*, and adult females 1·57–1·83 m *5 ft 3 in–6 ft* and 59–91 kg *130–200 lb*.

The smallest freshwater mammal is the European water shrew (*Neomys fodiens*) which has a head and body length of 72–96 mm *2·83–3·77 in*, a tail length of 47–77 mm *1·85–3·05 in* and weighs 10–23 g *0·35–0·81 oz* (Van der Brink, 1967).

The rarest land mammal found in Britain (excluding bats) is the Pine marten (*Martes martes*), which is now found only in the forest areas of NW Scotland (particularly Coille na Glas, Leitire and Ross and Cromarty), North Wales and – perhaps – the Lake District. This species was once very abundant, but later on it was heavily persecuted for its rich pelt. The last one to be killed near London was shot in Epping Forest in 1883. Another specimen was shot near Bristol in 1945, and one was trapped in Essex in 1950 (Hurrell, 1968). The largest specimens measure up to 86 cm *34 in* from nose to tip of tail (tail 152–228 mm *6–9 in*) and weigh up to 1·98 kg *4 lb 6 oz*.

The polecat (*Mustela putorius*), which was almost exterminated by the beginning of this century, is now recovering. The centre of abundance is central Wales, but its range now extends to the English border counties of Shropshire, Herefordshire and Cheshire. In 1937 a Caernarvonshire County Council roadworker complained that he had been trapped in his shelter for three hours by a ferocious 1·36 kg *3 lb* polecat which had come down from the mountains.

The rarest marine mammal is Longman's beaked whale (*Mesoplodon pacificus*), which is known only from two skulls. The type specimen was discovered on a beach near Mackay, Queensland, Australia, in 1926 (Longman, 1926), and the second skull near Mogadiscio, Somalia, East Africa, in 1955 (Azzaroli, 1968).

Until fairly recently Hose's Sarawak dolphin (*Lagendelphis hosei*), also called 'Fraser's dolphin', was known only from the type specimen collected at the mouth of the Lutong River,

Barama, Borneo, in 1895 in an advanced state of decomposition (Fraser, 1956). On 11 August 1966, however, a school of approximately 400 of these dolphins was sighted by an Australian cetologist NE of the Phoenix Islands, Central Pacific, and on 17 February 1971 about 25 were taken in the catch of the seiner MV *Larry Roe* about 800 km *500 miles* west of the Cocos Islands in the eastern Pacific. The species appears to be widely distributed in the tropical waters of the Pacific and Indian Oceans (Perring *et al*, 1973).

The fastest terrestrial animal over short distances (i.e. up to 400 m *437 yd*) is the Cheetah or 'Hunting Leopard' (*Acinonyx jubatus*) with a probable maximum velocity of 96 km/h *60 miles/h* over suitably level ground. Speeds up to 134 km/h *84 miles/h*, and even 129 km/h *90 miles/h*, have been claimed for this animal, but these figures must be considered exaggerated.

One individual chased by a car in Kenya recorded a speed of 82 km/h *51 miles/h* over a distance of 183 m *200 yd*, and Meinertzhagen (1955) says it was travelling flat out when it left the road.

The cheetah lives on the open plains and semi-arid savannahs of East Africa, Iran, Turkmenia and Afghanistan. It was also found in India until quite recently, but it became extinct in the wild state in the early 1950s. (The cheetahs used by Indian princes today for the hunt are imported from Kenya.)

In September 1937 eight cheetahs imported into England from Kenya by Mr K C Gandar Dower, the well-known animal-collector and writer, were matched singly against greyhounds in a series of races (after a mechanical hare) on the oval dog-track at Harringay, London. Unfortunately *cheetahs will never run flat out under domestic conditions*, and this was subsequently proved in the trials.

The fastest times were put up by a female cheetah named 'Helen' who recorded an average speed of 69·4 km/h *43·4 miles/h* (cf. 69·21 km/h *43·26 miles/h* for the fastest racehorse) over three 315 m *345 yd* runs, compared to 59 km/h *36·9 miles/h* for the fastest greyhound. This meant that a cheetah could give a greyhound a 18·28 m *20 yd* start over this distance and still win by about 4·57 m *5 yd*.

A few months earlier at Romford Stadium, Essex, it had been discovered that when (1) two cheetahs were raced against each other and one of them forged ahead the other would stop and refuse to finish the course; and (2) a cheetah running on its own would cover about half the course and then sit down waiting for the hare to come round again!

In 1960 Prof Milton Hildebrand of the Dept of Zoology, University of California, Los Angeles, carried out some studies on the locomotion of the cheetah by timing and photographing a specimen as it ran after a piece of meat attached to a cord pulled at speed by a *special* mechanism (an adapted bicycle wheel and crank!). The cheetah was a female named 'Ocala' who was used in an animal show at Ocala, Florida, and trained to run the length of a 59 m *65 yd* enclosure for rewards of food. The speed was determined for each of nine runs. The slowest velocity was 44·32 km/h *27·7 miles/h* the fastest 60 km/h *37·5 miles/h* and the average 52·8 km/h *33 miles/h*. 'Unfortunately', writes Prof Hildebrand, 'the cheetah never extended itself over any of the runs, and probably regarded the exercise as a game.'

If we are to believe Demmer (1966) the cheetah has been clocked at 114·6 km/h *71·6 miles/h* on a *racecourse*, but this figure is obviously based on the unsubstantiated claim by Bourlière (1954) that a cheetah had once been timed to cover more than 640 m *700 yd* in 20 s (= 114·54 km/h *71·59 miles/h*).

In 1964 Seraphino Antao, 26, said to be the fastest man in Africa, took part in a 91 m *100 yd* race against a three-year-old cheetah named 'Habash' in Kenya. At the half-way stage Antao led by 2·74 m *3 yd*, but eventually *lost the race by half a length*. The cheetah recorded a time of 9·1 s (= 36·08 km/h *22·55 miles/h*) and the Kenyan 9·2 s (= 35·68 km/h *22·30 miles/h*), which suggest the cat was only 'ambling'.

On 4 December 1971 a cheetah was observed chasing a hare in Serengeti National Park. After a hard pursuit both animals disappeared 'at speed' down a large pig hole and the pursuer was not seen again!

Incidentally, Prof Hildebrand has worked out that if the adult cheetah didn't have any legs it would still be able to propel its streamlined body along the ground at a speed of 8 km/h *5 miles/h* through sheer muscular contraction (cf. 6·4–8·0 km/h *4–5 miles/h* for a fast walking human).

The Caracal or 'Desert lynx' (*Caracal caracal*) of Africa, the Middle East and Central Asia is even swifter than the cheetah if we take size into consideration. Like the hunting leopard it is easily trained to hunt small deer, gazelle, hare, foxes and certain birds, and its agility is so extraordinary that it has been known to kill nine or ten of a flock of feeding pigeons before they could take to the air.

The fastest of all terrestrial animals over a sustained distance is the Pronghorn antelope (*Antilocapra americana*) of the western USA. On 10 October 1941 Mr Don Robins of the US Grazing Service paced four bucks for 6·4 km *4 miles* near the Tudor Ranch in Malheir County. The antelope had already run about 183 m *200 yd* before they came level with the car, and they travelled at a speed of 56 km/h *35 miles/h* for another 6·4 km *4 miles*. This was probably their *cruising* speed, because they showed no evidence of tiring (Einarsen, 1948).

Another small group of pronghorns which raced parallel with a car on a road near Rincon, New Mexico, in 1918 averaged 48 km/h *30 miles/h* for 11·2 km *7 miles* (Carr, 1927).

When it is hard-pressed, however, the pronghorn can attain speeds which compare favourably with those of the cheetah. On 14 August 1939 three agricultural scientists from Oregon University were driving across the dried bed of Spanish Lake in Lake County when they were challenged by a small group of pronghorns led by a magnificent buck. As they closed in from the right the buck took a lead of about 15·24 m *50 ft* and the men had to increase speed to keep up with the animal.

'The buck was now about 20 ft away and kept abreast of the car at 50 mph', writes Einarsen (1948). 'He gradually increased his gait, and with a tremendous burst of speed flattened out so that he appeared as lean and low as a greyhound. Then he turned towards us at about a 45 degree angle and disappeared in front of the car, to reappear on our left. He had gained enough to cross our course as the speedometer registered 61 mph. After the buck passed us he quickly slackened his pace, and when he reached a rounded knoll about 600 ft away he stood snorting, in graceful silhouette, against the sky as though enjoying the satisfaction of beating us in a fair race.'

If the car speedometer was *reliable* then this particular pronghorn must have been travelling at close to 120 km/h *75 miles/h* as it crossed in front of the vehicle, but this figure is so extreme that doubts must be cast on the accuracy of the report.

Other specimens have been observed to travel at 67 km/h *42 miles/h* for 1·6 km *1 mile* and 88 km/h *55 miles/h* for 0·8 km *½ mile*.

The fastest terrestrial wild mammal found in the British Isles is probably the Roe deer (*Capreolus capreolus*), which can cruise at 40–48 km/h *25–30 miles/h* for more than 32 km *20 miles*, with occasional bursts of up to 64 km/h *40 miles/h*. On 19 October 1970 a frightened runaway Red deer (*Cervus elephus*) registered a speed of 67 km/h *42 miles/h* on a police radar trap as it ran through a street in Stalybridge, Cheshire.

The fastest tree traveller in the mammalian world is the Fisher (*Martes pennanti*) of North America, which can exceed 16 km/h *10 miles/h* when pursuing squirrels.

The slowest moving terrestrial mammal is the Ai or Three-toed sloth (*Bradypus tridactylus*) of tropical America. The usual ground speed is 1·83–2·43 m/min *6–8 ft/min* (0·109–0·158 km/h *0·068–0·098 miles/min*), but one mother sloth speeded up by the calls of her infant was observed to cover 4·26 m *14 ft* in 1 min (0·249 km/h *0·155 miles/h*). In the trees this speed may be increased to 609 mm/s *2 ft/s* (2·19 km/h *1·36 miles/h*).

The longest-lived land mammal after man (up to 114 years) is the Asiatic elephant (*Elephas maximus*). The greatest age that has been verified with absolute certainty is 70 years in the case of a bull timber elephant named 'Kyaw Thee' (Tusker 1342), who died in the Taunggyi Forest Division, Southern Shan States, Burma in 1965. He was born and bred in captivity (Gale, 1974).

Like their heights and weights, exaggerated statements have also been made about the ages attained by elephants, and stories of individuals living 100, 150 and even 200 years have been published. These extreme claims, however, were largely based on the huge size and wrinkled skin of the elephant, for it was argued that the larger an animal was, the longer it could live. Take for example the case of the 'Napoleon elephant'. The Emperor was presented with the animal in Egypt in *c.* 1798 and brought it back to Paris with him. Later he presented it to his father-in-law, Francis, Emperor of Austria, who in turn sent the pachyderm to the Imperial Menagerie at Schönbrunn, Vienna. Here it turned dangerous, and eventually the beast was passed on to Budapest Zoological Gardens, Hungary, where it was said to be still living in 1930 at the age of about 150 years.

Fortunately Major Stanley S Flower, famous for his research on the longevity of animals, investigated this case thoroughly and discovered that the animal referred to was at the most 'only about 40'. Apparently 'Siam', as the elephant was known, arrived at the Imperial Menagerie at Schönbrunn on 11 November 1897 when a five-year-old calf and was moved from Vienna to the

The female Asiatic elephant 'Modoc', who may have been the oldest elephant on record (Ralph Helfer)

Budapest Zoo in 1900 (Flower, 1948). A pair of Asiatic elephants did in fact arrive at Schönbrunn in 1799, but the bull died in September 1810 aged approximately 17 years, and the cow died in July 1845 aged about 53 years (Fitzinger, 1863, cited by Flower).

Kyaw Thee's closest rival for age was probably a cow named 'Jessie' who was destroyed at Taronga Park Zoo, Sydney, Australia, on 26 September 1939 after developing abscesses on the soles of her feet which eventually made walking impossible (Le Souef, cited by Flower, 1947). She arrived at Moore Park Zoo, Sydney, in 1882 as a gift from the King of Siam (Thailand) and was transferred to Taronga Park Zoo just before the outbreak of the First World War. The exact age of Jessie on arrival in Australia was not recorded, but it was generally believed that she was about twelve years old. Patton (1940), on the other hand, thinks that she was about 20, in which case she would have been about 77 at the time of her death.

Mention should also be made of the magnificent Asiatic tusker 'Chandrasekharan' (see page 20), who died on 10 August 1940 aged 67 years. He was handed over by the Travancore Forest Dept to the Royal Elephant Stables on 15 August 1883 when aged about ten years. This elephant, incidentally, was renowned for his intelligence and gentleness, as demonstrated by Pillai (1941) in the following anecdote:

'On one occasion he refused to erect a pillar in one of the pits dug for the purpose in connection with one of the Murajapam festivals in Trivandrum. Usually the elephant was a very willing worker and his refusal to hoist the pillar, which he held still with his tusk and trunk, surprised the mahout who, on looking into the pit, found that a dog had strayed in and fallen asleep. It was only after the dog was rescued and driven away that Chandrasekharan lowered the pillar into the pit.'

On 17 July 1975 a former circus elephant named 'Modoc' died at Santa Clara, California, USA, reputedly aged 78 years, but this record has not yet been fully authenticated. According to one source the Asiatic cow arrived in the USA in 1898 and was purchased by the Barnum & Bailey Circus (later to become the Ringling Brothers-Barnum & Bailey Circus) six years later. By 1930 she was the largest elephant in the famous Ringling herd, measuring 2·38 m *7 ft 10 in* at the shoulder and weighing 5163 kg *9180 lb* (Benedict, 1936). In 1959 she came into the hands of Ralph Helfer, President and Head Trainer of Creative Animal Techniques of Saugus, California, after languishing at a roadside zoo in Tennessee for 20 years, and appeared in a number of television series including *Daktari* (wearing false ears!) before being semi-retired in 1968. Shortly before her death Mr Helfer (pers. comm. 23 February 1975) admitted to the compiler that he had no documentation to prove her date of birth, and that circus records were not always reliable, but he did say that a former trainer who had worked with the animal some 50 years previously when Modoc was already believed to be mature, had positively identified the animal in November 1974. This still doesn't explain, however, (1) why this matriarch's teeth were in excellent condition at the time of her demise, and (2) where she got the energy to rush about on film location at such an advanced age. One *possible* explanation is that two elephants were involved, and this is interesting because when Benedict measured this elephant he said her name was 'Big Modoc' (one of the elephants in the present Ringling herd is called 'Modoc'), but this still leaves the puzzle of the extended relationship between the pachyderm and her former trainer.

The greatest reliable age reported for an African bush elephant is *c.* 70 years for the outsized bull 'Dhlulamithi' (see page 16), who was probably fully mature (i.e. 25 years old) when Bvekenya Barnard tried to kill him in 1929.

The magnificent tusker 'Ahmed' of Marsabit Mountain Reserve, Kenya, who died in 1971, was believed to have been 50–55 years old, based on a count of seasonal and annual growth layers on the roots of the molars or grinding teeth, but this method of age allocation is only reliable up to 30 years.

Unlike man, a wild elephant's life-span is determined by the persistence of its last molar teeth. Once these are gone the animal is literally toothless, with the result that it cannot masticate its food and dies of starvation. Laws (1966) found that the absolute age limit based on molar material for a series of elephants shot in Uganda was 60 years; Sikes (1971), on the other hand, says there is *evidence* that elephants living in certain unrestricted parts of Africa sometimes live for 80 or even 100 years, but these readings are only approximate and therefore subjective.

The greatest reliable age reported for an African bush elephant in captivity is 'over 41 years' for a cow named 'Jumbina', received at the National Zoological Park in Washington, DC, USA, on 8 August 1913. She died there on 30

The Tibetan Yak, the world's highest living mammal (E.N.A.)

June 1952 after living in the zoo for 38 years 10 months 22 days. Her age on arrival was about three years (Mann, 1953).

The greatest reliable age reported for a forest elephant is 43+ years for a cow named 'Colonie' at the Gongala-na-Bodia training centre in the Congo, who was still alive in 1959. Another cow at the same centre named 'Bakela' was 40 at the time of her death (Bourliere & Verschuren, 1960).

The shortest-lived mammals are True shrews (family Soricidae). Some species have a life-span of less than two years, including the Common shrew (*Sorex araneus*) of Europe and Asia which usually dies before it is one year old (Crowcroft, 1956).

The highest living mammal is the Yak (*Bos grunniens*), the wild ox of Tibet and Szechwan. In 1899 Edgar Phelps shot a bull 35 km *22 miles* south of Horpu Cho at an altitude of 5486 m *18 500 ft*, but this animal may pass the 6096 m *20 000 ft* line when foraging.

The Bharal (*Pseudois nayaur*) and the Pika or Mouse hare (*Ochotona thibeta*) are also found above 5486 m *18 000 ft*, and the latter has been observed 'sunning' itself on a rock when the temperature was −17°C (= 65 degrees of frost!). There is also a reliable record of a Woolly hare (*Lepus oiostolus*) being seen at 6035 m *19 800 ft*.

The greatest altitude reported for a carnivore in the Himalayas is 5791 m *18 000 ft* for a Woolly wolf (*Canis lupus chanco*) and the Hill fox (*Vulpes vulpes montana*) and the Red Lynx (*Lynx lynx isabellinus*) have been sighted at 5638 m *18 500 ft* and 5486 m *18 000 ft* respectively (Napier, 1972). The Snow leopard (*Panthera uncia*) has allegedly been seen at an altitude of 5486 m *18 000 ft*, but the 'ceiling' for this species is probably nearer the 4267 m *14 000 ft* line.

In May 1954 two members of the *Daily Mail* Abominable Snowman Expedition to the Himalayas saw what appeared to be a Red bear (*Ursus arctos isabellinus*) at a height of 5486 m *18 000 ft* on the Reipimu Glacier and later found a fresh set of footprints. Most of the sightings and discovery of tracks of the so-called 'yeti' can be attributed to this creature – or bare-footed holy men (yogis) with congenital deformities of the feet!

Sir Harry Johnstone reports seeing African elephants at an altitude of 3962 m *13 000 ft*, and Abbot (1899) discovered tracks on Mt Kilimanjaro, Kenya, at 4572 m *15 000 ft*.

There is also a record of a frozen leopard carcass being found on the rim of the Kibo crater (5090 m *16 700 ft*) on the same mountain.

The largest herds ever recorded were those of the Springbok (*Antidorcas marsupialis*), also called the 'Springbuck', during migration across the high open plains of the western parts of southern Africa in the 19th century. These great *trekbokkens*, as the Boers called them, occurred at irregular intervals when overcrowding became acute and food and water supplies dwindled. The gazelles would join up to form a herd of almost inconceivable magnitude and then set off across the desert tracts and veld in search of fresh pastures. They moved in such densely packed masses that any unfortunate animals met on the way were either trampled to death or forced along with them, and the country was left completely devastated in their wake. Hardly any of the springbok survived the migration. Most of them died from starvation, drowning (attempting to cross the Orange River) or disease; others fell to predators (including man) or died from drinking salt water if they reached the sea.

Sometimes it would take days for the living wave of animals to pass a given point. In 1849, for instance, John Fraser (later Sir John Fraser) saw a *trekbokken* that took *three days* to pass through the settlement of Beaufort West, Cape Province.

According to an eyewitness account of the last great migration in July 1896 the enormous herd covered an area of 3472 km² *2170 miles²* (221 km *138 miles* long and 24 km *15 miles* wide) as it advanced towards Karree Kloof near the Orange River (Cronwright-Schreiner, 1925).

Fortunately these periodic movements are now a thing of the past. The migrations were so ruinous to crops and livestock that the farmers began destroying the springbok in great numbers, and this extermination, combined with outbreaks of rinderpest, reduced the enormous population drastically. Today springbok are found in reasonable abundance only in the Kalahari Desert (especially around Etosha Pan) and parts of SW Africa, and their numbers no longer threaten to disturb the delicate balance of nature.

The longest of all mammalian gestation periods is that of the Asiatic elephant (*Elephas maximus*) with an average of 608 days or just over 20 months.

Burne (1943) gives a record of a cow named 'Mai Mai', captured in the Mongpan Forest, Southern Shan States, Burma, who gave birth to a bull calf after a gestation period of 760 days – more than two and a half times that of a human. Another cow named 'Mai Myat' in the same stable gave birth to a bull calf after a gestation period of 743 days. The gestation period of the African bush elephant is reported to be 649–661 days (Lang, 1967).

The shortest of all mammalian gestation periods is probably that of the American opossum (*Didelphis marsupialis*), also called the 'Virginian opossum', which is normally 12/13 days, but may be as short as eight days (Reynolds, 1952). This species is born in a very immature state, and the young are immediately transferred to the ventral pouch, where they remain for ten weeks until embryonic development is completed. The gestation periods of the rare Water opossum or Yapok (*Chironectes minimus*) of central and northern South America (average 12/13 days) and the Eastern native cat (*Dasyurus quoll*) of Australia (average twelve days), may also be as short as eight days on occasion (Walker *et al*, 1964).

The shortest of all mammalian gestation periods in which the young are fully developed at birth is probably that of the Golden hamster (*Mesocricetus auratus*) of Syria with an average of 15/16 days. Gestations as low as 13 days have been reported for both the Common shrew (*Sorex araneus*) and the House mouse (*Mus musculus*), but the average periods are 16 and 17 days respectively (Burton, 1962).

The highest recorded number of young born to a wild mammal at a single birth is 32 (not all of which survived) in the case of the Common tenrec (*Centetes ecaudatus*) found in Malagasy and the Comoro Islands. The average litter size is 13–14.

The greatest number of litters produced by a mammal in a single year is 17 (average litter size 4–9) by the North American meadow mouse

The Common tenrec, producer of the largest litters among wild mammals (E Schuiling/F Lane)

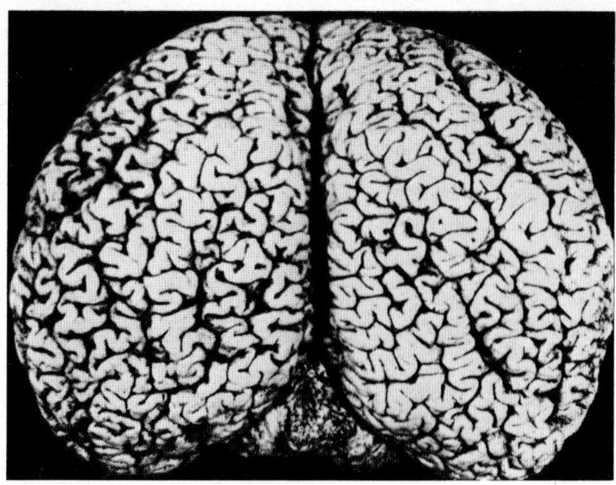

The massive brain of the Sperm whale $\frac{1}{6}$ nat. size (G. Pilleri)

(*Microtus pennsylvanicus*) (Burton, 1965). This animal is also the most prolific breeder in the mammalian world.

The fastest-developing mammal in the world is the Streaked tenrec (*Hemicentetes semispinosus*) of Malagasy. It is weaned after only five days, and females are capable of breeding three to four weeks after birth.

The heaviest mammalian brain is that of the Sperm whale (*Physeter catodon*). During the Japanese whaling expedition to the North Pacific in 1949/50 Dr Tokuzo Kojima (1951) of the Brain Institute, University of Tokyo, weighed the brains of 16 adult bulls brought aboard the factory ship *Nissin Marun No. 1*. The heaviest example, taken from a 49-footer *14·93 m*, weighed 9·2 kg *20 lb 4 oz* and the lightest 6·4 kg *14 lb 1½ oz* (from a very old individual). The average weight was 7·8 kg *17 lb 2½ oz*. Pilleri and Gihr (1971) say the maximum brain weight for the cachalot is about 10 kg *22 lb*.

The largest living terrestrial carnivore is the Kodiak bear (*Ursus arctos middendorffi*), which is found on Kodiak Island and adjacent Afognak and Shuyak Islands in the Gulf of Alaska, USA. The average adult male has a nose to tail length of 2·43 m *8 ft* (tail about 100 mm *4 in*), stands 1·23 m *52 in* at the shoulder and weighs between 476 kg *1050 lb* and 532 kg *1175 lb*. Females are about one-third smaller.

The greatest weight recorded for a Kodiak bear in the wild is 750 kg *1656 lb* for a male shot at English Bay, Kodiak Island, in 1894 by Mr J C Tolman, later Customs officer on Wrangel Island. The bear was killed by a single shot in the head from a Winchester rifle. The *stretched* skin (pegged to the side of a cabin on a frame and then weighed down with rocks at the bottom edge) measured 4·11 m *13 ft 6 in* from the tip of the nose to the root of the tail, and the hind-foot was 457 mm *18 in* long (Phillips-Woolley, 1894).

The largest Kodiak bear ever held in captivity was a male at Cheyenne Mountain Zoological Park, Colorado, USA, which scaled 757 kg *1670 lb* at the time of its death on 22 September 1955 (Crandall, 1964). Unfortunately nothing is known about the physical condition of this animal (or its nose to tail length) but it was probably 'cage-fat'. The bear was received at the zoo as a cub direct from Kodiak Island on 29 June 1940.

According to Couturier (1954) a weight of 1200 kg *2645 lb* was recorded for a Kodiak bear at Berlin Zoological Gardens, West Germany, in 1937, but this figure is incorrect. Dr Heinz-Georg Klos (pers. comm. 13 January 1968), the zoo's Director, said that the bear, killed in an air raid in 1943, scaled about 550 kg *1212 lb*.

'Sam' and 'Erskine', twin Kodiak bears at Chicago Zoological Park, Illinois, USA, were also very large. Sam, the *smaller* of the two – he could reach food suspended 3·20 m *10 ft 6 in* above the ground with his mouth! – recorded a posthumous weight of 640 kg *1412 lb* and Erskine weighed an estimated 726 kg *1600 lb* at his death (Robert Bean, pers. comm. 28 March 1974).

The largest Kodiak bear ever held captive in a British zoo was a male (one of twins) named 'Nick', born at Whipsnade Zoological Park on 29 January 1954. At the age of one year he weighed 270 kg *595 lb*, and at the time of his death in October 1963, an *estimated* 748 kg *1650 lb* (Tong, pers. comm. 27 March 1965). His father 'Kam' scaled 521 kg *1148 lb* at death.

Early one morning in May 1964 a rope-ladder and a bag of tools were discovered in the pit of three half-grown Kodiak bears at Griffith Park Zoo, Los Angeles, California. Worried officials immediately cleared the bears from their den, but no evidence of a human trespasser was found. The bears' expressions and actions also revealed nothing, although the keeper in charge of the Kodiaks later told a reporter that they all looked 'well fed'!

The Peninsula giant bear (*Ursus a. gyas*) of the Alaska Peninsula and other parts of the country has also been credited with the title of 'largest living land carnivore' by some authorities, but this race of brown bear is slightly smaller on the average than the Kodiak. Adult males measure about 2·36 m *7 ft 9 in* nose to tail and weigh about 499 kg *1100 lb*.

On 28 May 1948 Robert C Reeve of Anchorage shot an enormous specimen near Cold Bay which measured 3·04 m *10 ft* nose to tail (the skin weighed 88 kg *193 lb 8 oz*). Its weight was estimated at 725–771 kg *1600–1700 lb*, and judging from photographic evidence this was not exaggerated. Reeves said the bear had just come out of hibernation and carried little or no fat, and he believed that the animal would have weighed *c.* 839 kg *1850 lb* at the end of the summer (Couturier, 1954).

The largest Peninsula giant bear ever held in captivity was probably a male captured on 24 May 1901 near Douglas Settlement at the western entrance to Cook Inlet on the Alaska Peninsula and sent to the National Zoological Park, Washington, DC, where it was received on 9 January 1902. On 20 January 1911 it recorded a weight of 526 kg *1160 lb*, but Seton (1919) says the bear was heaviest about 1 December 1910 when it scaled an *estimated* 544 kg *1200 lb*. According to the same authority the diet of this bear was very carefully regulated, and at no time during its twelve-year stay at the zoo was it over-fat.

A large Peninsula giant bear (New York Zoological Society)

The only creatures that the large brown bears of Alaska have to fear are men – and perhaps the odd horse or so! Some years ago a surveying party tried to land a small number of horses near Geographic Bay on the Alaska mainland for transportation use. The animals were being towed in a barge, and as the shoreline came into sight a violent storm blew up. Somehow or other the two vessels became separated, with the result that the equine cargo was tipped into the rough sea. There was only one survivor, and this courageous beast managed to swim to the shore before collapsing.

During the next few months the stallion was often see on the open beach and passing fishermen would land there specially to feed the friendly animal. It must have helped considerably because three years later the steed, now a local celebrity, was still alive and flourishing, despite the fact that there were a number of giant brown bears in the area.

One day Bill Kvasnikoff, a professional hunter, was passing along this shoreline in his boat when he suddenly spotted two large brown bears making rapid tracks for the horse which, attracted by the noise of the outboard engine, was standing on the beach hoping for a titbit. The next moment he witnessed one of the most amazing battles in Nature.

'As the bears tore in, teeth grabbing and arms flailing', writes Clyde Ormond (1961), 'they were met with feet on each end of their adversary, and teeth almost equal to theirs. They were met also with a speed in whirling, ducking to knees and running . . . which baffled the bears and kept them off balance. Fur, hide and feet flew. Dust enveloped them all. The squeals of the furious nag mixed with the roars and bawls of the bears.'

The fight, which lasted about 5 min, eventually ended with the two bears running for their lives with the enraged stallion in hot pursuit, which suggests that the equine fury had been involved in battles with brown bears many times and had developed some expertise in this field.

Dr Sten Bergman (1936) of the State Museum of Natural History, Stockholm, Sweden, has described a giant variety of Kamchatka bear (*Ursus a. piscator*) from the southern part of the peninsula which he says exceeds even the Kodiak and Peninsula giant bears in size. He writes: 'In the autumn of 1920 I was shown in Ust-Kamchatsk a pelt which far surpassed in size every other bear-skin I have ever seen. It was perfectly black and short-haired. It is asserted generally by the hunters that the very largest bears always are quite black. Besides this they always are short-haired, in contrast to the animals of normal proportions, which in general are very long-haired. Malaise has told me that on one occasion he saw the skull of a gigantic bear of the black kind; its teeth were perfect, and hence it could not have been that of an aged individual. On another occasion he also measured and photographed a bear's foot-print that was 37 cm *14½ in* long and 25 cm *10 in* broad, so that the animal must have been a veritable giant. There is much, then, that speaks for the existence in Kamchatka of a quite black, gigantic bear, in addition to the ordinary brown type; but this question must remain an open one. . . .'

These claims are partially supported by Russian findings, although the maximum sizes recorded do not match up to the largest Kodiak and Peninsula giant bears. Baturin, who shot a great number of bears in Kamchatka, says his largest example weighed 40 poods 653 kg *1441 lb*, while Novikov (1969) mentions a small number of others which scaled between 500 kg *1102 lb* and 685 kg *1510 lb*.

This giant bear is now probably extinct, because no individuals remotely approaching this size have been killed in recent decades. In one series of 40 bears collected by Averin (1948) the largest individual weighed only 285 kg *628 lb*, and no bears killed in Kamchatka in the interim period have exceeded this poundage (Kistchinski, 1972).

Weights in excess of 453 kg *1000 lb* have also been reliably reported for the Grizzly bear (*Ursus a. horribilis*), although the average adult male weighs about 272 kg *600 lb* and stands about 1·06 m *42 in* at the shoulder. One huge male killed in the Okanogan Forest Reserve, Washington, USA, in early August 1924 'weighed over 1100 lb'. It was a notorious cattle-killer, and over a period of three years took nearly 50 head of cattle and more than 150 sheep. (*Annual Report Dept of Agriculture*, 1924.) Another grizzly killed in Idaho in the 19th century was sold to a butcher in Spokane, who claimed that he paid for 531 kg *1173 lb* of meat. This figure was placed on the carcass as it hung in front of his shop (Wright 1909).

In May 1920 archer Arthur Young of San Francisco killed a monstrous grizzly in the Yellowstone National Park, Wyoming, with a single arrow through the heart. Later Dr Saxton Pope, who accompanied Young on the trip, wrote: 'As we dismembered him, we weighed the parts. The veins were absolutely dry of blood, and without this substance, he was 916 lb. There was hardly an inch of fat on his back. At the end of the summer this adipose layer would be nearly 6 in thick. He would then have weighed over 1400 lb.'

If we make due allowance for over-enthusiasm, this bear probably scaled *c.* 521 kg *1150 lb* at its heaviest.

Although the largest brown bears are now found in Alaska, it has been claimed by some authorities (Lydekker, 1901; Hittell, 1926) that the grizzlies which formerly inhabited the high peaks of the Sierra Nevada Range were, in fact, the biggest of all, weights of 771 kg *1700 lb* and 816 kg *1800 lb* being regularly attained.

While these poundages must be considered somewhat excessive, it is interesting to note that *Rowland Ward* list a Nevada grizzly of 695 kg *1536 lb* in their 1907 edition. The bear, shot by W F Sheard in 1881, was stated to have measured 3.50 m *11 ft 6 in* in length (skin) and 3.09 m *10 ft 2 in* across the outstretched front paws.

In 1960 a race of king-sized grizzly bears long believed to be extinct was rediscovered 240 km *150 miles* north of Edmonton, Alberta, Canada. It was reported that 'average-sized males' measured 3·04 m *10 ft* from *nose tip to hind paw* and weighed about 454 kg *1000 lb*. Oil exploration teams surveying the little-known and inaccessible areas of northern Canada were the first to sight this super-bear in an area which has subsequently been named the 'Valley of the Giants'.

At least 400 of these outsized grizzlies live in this 12000 km² *8000 miles²* wilderness, and they are now protected by Government legislation.

The largest grizzly bear ever held in captivity was a male which lived for 18 years in Lincoln Park Menagerie, Chicago, Illinois. Shields (1899) says the bear 'was fed to suffocation by the thousands of visitors, and in his later years grew so fat that he could not walk, could only crawl around'. His weight was variously estimated at 816–997 kg *1800–2200 lb*, but he actually scaled 522 kg *1153 lb* at the time of his death. His nose to tail length was given as 2·35 m *7 ft 8½ in* and his shoulder height as 1·06 m *3 ft 4¾ in*.

'Grizzly Adams', alias James Capen Adams, an enterprising hunter and animal-collector who toured the country with his menagerie, claimed that his grizzly bear 'Sampson' (trapped in the Sierra Nevada Range in the winter of 1854/5) was 'the largest specimen of the Grizzly species perhaps that ever was taken alive . . .'. He said the bear had been weighed on a hay scale and tipped the beam at 545 kg *1510 lb*. Dr William T Hornaday, Director of the New York Zoological Society, who later saw the animal, estimated its weight at a surprisingly low 363 kg *800 lb* but this calculation may have been made in a moment of pique, because as stated earlier some exceptionally large grizzlies have been recorded from the Sierras.

Merriam (1918) lists seven full species and 15 sub-species of brown and grizzly bears in North America, but these identifications are largely based on variations in size and colours of pelts. The modern view is that they are all races of a single species, *Ursus arctos*.

The greatest reliable weight recorded for a European Brown bear (*Ursus a. arctos*) is 480 kg *1058 lb* for a 2·43 m *8 ft* long male shot in the Oural District, northern Russia (Kazeeff, 1878). The average adult male weighs 204–249 kg *450–550 lb*.

The Polar bear (*Ursus a. maritimus*), which is found in the Arctic regions of North America, Europe and Asia, is of necessity a more streamlined animal than the giant brown bears already mentioned and has a proportionately longer body, longer neck and less massive skull. The average adult male measures 2·36 m *7 ft 9 in* nose to tail (tail 76–203 mm *3–8 in*), stands 1·21 m *48 in* at the shoulder and weighs 385–408 kg *850–900 lb* (adult females are one-third smaller), but much bulkier individuals have been recorded in the same length category, and they may be members of a giant strain which, since the Second World War, has virtually been wiped out through over-shooting by wealthy North American *Sportsmen*.

Captain George F Lyon (1825), for instance, describes how members of his crew killed a large polar bear at the entrance to Hudson Strait, NW Territories, Canada, in July 1821 which measured 2·61 m *8 ft 7 in* nose to tip, 2·41 m *7 ft 11 in* in maximum girth and 1·45 m *4 ft 9 in* at the shoulder. 'On lifting him in, we were astonished to find that his weight exceeded 1600 pounds.'

Another huge male of similar proportions was shot by James Lamont (1876) in Deeve Bay, Spitzbergen. 'He was so large and heavy that we had to fix the ice-anchor, and drag him up with block and tackle as if he had been a walrus. This was an enormous old male bear and measured upwards of 8 ft in length, almost as much in circumference, and 4½ ft at the shoulder. . . . He was in very high condition, and produced nearly 400 pounds of fat; his skin weighed upwards of 100 pounds and the entire carcase of the animal cannot have been less than 1600 pounds.'

According to Perry (1966) the French-Canadian explorer and hunter Andrew Tremblay shot an enormous bear near Cape York, north Baffin Island, in the early part of this century which measured 3·35 m *11 ft* in length (skin), 1·37 m *4 ft 6 in* at the shoulder and weighed an estimated 816 kg *1800 lb*. The local Eskimos told him it was the largest bear they had ever seen. Perry also mentions a polar bear killed on Franz Josef Island by members of the Russian research station there in the winter of 1931 which measured 3·76 m *12 ft 4 in* in length and scaled between 589 kg *1300 lb* and 703 kg *1550 lb*, but the length quoted for this 'giant' must have been a skin measurement.

In October 1957 three Eskimo hunters returning to Point Barrow, northern Alaska, reported they had seen a polar bear measuring over 9·14 m *30 ft* (*sic*) roaming the ice-shelf along the Arctic sea coast. A few months later Tom Bolock, an American big-game hunter, shot a huge bear which could have been the same animal as it was making a beeline for Siberia. The animal measured just over 3·04 m *10 ft* in length and its weight was conservatively *estimated* at 816 kg *1800 lb*.

The greatest weight recorded for a polar bear in the wild is an incredible 1002 kg *2210 lb* for a white colossus shot by Arthur Dubs of Medford, Oregon, USA, at the polar entrance to Kotzebue Sound, NW Alaska, in 1960. In April 1962 the 3·39 m *11 ft 1½ in* tall mounted animal was put on display at the Seattle World Fair.

That it is possible for polar bears to exceed 907 kg *2000 lb* has also been confirmed by Ognev, who says that some of the old males killed on Spitzbergen have scaled as much as 60 poods 979 kg *2160 lb*, but even a weight of 725 kg *1600 lb* must be considered exceptional.

There are few records of captive polar bears exceeding 453 kg *1000 lb*. A large male which died at New York Zoological Gardens (Bronx Zoo) in 1960 scaled 467 kg *1030 lb* (Crandall, 1964). Another male named 'Harold' who died at Dudley Zoo in November 1965 reportedly measured 2·74 m *9 ft* nose to tip of tail and weighed ½ tonne/*ton* 508 kg *1120 lb*, but the compiler has not been able to confirm these figures. Even larger examples have probably been kept in Russian zoos, but no statistics are available.

On 21 February 1936 a posthumous weight of

The 1002 kg 2210 lb *Polar bear shot in Kotzebue Sound, Alaska, USA in 1960 (*Sports Illustrated*)*

526 kg *1160 lb* was reported for a male hybrid between a male polar bear and a female Kodiak bear at the National Zoological Park, Washington, DC, USA (Davis, 1950).

The polar bear is the only large land carnivore in the world which does not fear man – despite intensive hunting pressures. In fact, it will deliberately stalk and kill humans for food at every given opportunity, and when it is wounded this 'King of the North' is a deadly adversary. There is even one record of an old male covering more than 55 m *60 yd* in a charge despite the fact that it had already been *shot through the heart several times.*

On the other side of the coin, the polar bear also has an insatiable curiosity, and sometimes he can be quite a clown, as witness the following story. In 1969 a coastguard vessel in the Canadian Arctic received a visit from an adult male polar bear travelling atop a drifting ice-floe. The animal was obviously bent on a shopping expedition, and the crew obliged by throwing it a carton full of black molasses which the bear soon spread all over itself and the ice. This was followed by some jam, salt pork, two salami sausages, an apple which it spat out in disgust, and a jar of peanut butter which disappeared in about two seconds flat. It refused to touch bread or potatoes, but loved chocolate bars. Eventually the food supply ran out but the 363 kg *800 lb* bear, its appetite now thoroughly whetted, decided to investigate further and stuck its head through one of the port-holes in search of further nourishment. When nothing turned up it decided to climb aboard, much to the alarm of the crew, who decided to open up the hoses on it. This was a big mistake, however, because the bear absolutely loved the drenching and raised its paws in the air to get the jet of water under its armpits! In the end the coastguards were forced to fire a distress rocket rather close to the interloper before it reluctantly moved away.

The polar bear is also the most intelligent member of the bruin family. In the wild state it has been known to brain a sleeping walrus with a heavy lump of ice rather than risk an open battle, and it is a well-known fact that it will *cover its prominent black nose with a paw* when edging towards a snoozing seal on the ice.

In January 1976 a female of this sub-species decided to make a break for freedom at Lincoln Park Zoo, Chicago, Illinois, USA, and came up with a rather ingenious method of escape. The sub-zero temperature had apparently frozen the waterfall in her enclosure, and by dint of hard work and perseverence she had managed to haul her 272 kg *600 lb* body up the ice bridge which had been formed. The bear's freedom, however, was short-lived because 91 m *100 yd* away she was met by a keeper with a tranquilliser gun.

In captivity bears are long-lived animals. In September 1973 the death was reported of a female polar bear named 'Furba' at Frankfurt Zoo, West Germany, aged 35 years. Another polar bear received at Dublin Zoo, Ireland, in 1900 was also believed to be *c.* 35 years old at the time of its death in 1928. There is also a record of a Peninsula giant bear living 36 years 10 months 6 days and a Kodiak bear for 34 years 13 days, and Crandall (1964) says the director of the Druid Hill Park Zoo in Baltimore, Maryland, USA, told him that a grizzly bear had lived there for 33 years 8 months 7 days.

The greatest age recorded for a bear is 47 years for a European brown bear (*Ursus a. arctos*) at the Nordiska Museet och Skansen, Stockholm, Sweden (Kai Curry-Lindahl, quoted by Crandall). A figure of 47 years was also claimed for another European brown bear which died in the Menagerie du Jardin des Plantes, Paris, France, but Flower (1931) has queried the authenticity of this record.

The rarest species of bear is the Giant panda (*Ailuropoda melanoleuca*) which, until recently, was believed to be a member of the racoon family (Procyonidae). In 1973, however, Dr Vincent M Sarich, a biochemist at the University of California, Berkeley, declared that the animal was really a bear after studying blood and tissue samples taken from the famous 'Chi-Chi' after her death at London Zoo, in July 1972. He also revealed that the giant panda was more closely related to the American black bear (*Ursus a. americanus*) than it is to either the Panda (*Ailurus fulgens*) or the Raccoon (*Procyon lotor*). Further research is necessary, however, before it can be proved conclusively – the giant panda does not hibernate like the bear – and it has been suggested that the creature should be put in a separate family of its own.

The giant panda is found only in the mountains of eastern Tibet and the Szechwan Range, SW China, and only the inscrutable Chinese know how many are living in the wild state. It may be several thousands, but *c.* 600 is probably more realistic.

The largest land carnivore found in Britain is the Badger (*Meles meles*). The average adult boar measures 900 mm *3 ft* in length including a 100 mm *4 in* tail and weighs 12·3 kg *27 lb* in the early spring and 14·5 kg *32 lb* at the end of the summer when it is in 'grease'. Sows are a few pounds lighter.

One huge boar shot in Denbighshire tipped the scales at 19 kg *42 lb* (Davis, 1936), and Swayne (1908) mentions a 43-pounder *19·5 kg* killed in Hereford. Another outsized specimen taken at Ampleforth, North Yorkshire, on 10 February 1942 reportedly weighed 31 kg *68 lb*, but the compiler has not been able to authenticate this extreme poundage. Mortimer Batten (1923) claims that the heaviest badgers are found in the Scottish Highlands and says *estimated* weights up to 22·5 kg *50 lb* have been reliably reported.

The Common otter (*Lutra lutra*) is also a strong contender. In one series of twelve adult specimens the average weights for dogs and bitches worked out at 10·3 kg *23 lb* and 7·4 kg *16 lb* respectively (Stephens, 1957), but much larger individuals have been recorded. One dog killed by the Eastern Counties Otterhounds near Ispwich on 10 July 1907 scaled 15·4 kg *34 lb* (Clapham, 1922), and another one caught by the Darlington and Hurworth Otterhounds in 1840 weighed 15·9 kg *35 lb*. (This specimen is now on display at the Railway Tavern, Shincliffe, Co. Durham.) Daniel mentions a dog otter caught in the River Lea, Essex, in October 1794 which scaled 'upward of 40 pounds', and Millais (1905) cites a weight of 22·6 kg *50 lb* for another one killed in Carmarthenshire which measured 1·67 m *5 ft 6 in* in total length. The same poundage was also reported for an otter taken in Hampshire, but Stephens (1955) found that this animal only scaled 7·60 kg *16 lb 13 oz* without its pelt and feet. In 1873 an otter weighing '53 pounds and a few odd ounces' was caught in the River Avon near Ringwood, Hampshire (Corbin, 1873). In December 1952 a Scottish newspaper reported that a dog otter weighing 'about 60 pounds' had been killed in Caithness, but this figure probably owed something to journalistic licence.

The largest toothed mammal ever recorded is the Sperm whale (*Physeter catodon*), which has a world-wide distribution. The average adult bull measures 14·32 m *47 ft* in length and weighs 33 tonne/*ton*, and the average adult cow 10·05 m *33 ft* and 9·5 tonne/*ton*.

The largest accurately measured sperm whale on record was a 20·7 m *67 ft 11 in* bull captured near the Kurile Islands in the NW Pacific by the Third Soviet whaling fleet in 1950. The whale was not weighed but it must have scaled – in good condition – at least 83·5 tonne *82 ton*. Two other bulls captured (1) off the coast of Japan in 1946 and (2) off Kamchatka, NE USSR in 1964, both measured 19·8 m *65 ft* (*International Whaling Statistics*, 1945–66).

The largest sperm whale ever recorded in the Southern Hemisphere was a 19·5 m *64 ft* bull caught near South Georgia during the 1948/9 whaling season.

In the early days of whaling lengths up to 25·60 m *84 ft* and estimated weights up to and exceeding 102 tonne *100 ton* were claimed for bull sperm whales killed in the South Pacific (Beale, 1839), but as no individuals approaching anything like this size have been measured in the interim period by qualified biologists, these figures must be considered unreliable. (The bull sperm whales found in the Northern Hemisphere are generally larger than their southern counterparts.)

On 25 June 1903 a bull sperm whale reportedly measuring 20·72 m *68 ft* long was killed 96 km *60 miles* west of the Shetland Islands and landed at the Narrona whaling station (Millais, 1906), but this measurement may have been taken over the curve of the body instead of in a straight line.

In 1921 a length of 21·64 m *71 ft* was claimed for a sperm whale caught off the Aleutian coast and towed to the Akutan whaling station in Alaska (Young, pers. comm. 29 April 1965). This measurement, too, was probably taken over the curve of the body, because Mr Sidney Brown of the Whale Research Unit has calculated from the photographic evidence and data he obtained personally on the lower jaw/total length ratios of four sperm whales caught in Icelandic waters in 1973 that the cachalot in question had a 'straight' length of *c.* 18·59 m *61 ft* (pers. comm. 10 January 1974).

Fifteen sperm whales have been stranded on British coasts since 1913. The largest was probably a 18·72 m *61 ft 5 in* bull washed ashore at Birchington, Kent, on 18 October 1914 (Harmer, 1927). Another bull measuring *c.* 18·28 m *60 ft* stranded at North Roe, Scotland, on 30 May 1958. A huge sperm whale which stranded at Ferryloughan, Co. Galway, Ireland, on 2 January 1952 was credited with a length of 19·81 m *65 ft*, but Fraser (1974) says the body was so badly decomposed that there may have been some length extension.

The largest accurately measured cow sperm whale on record was a 12·2 m *40 ft 3 in* specimen caught in the Southern Ocean in 1940/41 (*International Whaling Statistics*, 1940/41). Another cow measured by Robert Clarke (1956) at the whaling station on Horta in the Azores was exactly 12·19 m *40 ft*, and the same length was also recorded for a cow caught off Kamchatka (Tomilin, 1936). In 1947 a 16·8 m *55 ft* female was reportedly caught off Norway, but this was a wrongly identified bull.

The heaviest animal ever weighed whole was a 13·34 m *43 ft 9 in* bull sperm whale caught in the Indian Ocean on 8 September 1969 and brought back to Durban Harbour. The following day the cachalot was loaded on to a 12·80 m *42 ft* long flat railway car, secured by chains and then shunted to the weighbridge at the whaling station (Union Whaling Co Ltd) some 4·8 km *3 miles* away. After deducting the known weight of the railway car, the whale scaled 31 433 kg *69 300 lb* or 30·93 tonne/*ton*.

'No account was taken of the weight of the chains (estimated at 60 lb) used to steady the carcase', writes Dr Ray Gambell (1970), 'since a nearly equivalent weight must have been lost from the whale when the tips of the tail flukes were cut off at sea as is usually done to facilitate towing; in any case, the total weight was correct to the nearest 100 lb.' (The grenade explosion caused only minor internal damage.)

The carcass was later weighed piecemeal on a 10 tonne/*ton* scale and the total weight in parts came to 27619 kg *60 890 lb* 27·18 tonne/*ton*; this means the loss of blood and other body fluids must have accounted for 3815 kg *8410 lb* 3·75 tonne/*ton* (12·13 per cent of the total weight).

The dental armament of the adult bull sperm whale is extremely formidable. The mandibular teeth can exceed 160 mm *6·29 in* in vertical height, and the compiler owns a specimen taken from a sperm whale killed in South African waters which has a vertical height of 203 mm *8 in* (245 mm *9·66 in* along the outside curve). The largest example in the Zoological Museum of the Academy of Sciences, Moscow, USSR, measures 205 mm *8·07 in* (Tomilin, 1967) and Berzin (1972) quotes a height of 230 mm *9·05 in* for another Russian ivory. There is also a record of a very large sperm whale caught by the British barque *Adam* off the Galapagos Islands in 1817 which yielded a tooth measuring 241 mm *9½ in* (weight 1·36 kg *3 lb*).

Several years ago an American collector offered the Union Whaling Co Ltd in Durban $5000 for a mandibular tooth measuring *over* 254 mm *10 in* in vertical height, but the money was not claimed. Andrew Hermansen, the platform foreman, whose whaling experience dates from 1926, owns a specimen measuring exactly 254 mm *10 in* and he told Mr J D MacDonald, the company's chemist (pers. comm. 29 May 1973) that he had *given away* two others which both measured 279 mm *11 in*. Unfortunately the compiler has not been able to track down the whereabouts of these latter ivories, so the measurements must remain unsubstantiated.

Probably the largest mandibular teeth on record are a pair in the collection of the New

The heaviest animal ever weighed whole – a 31 tonne/ton bull Sperm whale (National Institute of Oceanography)

Bedford Whaling Museum, Massachusetts, USA, which both measure 279 mm *11 in* in vertical height and together weigh 3·82 kg *8 lb 7 oz*. They were cut out of a sperm whale allegedly measuring over 27·43 m *90 ft* taken by the American barque *Desdemona* off the River Plate (between Uruguay and Argentina) in the late 1870s (Ashley, 1926).

Dr Robert Clarke informed Mr Sidney Brown of the Whale Research Unit of the National Institute of Oceanography (pers. comm. 9 March 1976) that he had once seen a tooth which measured 250 mm *9·84 in* in vertical height and weighed 1·84 kg *4 lb 1¼ oz*, but this specimen is not mentioned in his published reports on the hunting and biology of the sperm whale in the Azores.

Yablokov (1958) was told by Russian shore workers that they had cut out teeth measuring up to 500 mm *19·68 in* in length (*sic*), but the largest he knew of had a more reasonable vertical height of 270 mm *10·63 in*.

Finally, it should be pointed out that the size of a tooth does not necessarily equate with the length of the animal. Mr MacDonald said he had seen very respectable ivories taken from an elderly 14·63 m *48 ft* bull, while a 16·45 m *54 ft* individual had only average teeth for its size.

The sperm whale is the only cetacean that could swallow a man whole, and no discussion of this subject can be made without reference to the celebrated case of James Bartley, an English sailor, who was reportedly swallowed by a huge sperm whale and lived to tell the tale.

According to the widely publicised story – taken originally from the French publication *Journal des Debats*, 14 March 1898 – on the afternoon of 25 August 1891 the whaling vessel *Star of the East* was in the vicinity of the Falkland Islands, South Atlantic, when the lookout spotted a large sperm whale about 4·8 km *3 miles* away. Two longboats were launched and eventually one of the harpooners managed to lance-bomb the giant. The stricken whale then seized the offending boat in its massive jaws and crushed it in two, hurling the occupants into the sea. The men were quickly picked up by their comrades as the leviathan moved off, but the steersman James Bartley could not be found and was given up for dead.

A couple of hours later the cachalot was killed and brought alongside the ship and the crew started busying themselves with axes and spades removing the blubber. They worked for the rest of the day and part of the night. The following morning the stomach section was hoisted on deck, and when the whalers cut through the muscular walls they discovered the missing sailor inside – unconscious but still alive. His unprotected face, neck and hands had been bleached to a deathly whiteness by the action of the gastric juices and he was smeared with the whale's blood.

Bartley regained partial consciousness after being given a sea-water bath and brandy to revive him and was put in the Captain's quarters, where he remained for two weeks in a semi-delirious state. After that he rapidly improved, and by the end of the third week he was able to make a statement to the Captain. He told him:

'I remember very well from the moment that I fell from the boat and felt my feet strike some soft substance. I looked up and saw a big-ribbed canopy of light pink and white descending over me, and the next moment I felt myself drawn downward, feet first, and I realised that I was being swallowed by a whale. I was drawn lower and lower; a wall of flesh surrounded me and hemmed me in on every side, yet the pressure was not painful and the flesh easily gave way like soft india-rubber before my slightest movement.

'Suddenly I found myself in a sack much larger than my body, but completely dark. I felt about me; and my hands came in contact with several fishes, some of which seemed to be still alive, for they squirmed in my fingers, and slipped back to my feet. Soon I felt a great pain in my head and my breathing became more and more difficult. At the same time I felt a terrible heat; it seemed to consume me, growing hotter and hotter. My eyes became coals of fire in my head, and I believed every moment that I was condemned to perish in the belly of a whale. It tormented me beyond all endurance, while at the same time the awful silence of the terrible prison weighed me down. I tried to rise, to move my arms and legs, to cry out. All action was now impossible, but my brain seemed abnormally clear; and with a full comprehension of my awful fate, I finally lost all consciousness.'

This latter-day Jonah who was about 35 years of age at the time and of robust build, reportedly made a full recovery, and the only lasting effect of his terrible experience apart from his bleached skin seems to have been a recurrent nightmare in which he relived his sensations in the whale's stomach.

It is certainly an incredible piece of narrative if authentic; unfortunately, however, the whole thing was a complete fabrication. It was Dr Robert Cushman Murphy, a leading American zoologist and a man very familiar with whaling in Antarctic waters, who shot the story down in flames.

Writing in the April 1947 issue of *Natural History*, the journal of the American Museum of Natural History, about the feasibility of such an event taking place, he said: 'A man might well be swallowed by a sperm whale; indeed since sperm whales have been known to swallow seals, it is likely that they have engulfed sailors during some of the many fracases between the whales and Yankee whale-boats manned by six whalemen. Plenty of such individuals disappeared after the wrecking of their frail craft, and it is not to be assumed that they all merely sank. But if a whaler was swallowed by a sperm whale without being previously killed, he would hardly survive any longer inside the whale than if he had dived under water.'

This opinion was later corroborated by Dr Egerton Y Davis of Boston, Massachusetts, who as a young surgeon attached to a sealing fleet, witnessed a less fortunate accident in Newfoundland waters in February/March 1893/4. In a letter published in the same journal two months later (June 1947) he said:

'We sailed on the schooner *Toulinguet*, one of a considerable fleet of wooden ships bent on the winter's take of seal pups. One of the lads in another ship had the misfortune, in full view of his comrades, to become isolated from the others on an ice pan, from which he fell into the icy waters in the proximity of a huge sperm-whale. The whale was apparently as lost and out of season in those Arctic waters as he was confused and angered by the sudden appearance of a fleet of ships and men.

'Somehow the poor fellow was swallowed by the whale, which then made straight for one of the smaller sealers. A lucky shot from a small cannon mounted on her stern mortally wounded the huge mammal and served to change his course, though he travelled a full three miles out to sea before his final death thrashing.

'The next day he was found belly-up by one of the longboats as it was searching for seal; and although it was impossible under those conditions to bring him in, the men, by a valiant effort and many hours of hard labour, were able to hack their way through his abdomen below the diaphragm and isolate his huge gas-filled "upper stomach" which apparently contained their comrade. This was severed with some difficulty at the cardia and in the first portion of the duodenum. They brought it to me for inspection and also for preservation of the man's body, as it was hoped he could be returned to his native Argentina [Newfoundland] for burial.

'At first I attempted the dissection with my scalpel, but quickly gave it up in favour of the sharpest galley-knives. The stomach was finally opened and gave off an overpowering stench. A fearsome sight met our eyes. The young man had apparently been badly crushed in the region of his chest, which may have been enough to kill him outright (in any event, examination of his lungs revealed a general atalectasia with marked haemorrhage throughout).

'The most striking finds were external, however; the whale's gastric mucosa had encased his body (particularly the exposed parts) like the foot of a huge snail. His face, hands, and one of his legs, where a trouser leg had been pulled up or torn, were badly macerated and partially digested. It was my opinion that he had no consciousness of what happened to him. Curiously enough some lice on his head appeared still to be alive.

'The appearance and odour were so bad that all save I were forced to turn away, and we were obliged to consign him to the briny deep – the last resting place of many a good sealer – rather than carry him back to his rocky homeland.'

The smallest living carnivore is the Least weasel (*Mustela rixosa*), also called the 'Dwarf weasel', which is circumpolar in distribution. Four races are recognised (Hall, 1951), the smallest of which is *M. r. pygmaea* of Siberia. Mature specimens have a head and body length of 158–184 mm *6·22–7·24 in*, a tail length of 19–23 mm *0·74–0·90 in* and weigh between 35 and 70 g *1·25 and 2·5 oz*.

The least weasel probably packs more nervous energy into its tiny body than any other living mammal.

Nelson (1930) writes: 'Once when camping in spring among scattered snowbanks on the coast of the Bering Sea, I had an excellent opportunity to witness their almost incredible quickness. Early in the morning one suddenly appeared on the margin of a snowbank within a few feet. After craning its neck one way and the other as though to get a better view of me, it vanished, only to reappear so abruptly on a snowbank three or four yards away that it was almost impossible to follow it with the eye . . . certainly no other mammal can have such flashlike powers of movement.'

In Europe this family is represented by the larger Common weasel (*Mustela nivalis*), although a pygmy form has been found in Finland and eastern Europe which rivals the least weasel for diminutiveness. This latter animal has a head and body length of 130–195 mm *5·11–7·67 in* and a tail length of 28·52 mm *1·10–2·04 in* (Van den Brink, 1967).

The largest member of the cat family is the now very rare long-furred Siberian tiger (*Panthera tigris altaica*) also called the 'Amur' or 'Manchurian' tiger. Adult males average 3·12 m *10 ft 3 in* in length (nose to tip of extended tail), stand 99–107 cm *39–42 in* at the shoulder and weigh about 265 kg *585 lb*. Adult females are four-fifths this size.

The maximum size attained by this northern race of tiger is still a matter of some dispute, but if we are to believe the records of the old Russian hunters some of the tigers killed before the advent of 'over-shooting' were very much larger than those that exist today. George Yankovsky, for example, says he killed one with the aid of dogs in the T'u-men-Tzu region between NE Korea and SE Manchuria which measured 3·96 m *13 ft* in length (Taylor, 1956), while Barclay (1915) mentions another tiger killed near Vladivostock which was 4·09 m *13 ft 5 in* long. Both these measurements, however, were taken 'over the curves' or sportsman fashion, which means that the tape was stretched round the curves of the body – from the nose following a line between the ears and along the spine.

The *correct* way to measure a tiger is to lay the

The Least weasel – the smallest living carnivore (Ernest P Walker)

animal on its back and take a reading between pegs placed at the point of the nose and tip of the extended tail, the result being some 10–12 per cent less than the 'over the curve' measurement. If Yankovsky had used this method on his larger tiger it would have pegged *c.* 3·50 m *11 ft 6 in*, but even this length is totally unimaginable for a creature of such symmetrical build.

According to Baikov (1936) the man-eating tiger known as 'Great Van' which he shot in the Khailinkhe Forest, SE Manchuria, measured nearly 4 m *13 ft 1½ in* over the curves, but a published photograph showed a tiger of quite ordinary proportions. Another tiger shot by the same hunter in Kirin Province, central Manchuria, measured 3·60 m *11 ft 9¾ in* over the curves (*c.* 3·25 m *10 ft 8 in* between pegs), and weighed 254 kg *560 lb*.

There is also the question of skin measurements. Rowland Ward (1928) give details of five Mongolian/Manchurian dressed skins, the largest measuring 4·11 m *13 ft 6 in*. Cavendish (1894) speaks of a 4·41 m *14 ft 6 in* skin from Korea, and Burton (1928) says he saw skins of immense size at the Nijni Novgorod Fair in 1893. Such measurements, however, are quite valueless and bear no relationship to the true length of the tiger. (The skin of a 3·04 m *10 ft* tiger can be *stretched* to 3·94–4·26 m *13–14 ft*).

The tiger's length is also governed by its tail measurement. On an average this appendage is less than half the length of the head and body, but measurements up to 1·21 m *4 ft* have been recorded.

According to Perry (1964) all the Russian hunters, including Yankovsky and Baikov, claimed that there were two races of tiger, one large and one 'small' living in Manchuria, which would explain the wide variation in reported sizes. But although the larger-race Manchurian tigers are heavier and more powerful than their Indian counterparts, it would appear there is very little to choose between them in terms of total length.

Very little information has been published on the weights attained by large Siberian tigers. Filipek (1934) quotes a weight of 350 kg *771 lb* for a huge individual killed near the Amur River, and Baikov (1936) shot a tiger in Manchuria which tipped the scales at 325 kg *716 lb*. There is also another record of a tiger killed in central Manchuria which weighed 320 kg *705 lb* (Novikov, 1962).

During his visit to India in 1955 the Russian Premier, Nikita Kruschev, presented an adult pair of tigers captured in Ussuria, eastern Manchuria, to the Governor of West Bengal. Perry says the male was estimated to measure 3·20 m *10 ft 6 in* in length and weigh 317 kg *700 lb*, and its mate 3·04 m *10 ft* and 286 kg *630 lb*, but although photographic evidence confirms the great size of this pair, the male is much heavier than the female.

Guggisberg (1975) mentions an old Siberian tiger in Prague Zoo, Czechoslovakia, which weighed 192 kg *423 lb* at the time of its death, but probably weighed 250–260 kg *551–573 lb* when it was in its prime. This specimen measured 3·19 m *10 ft 5½ in* in total length.

Gary Smart of Windsor Safari Park, who went on a tour of Chinese zoos in February 1974, later told the compiler (pers. comm. 8 July 1974) that when he visited Shanghai Zoo he was shown a Korean–Chinese tiger (*Panthera tigris coreensis*) which weighed an astonishing 368 kg *813 lb*. The zoo's Director also said there was another captive specimen in the same area which was even larger! Although the Siberian and Korean–Chinese tigers are described as different subspecies this is solely on a geographical basis and physically they are practically identical.

Not surprisingly, lengths in excess of 3·65 m *12 ft* have also been reported for the Indian tiger (*Panthera tigris tigris*). Buffon (1778) speaks of one that measured 4·57 m *15 ft* and Blythe (1856) says another enormous tiger presented to the Nawob of Arcot measured an incredible 5·48 m *18 ft* (*sic*).

Another tiger killed on Cozzimbazar Island, Bengal, reportedly measured '13 ft and a few inches from the tip of his nose to the end of his tail' (Williams, 1807), and General W Rice (1857) states that the largest tigers shot by him and his friends in Rajputana and central India between 1850 and 1854 measured 3·84 m *12 ft 7½ in*, 3·83 m *12 ft 6¾ in* and 3·71 m *12 ft 2 in* respectively.

All these measurements, however, were (1) exaggerated; (2) taken from stretched skins; or (3) based on faulty tape readings, and in earlier times there appears to have been a direct relationship between the recorded size of a tiger and the hunter's social stature or position. Perry also points out that it was a common practice for shikaris to carry with them 'special' steel measures on which every inch *25 mm* was reduced by one-sixth. This meant that every 3·04 m *10 ft* tiger automatically measured out at 3·65 m *12 ft*, thus guaranteeing a handsome reward from a belted Earl or a high-ranking military man.

Col F T Pollock (1903) says he shot a heavy, loose-skinned tiger in southern India which

The 388 kg 857 lb *tiger shot in Uttar Pradesh, northern India in 1967 (David J Hasinger)*

measured 3·07 m *10 ft 1 in* between pegs when shot and yielded a 4·04 m *13 ft 4 in* long dried skin, and there is another record of an Indian tiger 25 mm *1 in* shorter yielding a 4·26 m *14 ft* skin when pegged out and dried.

Sir John Prescott Hewett (1938), who had records of 250 tigers shot in India (mostly in the swampy Terai), lists the largest specimen as a 3·19 m *10 ft 5½ in* male shot at Naini Tal, United Provinces, in 1893. Only seven of the others measured over 3·04 m *10ft*.

In reality the average adult male Indian tiger measures 2·82 m *9 ft 3 in* in total length, stands 91–96 cm *36–38 in* at the shoulder and weighs about 190 kg *420 lb*.

The longest accurately measured tiger on record was a 3·22 m *10 ft 7 in* specimen (tail 1·09 m *3 ft 7 in*) shot by Col Evans Gordon at Ramshai Hab, in the Duars, Bengal. It measured 101 cm *40 in* at the shoulder and weighed a surprisingly low 223 kg *491 lb* (Rowland Ward, 1907). Another huge male shot at Bhandai Pari, Central Provinces, by V A Herbert measured 3·21 m *10 ft 6¾ in* between pegs (Rowland Ward, 1907).

Until fairly recently the heaviest Indian tiger on record was a giant male shot in Nepal which scaled a colossal 320 kg *705 lb* (Smythies, 1942). Another tiger (length 3·05 m *9 ft 11½ in*) killed by Captain M D Goring-Jones in Central Provinces scaled exactly 317 kg *700 lb* (Rowland Ward, 1910).

Since then, however, Khan Seheb Jamshed Butt (1963) has claimed that tigers weighing up to 363 kg *800 lb* have been killed in India, and he was subsequently proved right in November 1967 when David J Hasinger of Jackson, Tennessee, USA, shot an enormous individual in northern Uttar Pradesh some 80 km *50 miles* south of the Chinese border. This tiger, killed by a single 300 gr soft-point bullet from a ·375 H & H Magnum, measured a record 3·22 m *10 ft 7 in* between pegs (3·38 m *11 ft 1 in* over the curves) and tipped the scales at an astonishing 388 kg *857 lb* (Hasinger, pers. comm. 3 July 1974). In 1969 the tiger was put on display in the rotunda of the United States Museum of Natural History, Smithsonian Institution, Washington, DC.

The enormous poundage was later confirmed by Dr Henry W Setzer, Curator of Mammals at the US Museum (NH), who told the compiler that the animal had been weighed on a scale at a sugar plantation (pers. comm. 25 March 1975); this avoirdupois, however, is somewhat deceptive inasmuch as the tiger had killed a buffalo the previous evening and had probably eaten heavily. George B Schaller (1967) says the amount of food ingested by a tiger at one sitting can total as much as 20 per cent of its bodyweight, which means a very hungry 317 kg *700 lb* tiger could – in theory – dispose of 63 kg *140 lb* of meat. On the other hand, it also proves that Hasinger's monstrous beast would have scaled at least 324 kg *715 lb* – even with an empty stomach!

In 1941 an enormously fat hermaphrodite tiger/tigress measuring exactly 3·04 m *10 ft* between pegs was shot in the Nilgiri Hills, Madras Province (Fraser 1942). The animal was not weighed, but it must have scaled at least 317 kg *700 lb*.

E H Morbey shot an exceptionally large male (length 3·20 m *10 ft 6 in*) in Kumaon, United Provinces, which weighed 292 kg *645 lb*, and a 608-pounder *276 kg* (length 3·10 m *10 ft 2 in*) was killed by the Kumar of Bikaner in Gwalior (Rowland Ward, 1928).

All of these exceptionally heavy tigers were confirmed cattle-killers, and consequently very bulky.

Although the African lion (*Panthera leo*) is smaller than the Siberian tiger, it rivals the Indian tiger for size. The average adult male measures 2·74 m *9 ft* in length, stands 91–96 cm *36–38 in* at the shoulder and weighs about 181 kg *400 lb*. Adult females are one-fifth smaller.

The longest accurately measured African lion on record was a 3·33 m *10 ft 11 in* black-maned giant shot by G Prud'homme in Uganda (Rowland Ward, 1969). Another lion shot by J K Roberts in the Sudan also measured 3·33 m *10 ft 11 in* (Rowland Ward, 1969). Both these measurements, however, must be considered freakish because the lion is on average a shorter-bodied animal than the tiger.

Edouard Foa (1899) shot a lion on the Zambesi which he said measured 3·57 m *11 ft 8¾ in*, and Wells (1933) cites a length of 3·81 m *12 ft 6 in* for another lion killed on the Klaserie River in northern Transvaal, but these figures were skin measurements.

Unlike the tiger, very few lions exceed a weight of 227 kg *500 lb* in the wild state. Col Richard Meinertzhagen (1938) shot a 2·84 m *9 ft 4 in* individual in Kenya which weighed 229 kg *506 lb*, and another male measuring 3·01 m *9 ft 10½ in* killed in the same country by Rear-Admiral R Montgomerie tipped the scales at 235 kg *516 lb* (Rowland Ward, 1928). Roberts (1951) mentions a 2·99 m *9 ft 10 in* lion shot in the Sabi District, Transvaal, South Africa, which scaled 251 kg *553 lb* and a weight of 264 kg *585 lb* was reported for a lion (length 2·87 m *9 ft 5 in*) taken in the Orange River Colony, South Africa, in 1865. Another specimen shot by White (1912) was said to have measured 3·01 m *9 ft 11 in* in length and scaled just under 272 kg *600 lb*.

The heaviest wild African lion on record was a man-eater shot just outside Hectorspruit, eastern Transvaal, South Africa, in 1936 by Lennox Anderson, which scaled 313 kg *690 lb*. This weight was so extreme – taken on the local railway scale – that it was checked by several people before being officially accepted (Campbell, 1937).

The protected Asiatic or Indian lion (*Panthera leo persica*), which is now confined to the Gir Forest in the SW of Kathiawar Peninsula, western India, and the Chandraprabha Sanctuary in Uttar Pradesh, United Provinces (Fitter, 1968), is about the same size as its African cousin, although a little stockier in build. A 2·68 m *8 ft 9½ in* male shot by Captain Smeel weighed 222 kg *490 lb* excluding entrails (Sterndale, 1884).

In *c*. 1620 Emperor Jehangir reportedly speared a very large lion in the neighbourhood of Rahimabad, western India, which measured 3½ cubits 2 tassu (= *c*. 3·12 m *10 ft 3 in*) in length and weighed 8½ Jehangiri maunds (= *c*. 308 kg *680 lb*) (Ali, 1927).

The largest lion ever held in captivity was a 3·20 m *10 ft 6 in* long black-maned giant named 'Simba' who weighed 375 kg *826 lb* in July 1970. Other statistics included a shoulder height of 112 cm *44 in* and a 132 cm *52 in* neck. Simba's owner, professional wild-animal trainer Adrian Nyoka, purchased the lion as a six-month-old cub from Dublin Zoo in October 1958. Up to the age of three years the animal grew along normal lines. Then suddenly, for some unknown reason, he started growing at an alarming rate. At four and a half years he tipped the scales at 227 kg *500 lb*, and at six years he was 281 kg *620 lb*. Two years later he weighed in at 319 kg *704 lb*, and by 1969 his weight had increased to 343 kg *756 lb*. Simba died on 16 January 1973 and his stuffed body is now on display at Knaresborough Zoo, Yorkshire.

In February 1973 a weight of 408 kg *900 lb* was reported for a very obese lion named 'Ali' owned by Mr Charles Mason of Kingsbridge, Devon, but this poundage was estimated. In January 1976 the diet-reduced 13-year-old animal was sold to an unnamed buyer, but died shortly afterwards.

In 1953 a weight of 340 kg *750 lb* was recorded for an 18-year-old male 'liger' (a lion-tigress hybrid) living in Bloemfontein Zoological Gardens, South Africa (Crandall, 1964).

The largest feline found in Britain (excluding the domestic variety) is the Wild cat (*Felis silvestris*) which is now confined to the remoter parts of Scotland. Specimens measuring up to 1·29 m *51 in* from nose to tip of tail and weighing up to 7·07 kg *15 lb 10 oz* have been officially recorded.

The smallest member of the cat family (Felidae) is the Rusty-spotted cat (*Felis rubiginosa*). Two races are recognised: *Felis r. rubiginosa* of southern India and *Felis r. phillipsi* of Sri Lanka. The average adult male has a head and body length of 406–457 m *16–18 in*, a tail length of 228–254 mm *9–10 in* and weighs about 1·36 kg *3 lb*, while adult females are slightly smaller (cf. 4·53 kg *10 lb* for an adult domestic cat).

According to Guggisberg (1975) the smallest wild cat in the world is the Black-footed cat (*Felis nigripes*) of southern Africa, but although adult males are slightly shorter in over-all length (575–700 mm *22·63–27·55 in*) than *F. rubiginosa*, specimens have been weighed up to 2·75 kg *6·06 lb*.

The rarest large member of the cat family is the Javan tiger (*Panthera tigris sondaica*). Up to the early 1930s this tiger was still fairly abundant, but in the years that followed excessive killing by man virtually wiped out the population. By 1955 there were only an estimated 20–25 tigers left in Java, of which 10–12 were in the Udjung Kulon Reserve at the western tip of the island (Talbot, 1959). During the 1960s the tigers in Udjung Kulon and Baluran reserves were exterminated, and by 1972 there were only three to seven in the Betiri Forest Reserve, and another five in the Sukamati Reserve, SE Java. The following year villagers in the central area reported sighting one tiger, but when two Indonesian zoologists searched central and western Java in 1974 they could find no trace of the beast. Since then the remains of animals killed by predators have been found in NW Java, but there is no definite proof that tigers were involved. (The leopard is also found on this island.)

The Bali tiger (*Panthera tigris balica*), which is similar in appearance to the Javan tiger but smaller, was also fairly common at the turn of this century, but indiscriminate hunting eradicated the population even faster than that of the Javan tiger. The last individual killed in the wild was shot at Sumbar Kima, West Bali, on 27 September 1937, and by 1963 there were only three to four left in a reserve in western Bali. Since then, however, there have been no sightings and Curry-Lindahl (1972) says it is almost certain that this race of tiger is now extinct.

Despite legal protection there are now only about 130 Siberian tigers left in the Soviet Far East today, and most of these live in the Khabarovsk and Primorye territories. There may also be another 40–50 in Korea, 50 in northern China and possibly a few in Mongolia.

In 1973 there were *c.* 200 Asiatic lions left in the Gir Forest.

The greatest reliable age recorded for a 'big cat' under zoo conditions is *c.* 29 years for a lion named 'Nero' which died in Cologne Zoological Gardens, West Germany, in May 1907 or May 1908 (Flower, 1931).

Another lion named 'Brutus' owned by Mrs Joan Temple of Boreham, Kenya, died at the Orphanage in Nairobi National Park in March 1972 aged 24 years.

The famous lion 'Pompey' was reputedly 70 years old when he died in the Tower of London menagerie in 1760, and his replacement was said to have lived for 63 years (Nott, 1886), but both these figures belong in the fantasy category. Lions rarely live longer than 20 years in the wild state, and Guggisberg (1961) gives the normal expectation of life as 12–14 years.

According to Russian claims the Siberian tiger lives up to 50 years in the wild, but this figure is not supported by zoo records. A Siberian tigress which died in Cologne Zoological Gardens in October 1930 from *extreme senile debility* was only 19 years old. Perry puts the maximum life potential in the wild at *c.* 25 years, but he says few individuals live longer than 15 years.

R Morris says the famous 'Chendragiri maneater' was at least 20 years old at the time of its presumed death in 1929, having terrorised the neighbourhood since 1912 when it must have been about four years of age. The 'Bargur maneater', which roamed the Ramapuram-Bargur Ranges (Kollegal Division, Coimbatore District) for 15–16 years, was also at least 20 years old when it was finally shot in 1928.

The greatest reliable age recorded for a tiger under zoo conditions is 22 years 1 month 4 days for a female Sumatran-Bengal cross named 'Sandra' at Racine Zoological Park, Wisconsin, USA. At the time of her death on 17 June 1968 she weighed an incredibly low 63 kg *138 lb*, compared to 204 kg *530 lb* in her prime (Christensen, pers. comm. 17 June 1968).

The previous record-holder was a Sumatra tiger (*Panthera tigris sumatrae*) named 'Slome' who died in Rotterdam Zoological Gardens in the Netherlands on 5 January 1966 aged 21 years. He was captured in Deli, Sumatra, and arrived at Rotterdam Zoo in 1948 when his age was estimated to be not less than three years.

Guggisberg (1975) says one Siberian tiger which died at the age of 26 years 'showed clear signs of advanced senility, and was found to have suffered from debility of the heart and lungs', but he does not quote the source of his information.

On 31 December 1954 a tigress at Wellington Zoological Gardens, New Zealand, allegedly celebrated her 36th birthday (*International Zoo News*, No. 1, January 1955), but the Manager, K Kuiper, later told the compiler (pers. comm. 3 February 1973) that there was no truth whatsoever in this claim.

In July 1972 the death was reported of 'Shasta', a liger at Hogle Zoological Garden, Salt Lake City, Utah, USA, aged 24 years. She was born there on 6 May 1948, the product of an African lion and a Bengal tigress.

The only true man-eaters of the cat family are the tiger, lion, leopard and jaguar which, between them, probably account for close on 1000 deaths annually.

The individual man-eating record is held by a notorious tigress known as the 'Champawat man-eater' which operated first in Nepal and then in Kumaon. She killed an incredible *438 people in eight years* before being shot by Jim Corbett in 1911.

In January 1966 a man-eating tiger said to have killed and eaten 500 people in six years was shot at Ramgiri Udaygiri by Mrs Alida Sverdsten of Idaho, USA, but Clark (1969) believes this figure was grossly exaggerated.

In 1869 a tigress reportedly killed 129 people in the Sunderbunds, a great swampy region in the Ganges Delta (Nott, 1886).

Man-eating lions prefer to do their killing in prides rather than singly, but there is a record from Malawi of a rogue lion which killed 14 people in the space of one month (Guggisberg, 1963). Another man-eater shot in the Numgari District of Portuguese East Africa in September 1938 was responsible for the deaths of 22 natives in eight weeks. There is also a reliable record of a lion killing 40 people in the Kasama District of northern Rhodesia before it was shot in October 1943.

The Leopard (*Panthera pardus*) has an even more unsavoury reputation. The infamous man-eater of Panar, for instance, accounted for 400 victims before it was shot by Corbett in 1910, and as recently as October 1972 a leopard attacked and killed three boys aged four, seven and twelve respectively within the space of eight hours in villages near Junagadh, western India.

Sometimes, however, the boot is on the other foot. . . .

In November 1934 a black panther (leopard) escaped from Zürich Zoological Gardens, Switzerland, and immediately ran to ground. A few days before Christmas a peasant who lived in a remote woodland village some miles from Zürich discovered the animal in a barn and promptly despatched it with a wood axe. He said later he had no idea what the creature was, and could think of nothing better than *to cook and eat it*!

The Jaguar (*Panthera onca*), which ranges from the SW USA through Central and South America to Patagonia, is more a 'man-killer' than a 'man-eater', although Sasha Siemel (1952) makes several references to man-eating jaguars in Brazil. Maw (1829) says another man-eater in Peru accounted for nearly 50 victims before it was shot, but this statement was based on hearsay information.

The greatest number of live cubs produced by a 'big cat' in a single litter is seven by a lioness named 'Maire' (b 1956) at Dublin Zoo in February 1964. She has also had at least three other litters of six live cubs by the same breeding male 'Cormac' (b 1959). (Terence Murphy, pers. comm. 15 November 1972). Earlier another lioness in the same zoo had given birth to four living cubs and was later found to have retained three dead ones (Steyn, 1951).

In 1971 a pride of eleven young lionesses was introduced to a game park at Laguna near Long Beach, California, USA. To establish the group and keep the ladies happy a young, virile lion was brought in. Unfortunately they did not take to him at all, and the next morning keepers found the badly mauled animal licking its wounds. For the next five nights the keepers introduced a different young lion – and each morning the result was the same.

Finally, in desperation, they called upon the services of an arthritic 17-year-old lion named 'Frasier'. He was unimpressive to say the least, being short of breath, teeth, strength and just about everything else. In fact he was so decrepit that when his tongue lolled out he didn't have

'Simba', the largest lion ever held in captivity, with his owner Adrian Nyoka (J Arthur Dixon Ltd)

The lioness 'Maire', who produced a record litter of seven live cubs in 1964 (Royal Zoological Society of Ireland)

enough muscular power to pull it back in again, and if he took two solitary steps by himself he would collapse in an exhausted heap . . . but when it came to loving this moth-eaten Casanova had no peer, despite his infirmities.

As soon as he was put in the enclosure the lionesses started pestering him for his favours, obviously anxious not to let the rheumy-eyed pensioner waste a lifetime of experience, and during the next 18 months Frasier kept his harem almost constantly pregnant. In return they chewed his meat for him, fed him and licked his coat clean, and when he wanted to walk a lioness would position herself each side of him and act as supports. They also protected him from the other lions, which were obviously jealous of his romantic prowess. When the love-king, whose name by now was a household word in the USA, eventually died in September 1972 he had fathered 35 cubs.

His finest tribute, however, came after his demise when an enterprising whisky company announced a new drink called 'The Frasier' – a combination of whisky and passion fruit.

The greatest jumper (horizontally or vertically) among the Felidae is the Puma (*Felis concolor*) of the Americas. Barnes (1960) writes: '. . . probably no other cat in the world can equal it [the puma] in ease and resilience of spring. It can vault over the highest fences, bound safely down from incredibly high cliffs, and even plunge, as if it were some mammoth flying squirrel, into a labyrinth of tree tops; for, not being so heavy as either the African lion or the Indian tiger, nor so stocky as the leopard, it can excel them all in the grace and distance of its magnificent leaps. . . . I have . . . seen it jump from a branch to the earth 50 or 60 feet below and light on its feet apparently unhurt.'

Merriam (1884) mentions a puma which made a measured leap of 18·28 m *60 ft* from a ledge of rocks *c.* 6·09 m *20 ft* above the ground, and the same writer also quotes an instance of a 'measured leap over snow of nearly 40 ft'. Tinsley (1920) says he once saw a puma bound into a tree fork 3·65 m *12 ft* from the ground with the carcass of a *deer* in its mouth, and Barnes cites a height of 5·48 m *18 ft* for another puma which was not encumbered in any way.

The leaping powers of the Snow leopard (*Panthera uncia*) of Central Asia, which is very similar to the puma in shape and size, are also of comparable ability. 'The Siberian leopard', writes Ognev (1962), 'makes unbelievable leaps and I would never have accepted stories about them if I had not seen them myself. I have seen a snow leopard leap not less than 15 m *49 ft 2½ in* uphill over a ditch.'

Turnbull-Kemp (1967) once saw an African leopard leap into the fork of a tree 5·48 m *18 ft* above the ground, and on another occasion he witnessed a frightened leopard leap 9·14 m *30 ft* out from a tree.

Despite their much greater size the jumping abilities of the tiger and lion are also impressive.

According to Crandall (1964) the water moat surrounding the tiger enclosure at Detroit Zoological Park, Michigan, USA, is 7·62 m *25 ft* wide and 4·87 m *16 ft* deep, and that for the lions 6·40 m *21 ft* wide and 3·65 m *12 ft* deep, which suggests that the tiger is the superior jumper of the two. This is also confirmed by Perry (1964) who says a healthy adult tiger can leap a 5·79 m *19 ft* wide gulley from a level surface without even gathering itself for a spring.

Yankovsky once saw a Siberian tiger chased by dogs clear a 9·14 m *30 ft* wide ditch in a downhill leap, and there is also a record of a 11·88 m *39 ft* downhill leap by a lion, but both these measurements are completely valueless because they involve other complexities like the steepness of the slope and the condition of the ground.

The largest of the 32 known species of pinniped is the Southern elephant seal (*Mirounga leonina*), which inhabits the sub-Antarctic islands. Adult bulls average 5 m *16 ft 6 in* in length (tip of inflated snout to the extremities of the outstretched tail flippers), 3·65 m *12 ft* in maximum bodily girth and weigh about 2267 kg *5000 lb* 2·18 tonne/*ton*. Adult cows are much smaller, averaging 2·74–3·04 m *9–10 ft* in total length, 1·83 m *6 ft* in maximum bodily girth and weighing *c.* 453 kg *1000 lb*.

The largest accurately measured southern elephant seal on record was probably a bull killed in Possession Bay, South Georgia, on 28 February 1913 and examined by Dr Robert C Murphy (1914) of the American Museum of Natural History, New York. It measured 6·50 m *21 ft 4 in* after flensing (original length *c.* 6·85 m *22 ft 6 in*), and must have weighed in the region of 4 tonne/*ton*. The girth of this animal was not taken, but the maximum circumference must have been about 5·48 m *18 ft*.

According to Murphy the fattest bull seen by him at South Georgia was a specimen measuring 5·58 m *18 ft 4 in* in length. 'It was so round and distended that it had the appearance of being pneumatic and inflated under pressure. Seven men could barely turn its body over with the aid of ropes and hand holes in the blubber, even after half the blubber had been removed, and a trench had been scooped under one side of its carcase.'

Another exceptionally large bull shot by Herbert Mansel 72 km *45 miles* west of the Falkland Islands in 1879 measured just over 6·40 m *21 ft* in length and 'must have weighed several tons'. The skeleton of this giant is preserved in the Museum of the Royal College of Surgeons in London (Flower, 1881).

Of 226 elephant seals shot at South Georgia and examined by Laws (1953) the largest bull measured 5·51 m *18 ft 1 in*, but this measurement was taken over the curve of the back. Another bull measuring in excess of 6·09 m *20 ft* was observed, but it was not killed.

There are old records of bulls measuring 7·62 m *25 ft*, 9·14 m *30 ft* and even 10·05 m *33 ft*, but these figures must be considered exaggerated.

According to Hamilton (1949) a 4·06 m *13 ft 4 in* bull weighed in pieces by Messrs Christian Salvesen and Co at Leith Harbour, South Georgia, tipped the scales at 1976 kg *4357 lb* 1·94 tonne/*ton*. The skin weighed 115 kg *254 lb*, the blubber 667 kg *1469 lb*, the heart 42 kg *92 lb*, the head 52 kg *114 lb* and the blood (estimated) 99 kg *218 lb*. 'It was a medium-sized bull', writes Hamilton, 'and in the absence of further data it is reasonable to believe that a large bull may weigh anything up to four or five tons.'

In 1924 a large southern elephant seal hauled up on the beach at Simonstown, Cape Province, South Africa, and caused a great deal of panic among the local population before it was shot. It measured 7·25 m *16 ft* in length and weighed just over 2 tonne/*ton*.

The largest accurately measured cow southern elephant seal on record was an 3·48 m *11 ft 5 in* specimen obtained by Harris in the Falkland Islands in 1909. The weight of this animal was not recorded, but it probably scaled about 771 kg *1700 lb*. Another cow shot in South Georgia in *c.* 1952 measured 3·50 m *11 ft 6 in*, but the length was taken along the curve of the back (Laws, 1953).

Most of the cows measured by Murphy at South Georgia were under 2·59 m *8 ft 6 in*.

The much rarer Northern elephant seal (*Mirounga angustirostris*), now restricted to the islands off the Pacific coast of Mexico and southern California, is slightly smaller than *M. leonina*, adult bulls *now* measuring 4·26–5·18 m *14–17 ft*, but the proboscis in this species is much larger. In former times before they were commercially exploited the northern beachmasters reached much greater size. One bull examined by Captain Charles M Scammon (1870) measured 6·70 m *22 ft* over the curve, which means it must have had a straight-line length of *c.* 6·40 m *21 ft*.

The largest accurately measured cow of the northern race on record was probably an 3·55 m *11 ft* example collected by Townsend (1912) on Santa Barbara Island, California. Another one measuring 3·48 m *11 ft 5 in* was reportedly killed in the same area in *c.* 1909 (Scheffer, 1958), but this may have been a curve length for the same specimen.

The largest elephant seal ever held in captivity was a bull of the southern race named 'Goliath' (one of several of that name) received at Carl Hagenbeck's Tierpark, Hamburg-Stellingen, Germany, in 1928 from South Georgia, who measured 6·24 m *20 ft 6 in* in length and weighed over 3 tonne/*ton* at the time of his death in 1930 (Hagenbeck, pers. comm. 20 May 1971). This colossus died from severe lacerations after a sadistic member of the public had thrown the jagged neck of a broken beer-bottle into his open jaws (Hediger, 1969).

In 1968 a length of 6·09 m *20 ft* and a weight of 4 tonne/*ton* were reported for another southern bull named 'Roland' at West Berlin Zoo (received 1962), but Dr Heinz-Georg Klos (pers. comm. 18 September 1970) said that this seal measured *c.* 5 m *16 ft 5 in* in length and weighed 1540 kg *3395 lb* at the time of his death on 21 August 1969. In 1966 the same animal weighed 2199 kg *4850 lb*, and his mate 'Bollie' 708 kg *1562 lb*.

In 1970 a posthumous weight of 2851 kg *6287 lb* 2·80 tonne/*ton* excluding blood was recorded for a 4·47 m *14 ft 8 in* long southern bull called 'Spot' at Edinburgh Zoo. As the blood usually accounts for about 8 per cent of the total body-weight in pinnipeds this heavyweight, who measured 4·34 m *14 ft 3 in* in circumference, must have scaled about 3084 kg *6800 lb* 3·03 tonne/*ton* when alive (Rushton, pers. comm. 8 April 1971).

The largest elephant seal living in captivity today is another bull of the southern race called 'Daikichi' at Enoshima Marineland, Fujisawa, Kanagawa Prefecture, Japan. In September 1972 this individual measured 5·3 m *17 ft 4½ in* in length and weighed 2267 kg *5000 lb* 2·23 tonne/*ton* (Tetsunosuke, pers. comm. 17 June 1973), but since then he has grown to *c.* 5·48 m *18 ft* and 2948 kg *6500 lb* 2·90 tonne/*ton*. The bull arrived at the marineland in April 1964 from South Georgia.

The largest northern elephant seal ever held in captivity was a bull received at San Diego Zoo, California, USA, in 1929 which measured 5·02 m *16 ft 6 in* in length and weighed 'nearly five thousand pounds' (Benchley, 1930).

Apart from being the largest member of the Pinnipedia, the elephant seal also has the longest intestinal tract of any living mammal, although the reasons for this are obscure. One 4·80 m *15 ft 9 in* southern bull had a gut measuring 202 m *662 ft* or 41½ times its own body length (cf. 9·14 m *30 ft* for a man, 24 m *80 ft* for a horse and 76 m *250 ft* for a large sperm whale).

And, if that's not enough, the elephant seal also has the most flexible spine in the mammalian world. In this case, however, *M. leonina* is the more supple of the two and can bend itself far back over the vertical angle into a remarkable U or even V shape. This fact is often demonstrated by zoo-keepers to amuse visitors. A fish is placed on the animal's back near its hind-flippers, and the huge creature then bends over backwards to seize its reward.

The northern elephant seal, on the other hand, cannot bend back much over the vertical.

The largest pinniped among british fauna is the Grey seal (*Halichoerus grypus*), also called the 'Atlantic seal'. It is found mainly on the western coasts of Britain, but the main centre of abundance is the Farne Islands off the coast of Northumberland. Adult bulls average 2·43 m *8 ft* in length and weigh about 227 kg *500 lb* and adult cows 1·98 m *6 ft 6 in* and 159 kg *350 lb* (lactating cows can be 45 kg *100 lb* heavier).

According to Hickling (1962) one of the largest grey seals killed by Blacket on the Farne Islands in *c.* 1772 measured 2·7m *9 ft* in length, 2·28 m *7 ft 6 in* in maximum girth and weighed 298 kg *658 lb*. Edmondston (1837) says the largest bull collected by him on Shetland Island measured 2·43 m *8 ft* in length and weighed 305 kg *672 lb*, but as the maximum girth of this creature was only 1·83 m *6 ft* this poundage must be considered suspect. The largest of the 27 adult bulls measured by Millais (1906) was 2·89 m *9 ft 6 in*. He also mentions another bull shot by Sir Reginald Cathcart on South Uist, Outer Hebrides, which scaled 317 kg *700 lb*. Unconfirmed lengths up to 3·96 m *13 ft* have been claimed.

In November 1973 a giant grey seal said to measure over 4·57 m *15 ft* in length (?) was reported to be coming ashore regularly on the beach at Kingsdown between Deal and Dover where children fed it with mackerel. It was thought to be living on the Goodwin Sands.

Adult cows have been reliably measured up to 2·31 m *7 ft 7 in* and 249 kg *550 lb*.

The smallest pinnipeds are the Baikal seal (*Pusa sibirica*) of Lake Baikal, a large freshwater basin in southern Siberia, USSR, and the Ringed seal (*Pusa hispida*) of the circumpolar Arctic coasts. Both animals have their supporters, and in reality there is very little to choose between them in terms of size.

According to King (1964) adult examples (both sexes) of *P. sibirica* grow to a length of *c.* 1·37 m *4 ft 6 in* and a weight of about 63 kg *140 lb*, but Koshov (1963) says the maximum size reached by this species is 1·65 m *5 ft 5 in* and 130 kg *286 lb*.

The ringed seal has been credited with a length of 1·47 m *4 ft 10 in* and a weight of 91 kg *200 lb* (King, 1964), but specimens have been recorded up to 1·67 m *5 ft 6 in* and 113 kg *250 lb*.

Although these two pinnipeds are classified as different species, it has been suggested that *P. sibirica* may be a land-locked race of ringed seal because of the similarities in size and appearance.

Some species of seal are among the swiftest mammals in the sea.

The fastest-swimming speed recorded for a pinniped is 40 km/h *25 miles/h* for a Californian sea lion (*Zalophus californianus*) (MacGinitie & MacGinitie, 1949). This velocity may be matched by the Leopard seal (*Hydrurga leptonyx*) of the sub-Antarctic islands which chases and frequently catches penguins (maximum speed *c.* 37 km/h *23 miles/h*. The Alaska fur seal (*Callorhinus ursinus*) has been timed at 24 km/h *15 miles/h* for short bursts (Caldwell & Caldwell, 1972).

The deepest-diving pinniped is the Weddell seal (*Laptonychotes weddelli*), the world's most southerly mammal, which is found along the Antarctic mainland and neighbouring islands. Adult bulls regularly descend to 274–305 m *900–1000 ft* in search of food like fish, squid and crustaceans, and dives can last from 20 min to an hour or more.

In March 1966 an elderly bull with a depth-gauge attached to it recorded a dive of 600 m *1968 ft 6 in* in McMurdoe Sound and remained below for 43 min 20 s. At this depth the seal withstood a pressure of 397 kg/cm^2 *875 lb/in^2* of body (Kooyman, 1969).

Ray (1966) says the Weddell seal survives these tremendous dives by (1) constricting surface blood-vessels to a minimum to ensure a steady supply of blood to the heart and brain; and (2) tolerating large amounts of carbon dioxide in the blood.

The Harp seal (*Pagophilus groenlandicus*), also called the 'Greenland seal', is another spectacular diver and Nansen (1925) describes how he and his men caught one in a net which had been dropped to a depth of 275 m *902 ft*.

Scholander (1940) once forcibly dived a young Hooded seal (*Cystophora cristata*) to 300 m *984 ft* in 3 min, but the unfortunate creature died soon after being brought back to the surface. He came to the conclusion that the pinniped was probably unprepared for the dive and went down with excess air in its lungs.

More recently a young female Californian sea lion captured off the coast of California in 1960 has been trained to retrieve a ring at a depth of 170 m *558 ft* (Evans, 1968), which suggests that large bulls of this species may be capable of diving to the 305 m *1000 ft* mark.

The elephant seal – rather surprisingly – does not seem to go below 200 m *656 ft* when diving naturally, although the exceptionally large eyes of this animal point to a deep-diving ability.

The greatest reliable age recorded for a pinniped is 'at least 46 years' based on a count of dental annuli for a female grey seal shot at Shunni Wick in Shetland on 23 April 1969 (Bonner, 1971). A ringed seal collected on the SW coast of Baffin Island in the eastern Canadian Arctic in 1954 by McLaren (1958) was believed to be over 43 years old on a similar count.

The greatest reliable age reported for a captive pinniped is 'at least 43 years' for a bull grey seal named 'Jacob'. He was captured off the Stockholm Archipelago in the Baltic Sea and arrived at Skansen Zoo, Stockholm, Sweden, on 28 October 1901 when he was an estimated two years of age. He died there on 30 January 1942 (Kai Curry-Lindahl, pers. comm. 2 February 1966).

The life expectation of the elephant seal under natural conditions is quite short by comparison. In one series of tooth counts of *M. leonina* by Laws (1953) the maximum ages recorded for bulls and cows were 20 years and 18 years respectively. In 1940 a cow of the southern race named 'Nixe' died at Carl Hagenbeck's Tierpark aged about 16 years (Steinmetz, 1954).

The rarest of all pinnipeds is the Caribbean or West Indian monk seal (*Monachus tropicalis*). At one time this animal was the basis of a profitable seal fishery in both the Caribbean and the Gulf of Mexico, but it was so persistently slaughtered for its oil, meat and skin that by the end of the 19th century the seal was virtually extinct.

On 14 June 1909 an adult bull and two yearlings were received at New York Aquarium (Townsend, 1909). In March 1911 one of the young seals was still living, but Crandall says 'no further information concerning it is now available'.

In January 1911 some fishermen visited the Triangle Keys, a group of islets to the west of Yucatan and killed about 200 seals (Allen, 1942), and on 15 March 1922 one was killed near Key West, Florida (Townsend, 1923). In 1949 two specimens were reportedly seen in the waters south of Kingston, Jamaica, and another one was sighted in the same area in 1952. Ten years later a Caribbean monk seal was seen on the beach at Isla Mujueres off the Yucatan Peninsula, Mexico (Fisher *et al*, 1969). Some monk seals may still survive in the remoter areas of the Caribbean, but the total population must be very small. Incredibly, it has no protection whatsoever.

Of sub-species the Japanese sea lion (*Zalophus californianus japonicus*), formerly widespread in the Japanese Archipelago, probably became extinct in the early 1950s.

The Juan Fernandez fur seal (*Arctocephalus philippii phillippii*), which was believed to have become extinct in 1917, was rediscovered on the Juan Fernandez Islands some 640 km *400 miles* west of Chile in November 1968. The sub-species is now protected by the Chilean Government, and in 1972 there were believed to be about 450 of these animals living on Mas a Tierra and Mas Afuera islands.

The only flying mammals are Bats (Chiroptera) of which there are *c.* 950 living species. They are found throughout the world, with the exception of the polar regions.

The largest known bat in terms of wing expanse is probably the Kalong (*Pteropus vampyrus*), a fruit bat (= flying fox) found in Malaysia and Indonesia. It has a wing-span measurement of up to 1·70 m *5 ft 7 in*, a head and body length of about 400 mm *15¾ in* and weighs up to 900 g *31·8 oz* (Walker *et al*, 1968).

According to Peterson (1964) a huge example of *P. neohibernicus* from New Guinea preserved in the American Museum of Natural History, New York, has a head and body length of 455 mm *17·91 in* and a span of 1·65 m *5 ft 5 in*. He thinks that some unmeasured specimens may reach 1·83 m *6 ft*, but this figure is unconfirmed.

A span of more than 1·52 m *5 ft* has also been

The fruit bat has been measured up to 170 cm 5 ft 7 in *across the wings (Otho Webb)*

reliably reported for *P. giganteus* (= *P. medius*) of India. This is a bulkier species than *P. vampyrus* and *P. neohibernicus*, and weights up to 1548 g *3 lb 8 oz* have been recorded.

The largest bat found in Britain is the very rare Large mouse-eared bat (*Myotis myotis*). Mature specimens have a wing span of 355–450 mm *13·97–17·71 in*, a head and body length of 68–80 mm *2·67–3·1 in* and weigh up to 45 g *1·59 oz* (Blackmore, 1964).

The smallest known mammal is the rare Kitti's hog-nosed bat (*Craseonycteris thonglongyai*), also called the 'Bumblebee bat', of southern Thailand. This species, which was not discovered until 1973 and is the basis of a new family and genus (Hill, 1974), is restricted to two caves near the forestry station at Ban Sai Yoke on the Kwae Noi River, Kanchanaburi. Mature specimens (both sexes) have a wing span of *c.* 160 mm *6·29 in*, a head and body length of 29–33 mm *1·14–1·29 in* and weigh 1·75–2 g *0·062–0·071 oz*. The Tiny pipistrelle (*Pipistrellus nanulus*) of West Africa has a smaller wing span (152 mm *6 in*) but this species has a larger head and body (38 mm *1½ in*) than *C. thonglongyai* and scales about 2·5 g *0·088 oz* (Koopman, pers. comm. 24 April 1967).

The smallest British bat is the Pipistrelle (*Pipistrellus pipistrellus*). Adult examples have a wing span of 200–230 mm *7·87–9·05 in*, a head and body length of 45–52 mm *1·65–2·04 in* and weigh between 5·5 and 7·5 g *0·193 and 0·264 oz* (Lovett, 1961).

At least three species of bat are known only from the type specimen: *Neopteryx frosti* (collected at Tamalanti, West Celebes, in 1938/9); *Paracoelops megalotis* (collected at Vinh, Annam, Indochina, in 1945) (Walker *et al*, 1968), and *Lakidens salimalii* (collected in the High Wavy Mountains, southern India, in 1948). John Edwards Hill, however, a chiropterologist at the British Museum (Natural History), thinks that this information is misleading. He told the compiler (pers. comm. 21 January 1973): 'Because bat species are known only from one specimen does not mean that they are necessarily rare in occurrence: it means merely that no more have been collected, and it is my experience that changes in collecting techniques can often produce many more specimens.'

The rarest bat in Britain is Bechstein's bat (*Myotis bechsteini*), which is confined to a small area in southern England, with the New Forest as the main centre of population. Up to 1886 there was only one English record of this species (New Forest, before 1837), but that year a small colony was discovered in a woodpecker's hole near Burley, Hampshire. Ten years later an adult male was shot near Battle, Sussex, and in 1901 one was found asleep in a chalk tunnel on the Berkshire side of the river at Henley-on-Thames. Two more specimens were shot at Newport, Isle of Wight, in 1909, and there have been about half a dozen records since.

In January 1965 about 15 specimens of the Grey long-eared bat (*Plecotus austriacus*) were discovered in the roof of the Nature

Conservancy's research station at Furzebrook, Dorset, by R E Stebbings. Up to then this species, which is found all over Europe, had only been recorded once in Britain (Hampshire, 1875).

The Parti-coloured bat (*Vespertilio murinus*) of eastern Europe has only been recorded three times in Britain, but this species is considered a rare vagrant.

The greatest reliable age reported for a bat under natural conditions is at least 24 years for a female Little brown bat (*Myotis lucifugus*) found on 30 April 1960 in a cave on Mt Aeolus, Vermont, USA. It had been banded at a summer colony in Mashpee, Massachusetts, on 22 June 1937, at which time it was already fully adult. Another specimen found in a decomposed state in the same cave on 26 December 1964 was also believed to be 24 years old at the time of its death (Griffin & Hitchcock, 1965). There is also an unconfirmed record of a Greater horseshoe bat (*Rhinolophus ferrumequinum*) living 26–27 years in France (Yalden, pers. comm. 6 May 1974). This may be a reference to a fully grown bat caught by Norbet Casteret in the Grotto of Labastide-de-Neste in the French Pyrennes on 31 December 1936 which he reportedly kept in captivity for 25 years before finally setting it free.

The greatest reliable age reported for a British bat under natural conditions is 19 years 3 months for a Greater horseshoe bat banded in March 1949 which was still alive in Devon in October 1967 (Hooper & Hooper, 1967). A Daubenton's bat (*Myotis daubentoni*) banded in 1949 was still alive in 1967 aged 18+ years (Yalden & Morris, 1975).

The greatest reliable age recorded for a captive bat is 22 years 11 months for a Rousette fruit bat (*Rousettus leachii*) which died in Giza Zoo, Cairo, Egypt, in 1918. In 1968 a Straw-coloured fruit bat (*Eidolon helvum*) died at London Zoo aged 21 years 1 month (Jones, 1972). The smaller species of bat rarely live long in captivity.

The fastest-flying bats are the Noctule bat (*Nyctalus noctula*) and the Long-winged bat (*Miniopterus schreibersi*), both of which have been timed at 49·6 km/h *31 miles/h* in the open (Kolb, 1955; Constant & Cannonge, 1957). According to Yalden & Morris (1975) the Big brown bat (*Eptesicus fuscus*) of North America has been clocked at 21 m/s *69 ft/s* (=75·6 km/h *47¼ miles/h*) flying from a maternity colony in Kentucky, but this timing was misread. The original speed given by Patterson & Hardin (1969) for this species was in fact 33·28 km/h *20·8 miles/h*.

During migration bats fly considerable distances. Adam Krzanowski (1964), citing Buresch & Beron (1962) mentions three bats which flew 1697 km *1060 miles*, 1950 km *1219 miles* and 2347 km *1467 miles* respectively from the interior of the USSR to Bulgaria. The species concerned were *P. pipistrellus* and *M. mystacinus*, both small, mediocre flyers, and *N. noctula*, a strong, vigorous flyer. It is interesting to note that both *M. lucifugus* and *Lasiurus cinereus* have been recorded in Iceland. In the case of the former, however, the bat probably covered the distance as an accidental passenger aboard a ship, but *L. cinereus* was probably blown across by prolonged westerly winds.

Bats have the most highly developed sense of hearing of any terrestrial mammal. The sound frequencies used by these creatures normally range between 20 kHz (kiloherz) and 130 kHz (Yalden & Morris, 1975), but vampire bats (Desmodontidae) and fruit bats (Pteropodidae) can hear frequencies as high as 160 kHz (cf. 15–20 kHz for the range of human hearing).

The most dangerous bat is the Common vampire (*Desmodus rotundus*) of tropical and subtropical America, which not only drinks the blood of its victims but also transmits disease, including the paralytic rabies virus (hydrophobia) which is almost 100 per cent fatal to livestock and man.

According to reliable estimates over 1 000 000 head of cattle die every year in Latin America from rabies carried by vampire bats and there are a number of human fatalities. In 1933 over 40 people died of vampire bites in Trinidad, and there were five deaths in the Mexican State of Sinaloa in 1951.

Vampires usually alight on their sleeping or quiet victim and then crawl softly over its body until they find an attractive piece of bare skin, i.e. on the neck, ears or back. They then make a slight cut with their minute, razor-sharp incisor teeth and greedily lap up the blood oozing from the wound. Their saliva contains an anti-coagulant, which means the blood does not clot properly, thus allowing the nightmarish creatures to obtain full sustenance from the tiny puncture marks they have made. Sometimes the bats gorge themselves with blood to such an extent that they look like furry balls, and Eisentrait (1936) has described how one specimen which fed on the blood of a domestic goat for ten minutes became so bloated that it could not fly. (A vampire can consume about 26¼ litres *46 pints* of blood a year.)

When these mainstays of horror fiction attack sleeping humans they usually go for the toes, sometimes creeping under the bedclothes. During a collecting expedition to Brazil in 1927 on behalf of the New York Zoological Society Dr William Beebe carried out some experiments to determine the puncturing properties of this animal. He crept up on his sleeping companions and attempted to pierce the skin of their toes with the finest needles he could find without waking them. In each case they were aroused immediately, yet each of these men was robbed of quantities of blood by vampires without being aware at the time of the attack.

In recent years American Government scientists have discovered that an anticoagulant drug called 'diphenadione' provokes fatal haemorrhaging in vampires, and they have since come up with two satisfactory ways of controlling these bats. The most effective method is the application of small amounts of the drug to the backs of captive bats with a brush. The vampires are then released and make their way back to their colonies where they die shortly afterwards. As vampires have a habit of grooming each other like cats the drug has a multiplier effect, and in one series of nine individuals treated with the drug under laboratory conditions the vampires accounted for 310 of their companions. The other method is to inject the drug into the stomachs of cattle or simply spray them with the anticoagulant. The compound is harmless to cattle but renders the blood lethal to vampires for at least three days and nights.

The largest living primate is the Eastern lowland gorilla (*Gorilla gorilla graueri*), which inhabits the lowlands of the eastern part of Zaire and SW Uganda. The average adult bull 'stands' 1·76 m *5 ft 9 in* tall (because the erect position is unnatural for a gorilla the measurement is taken between sticks placed at the crown and heel); has a chest circumference of 147–152 cm *58–60 in* and weighs about 163 kg *360 lb*. The average adult female is much smaller, standing about 1·40 m *4 ft 7 in* and weighing 77–95 kg *170–210 lb*.

The two other races of gorilla, the Mountain gorilla (*Gorilla g. beringei*), which is found on the volcanic Virunga Mountains, Zaire, and SW Uganda, and the Western lowland gorilla (*Gorilla g. gorilla*), which lives in the lowland rain forests of the Congo, the Cameroons and Gabon, are both slightly shorter in stature than *Gorilla g. graueri*, adult bulls averaging 1·73 m *5 ft 8 in* and 1·68 m *5 ft 6 in* respectively (Groves, 1971).

If we are to believe the claims of some of the early gorilla-hunters a small number of really colossal bulls have been killed in the past, but these assertions were either grossly exaggerated or the measurements taken in such a way as to considerably enhance the true size of the primate.

For instance, one western lowland gorilla killed in eastern Cameroon allegedly measured 2·29 m *7 ft 6½ in* in height, 1·08 m *3 ft 7 in* across the shoulder and weighed 349 kg *770 lb*, but Willoughby (1950) says a photograph published in the French journal *La Nature* on 29 July 1905 showed a gorilla of quite ordinary size.

Another bull of the same race shot by M Villars-Darasse in the Forest of Bambio, Haute-Lobaze, former French Equatorial Africa, in 1919 was stated to have measured 2·84 m *9 ft 4 in* (*sic*), but once again photographic evidence (*L'Illustration*, Paris, 14 February 1920) failed to substantiate this figure, although the primate was quite tall. Don Cousins (1972) thinks this particular specimen was 'a little nearer six feet or slightly over as the animal had particularly long legs for a gorilla'.

In the first case the hunter probably obtained his measurement by running the tape over the contours of the body instead of in a straight line between sticks, and in the second the 'height' must have been taken from the tip of the uplifted arms to the end of the longest toe with the foot bent downwards. Either (or both) of these methods were favoured by gorilla-hunters in the past who were anxious to make the total length as great as possible. The *correct* way to measure a gorilla in the field is to take the distance from the crown (including the crest in bulls) to the base of the heel in the prone position.

In December 1930 an Italian scientific expedition which had been granted a special permit to explore the Virunga Mountains, reportedly shot a 2·13 m *7 ft* bull in the Albert National Park. Other measurements included a chest circumference of 185 cm *72·8 in* and an arm-span of 3·09 m *10 ft 1½ in*, and the weight was said to be 325 kg *716 lb 8 oz*.

The only Italian expedition in the Virunga Mountains that year, however, was one led by Commander Attilio Gatti (1932) for the Royal Museum of National History, Florence, and he said his largest bull – shot in Tchibinda Forest, near Lake Kivu (Republic of the Congo) – measured 2·87 m *8 ft 9 in* from the bottom of its feet to the tips of its raised arms and 2·06 m *6 ft 9 in* from the crown to the tip of the foot bent downwards. From these statistics (and photographic evidence) Willoughby was able to calculate that the gorilla must have stood about 1·78 m *5 ft 10 in* in life.

Paul Belloni du Chaillu (1861), the French–American explorer-naturalist, and father of gorilla-hunters, said adult bull gorillas ranged in height from 1·57 m *5 ft 2 in* to 1·88 m *6 ft 2 in*. The largest of nine specimens collected by him between 1856 and 1859 in Gabon, Equatorial Africa, measured 1·73 m *5 ft 8 in*.

According to Bourgoin (1955) the *record gorilla* was a bull of the mountain race shot in the Angumu Forest in the eastern Congo in March 1948 which measured 1·96 m *6 ft 5 in* in height and 166 cm *61 in* round the chest, but as the arm-span of this individual was only 2·49 m *8 ft 2 in*, the stature could not have been much above 1·70 m *5 ft 7 in*. (The height/arm-span ratio is normally 50:75.)

The largest accurately measured gorilla on record was probably a bull of the mountain race shot by T Alexander Barns (1923) on Mt Karisimbi, eastern Congo, which measured 1·88 m *6 ft 2 in* from the crown of the head to the heel and had a chest measurement of 152 cm *60 in*.

In May 1900 Henry Paschen, the German animal-trader, shot an eastern lowland gorilla near Taounde, in the Cameroons, which he said measured 2·07 m *6 ft 9½ in* in length, but the measurement was taken from the crown of the head to the tip of the extended foot. (The mounted specimen in the Rothschild Museum, Tring, Hertfordshire, has a standing height of 1·70 m *5 ft 7 in*.)

The greatest reliable weight recorded for a gorilla in the field is 219 kg *482 lb* for the large eastern lowland gorilla shot by Commander Gatti in the Tchibinda Forest. (Willoughby quotes a weight of 241 kg *531 lb* for the same specimen.) Another bull shot by Henry Raven in the same area in July 1929 stood 1·78 m *5 ft 10 in* (chest 142 cm *56 in*) and must have weighed at least 204 kg *450 lb*; he also killed another bull of the same race in the Belgian Congo which weighed nearly 209 kg *460 lb* (height 1·74 m *5 ft 8½ in*; chest 152 cm *60 in*).

In 1934 the George Vanderbilt African Expedition collected a huge bull of the eastern lowland race which had been killed by natives in the neighbourhood of Aboghi in the Sanga River area, French Equatorial Africa. This gorilla measured exactly 1·83 m *6 ft* between sticks and boasted a 140 cm *55 in* chest (Coolidge, 1937). Unfortunately there were no facilities for weighing this giant, but it must have scaled at least 227 kg *500 lb*.

Another large bull killed on Mt Sabini by Edmund Heller's American expedition in 1925 measured 1·82 m *5 ft 11½ in* between sticks and was estimated to weigh over 227 kg *500 lb*, and Ben Burbridge (1928) shot a gorilla on Mt Kikeno in the Kivu area which measured 1·83 m *5 ft 11¾ in* and was extremely bulky.

Fred G Merfield (1956), a very reliable English observer, who collected 115 gorillas for Euro-

The famous Western gorilla 'Phil', who reputedly weighed 352 kg 776 lb *at the time of his death (St Louis Zoological Park)*

pean museums over a four-year period (*c.* 1918–22) while he was a planter in the Mendjim Mey, French Cameroons, says he only shot one bull standing over 1·83 m *6 ft*. This giant, collected in the Ambam district after a fierce battle in which the hunter received a deep thigh wound was estimated to have weighed between 260–267 kg *574* and *588 lb*, although at the time he made this statement Merfield admitted that 'the abdomen was enormously extended, partly by a great quantity of vegetable matter in the intestines and partly by putrefaction'.

The skeleton of this gorilla was later presented to the Science Museum at the University of Texas, where primatologists estimated that the animal must have stood just over 1·83 m *6 ft* in life. Cousins, who later saw a photograph of this gorilla in an advanced state of decomposition, said it was clearly 'a monstrous animal'.

Very few adult female gorillas have been measured in the field and information is therefore scanty. In 1924 the following measurements were reported for a large female of the mountain race killed, very unusually, in open country on the Uganda–Congo border: arm-span 2·37 m *7 ft 9½ in*; crown of head to fork in sitting position 1·27 m *4 ft 2 in*; hip to heel 81 cm *2 ft 8 in*; neck 66 cm *26¼ in*; forearm 40 cm *16 in*; thigh 73 cm *28¾ in*; chest 162 cm *64 in* (post-mortem inflation), hand length 25 cm *9¾ in* and foot length 30 cm *11¾ in*. This specimen was stated to have stood 2·08 m *6 ft 10 in* tall in life (Pitman, 1931), but its true height was probably nearer 1·57 m *5 ft 2 in*.

The largest female (*Gorilla g. gorilla*) in a series of 60 skeletons examined by Schultz (1931) measured about 1·45 m *4 ft 9 in* in life.

A female mountain gorilla collected by Prince Wilhelm's Swedish Expedition to the Virunga Mountains in 1921 weighed 97 kg *124 lb 8 oz* (Gyldenstople, 1928), and Grzimek (1957) quotes a weight of 72 kg *159 lb 8 oz* for a female eastern lowland gorilla.

The heaviest gorilla ever held in captivity was probably a bull of the eastern lowland race named 'M'bongo', who died in San Diego Zoological Gardens, California, USA, on 15 March 1942. On 1 June 1941 he scaled 281 kg *619 lb* and, during an attempt to weigh him shortly before his death, the platform scales 'fluctuated from 645 pounds to nearly 670' (Benchley, 1942). M'bongo's posthumous weight was given as 264 kg *582 lb*, but this marked decrease in poundage was due to the fact that the gorilla took very little food during his terminal illness, which lasted 45 days. His posthumous measurements were: height 1·71 m *5 ft 7½ in*; chest 175 cm *69 in*; waist 183 cm *72 in*; wrist 36 cm *14⅓ in*; thigh 69 cm *27¼ in* and calf 39 cm *15⅓ in*.

M'bongo and another eastern lowland gorilla named 'N'gagi' arrived at San Diego Zoo on 5 October 1931. They had been captured the previous year by Martin and Osa Johnson in the Alumbongo Mts in the Kivu region of the Congo. At the time of their arrival they were four to five years of age, and their combined weight was 123 kg *269 lb*. M'bongo was about six months younger than N'gagi and his weight was 57 kg *125 lb*. Throughout their life in captivity N'gagi was always the dominant animal, although when the gorillas did actually fight M'bongo always proved the fiercer and more crafty of the two. In February 1940 M'bongo overtook his big companion in weight for the first time, tipping the scales at 234 kg *517 lb*, compared to N'gagi's 212 kg *468 lb*, but this was only because the latter animal had been unwell for six months and had lost considerable weight. (In September 1939 N'gagi scaled 227 kg *501 lb*.)

On 23 April the same year they were weighed again and scaled 268 kg *592 lb* and 238 kg *525 lb* respectively. But whereas M'bongo was 'round and paunchy with an enormously fat abdomen', N'gagi was 'broad of shoulder and very trim and slender of waist and hips' (Benchley, 1940). Fifteen days later M'bongo tipped the scales at 273 kg *602 lb* and N'gagi 244 kg *539 lb*. N'gagi weighed 287 kg *633 lb* shortly before his death on 12 January 1944 (Anon, 1944).

According to Cousins (1972) the heaviest bull ever kept in captivity was the famous western lowland gorilla 'Phil' of St Louis Zoological Park, Missouri, USA, who reportedly weighed 352 kg *776 lb* shortly after his death in 1958. This colossal poundage, had previously been queried by Dr R Marlin Perkins, Director of St Louis Zoo who told the compiler (pers. comm. 25 July 1966): 'Since coming to the St Louis Zoo four years ago, I have made inquiries of the staff about the weight of Phil the gorilla. They assure me that the animal was placed in a truck, taken to a public scale, weighed, returned to the zoo for autopsy, and that the figure of 776 pounds is the weight of the gorilla. I, too, find this very difficult to believe. I was not in St Louis at the time and had no part in the weighing of the animal. I had been at the zoo a short time before Phil's death, however, and had seen him alive. Because of this, and because of my knowledge of Bushman at Lincoln Park Zoo in Chicago and M'bongo at the San Diego Zoo (whom I also saw a few months before death), I still find it difficult to accept 776 pounds for Phil. It must be remembered that this magnificent specimen had been ill for several months and had lost considerable weight. The keepers tell me that he ate practically nothing for a period of ten days to two weeks prior to death.'

Phil's posthumous measurements were given as: height 1·81 m *5 ft 11 in*; chest 183 cm *72¼ in*; waist 162 cm *64 in*; neck 91 cm *36 in* and wrist 38 cm *15 in*; and as the weight can be calculated reasonably accurately when the height and maximum girth are known, this creature must have scaled about 279 kg *615 lb* at the time of his death.

Another western lowland gorilla called 'Gargantua' owned by Ringling Bros Circus, was billed as standing 1·98 m *6 ft 6 in* in height and weighing 340 kg *750 lb*. 'Gargantua' – or 'Buddy' as he was originally called – was purchased by Mrs Gertrude Lintz of Brooklyn, New York, in 1931 when he was a year old, and she kept the gorilla for six years before selling him to the circus for $10 000. According to Riess (1949) Gargantua weighed 249 kg *550 lb* at the time of his death on 25 November 1949, but Dr S Dillon Ripley, Curator of Vertebrate Zoology at the Peabody Museum, Yale University, where the gorilla's skeleton is on display, told Plowden (1972) that the primate only scaled 141 kg *312 lb*. This is rather surprising because Gargantua's posthumous measurements – height 1·70 m *5 ft 7½ in*; chest 170 cm *67 in*; upper arm 61 cm *24 in* – point to a weight in excess of 227 kg *500 lb if the figures were reliable*.

Ringling Bros later acquired another western lowland gorilla, 'Gargantua II', which they said was the world's heaviest captive gorilla. In 1972 he allegedly scaled 338 kg *745 lb*, but his actual weight was believed to be nearer 202 kg *445 lb* (Gordon Hull, pers. comm. 11 June 1972).

The largest gorilla ever held captive in a British zoo was the western lowland bull 'Congo'. He arrived at Bristol Zoo from the French Cameroons on 30 August 1954 and weighed 216 kg *476 lb* in February 1966. At the time of his death in December 1968 he scaled an estimated 254 kg *560 lb*. 'When I saw him a few months before his demise', writes Cousins (1972), 'I was struck by the obesity of this animal. His abdomen was so vast that it almost touched the ground when he walked. His height was estimated at 1·73 m *5 ft 8 in*, but I believe that he was smaller than this.'

The heaviest gorilla living in captivity anywhere in the world today is believed to be the western lowland bull 'Samson' of Milwaukee Zoological Park, Wisconsin, USA. In April 1971 he scaled a very obese 296 kg *652 lb* (Hull, pers. comm. 28 January 1975), but by May 1974 he had reduced to 264 kg *582 lb*. Today his weight fluctuates between 267 kg *585 lb* and 274 kg *605 lb*.

'Samson', the world's heaviest captive gorilla (Milwaukee Zoological Park)

Another western lowland bull called 'Casey' at Como Zoo, St Paul's, Minnesota, USA (received 14 May 1959) has scaled as much as 261 kg *576 lb*, but his best weight is about 247 kg *545 lb*. This gorilla, named after Casey Stengle, the American baseball-player because he has a wicked side-arm throw with a rock and has been known to hit visitors on occasion, was loaned to the Henry Doorly Zoo in Omaha, Nebraska, in 1970 for 'family purposes'. He was anaesthetised, strapped to a litter, and then placed in a small aircraft uncaged but attended by a veterinarian. This was the *first time* this technique had ever been used on an adult gorilla (Fletcher, pers. comm. 14 April 1971). In March 1973 Casey tried to grab one of his girl-friends, 'Brigitte', through the bars of the adjoining cage, and had one of his fingers bitten off for his trouble!

The heaviest gorilla held captive in a British zoo today is probably the celebrated western lowland gorilla 'Guy' of London Zoo (received 5

November 1947). In February 1966 he scaled 212 kg *468 lb*, and on 31 July 1967 an obese 222 kg *488 lb 8 oz*. Shortly afterwards he was put on a strict diet and his poundage now fluctuates between 168 kg *370 lb* and 177 kg *390 lb*. This moody gorilla is rather short for an adult bull, standing only 1·61 m *5 ft 3½ in* tall.

The heaviest captive female gorilla on record was probably 'M'Toto' (b November 1931), a member of the western lowland race, who came to Ringling Bros Circus via Mrs Maria Hoyt of Havana, Cuba in 1941. In 1947 this massively built animal weighed 199 kg *438 lb* (Riess, 1949), and probably exceeded 227 kg *500 lb* at her heaviest, but the statement by Ringling Bros that she scaled 295 kg *650 lb* in 1953 can be disregarded. It is interesting to note, however, that José Tomas, her Spanish-born trainer, has been quoted as saying that she 'later got up to 575 lb' (Plowden, 1974). M'Toto died in Venice, Florida, on 17 July 1968.

The heaviest female gorilla held captive in a British zoo was 'Josephine', a member of the western lowland race, who arrived at Bristol Zoo on 30 August 1954 along with Congo from the French Cameroons. At the time of her death on 3 March 1966 she weighed 161 kg *355 lb* (Cousins, 1972).

At the other end of the scale a creature called the 'Pigmy gorilla' (*Pseudogorilla mayema*) of the Gabon coastal regions, West Africa, has also been described (Aix & Bouvier, 1877), but Dr Colin Groves (1970) says the small number of specimens preserved in various museums have since been reidentified as either young female gorillas or chimpanzees! This still doesn't explain, however, why a female gorilla named 'Pussi' at Breslau Zoo, Germany, weighed only 30 kg *66 lb* at the time of her death in August 1904 aged eleven years, or why another female gorilla named 'Empress' (received Dublin Zoo from Gabon in January 1914) measured only just over 61 cm *2 ft* tall and weighed 14 kg *31 lb* at the age of *c.* four years – half the normal poundage (Cousins, 1972).

The greatest age recorded for a gorilla is 45+ years for the western lowland bull 'Massa' (b December 1930) of Philadelphia Zoological Garden, Pennsylvania, USA, who was still alive in March 1976. He was purchased by Mrs Gertrude Lintz in September 1931 and arrived at Philadelphia Zoo on 30 December 1935.

Although the ferocity of the gorilla has been greatly exaggerated, most of the stories about its incredible strength are true. In one strength experiment carried out by Yerkes (1927) on 'Congo', a five-year-old female gorilla, which involved the pulling of a rope attached to a spring-balance in an effort to obtain food, the 58 kg *128 lb* subject pulled 109 kg *240 lb* working with both hands and feet braced, but she was only half trying. Yerkes concluded from this test that the arm strength of the young gorilla was two or three times greater than that of a human of comparative size.

During her stay in Havana the famous M'Toto went into Mrs Hoyt's garage one day and found a station wagon blocking her way. The young gorilla seized the rear axle with *one hand* and pushed the car – the emergency brake was on – with tremendous force against the wall of the building.

Despite their fierce appearance gorillas are generally timid and peaceful creatures that shun human contact, but they can become extremely dangerous when wounded.

During the ten-year period 1956–65 Jorge Sabater Pi (1966) collected records of all the authentic cases of gorilla attacks against humans in Rio Muni, Spanish Guinea. Most of the victims were injured – some seriously – but no fatalities were reported.

One of them was a hunter named Pedro Beyeme who wounded a large solitary bull in the Mikomeseng Forest. 'The injured animal attacked the hunter vigorously, and caused severe injuries with his hands, completely ripping the buttocks and the ankles of both legs. A companion who was nearby when the attack occurred came quickly to his rescue and killed the animal.'

The only case not involving a wounded gorilla occurred in the Forest of Mosumu (Group Mobumuom-Monte Mitra): 'On 26 January 1964 the natives Manuel Nsue and Jesus Abeso, both of Mabeuolo (Niafang), departed from Mosum early in the morning, crossed the Benito river in a dug-out canoe, and entered the large forest known by the name of Mobumuom-Monte Mitra for the purpose of hunting monkeys (*Cercopithecus*) and small forest antelopes (*Cephalophus*). Manuel carried only a machete and Jesus an old single-barrelled 12-gauge shotgun with a few cartridges of No. 3 shot and only one cartridge with a ball. About 3 pm they started to return and Manuel Nsue went on ahead to prepare the canoe. At a sharp turn in the forest he suddenly encountered at a short distance a family of gorillas composed of four individuals (one adult male, two females and a small juvenile) seated, silently eating the fruit of the "Afrafmonum" shrub. They were surprised and very frightened at the sudden appearance of the man.

'Manuel, amazed by the proximity of the group [3–4 m *9¾–13 ft*] attempted a prudent retreat, facing the animals (it is a very strong belief among the natives that the gorillas will not attack if one always keeps his face toward them), but it availed him nothing. The dominant male, recovering from the fright, gave a loud shout [*sic*] and charged furiously, knocking Manuel down and biting him in the legs. The victim fell face down, losing his machete. The enraged animal pressed its attack on Manuel's hands, who used them to ward off the animal's blows. Jesus, who was left behind, came to the rescue when he heard the shouts of Manuel and the gorilla, and shot the animal with the ball cartridge that he had saved. The gorilla attempted another charge but fell mortally wounded at the feet of Jesus. Manuel Nsue came out of the encounter with various fingers on both hands amputated and with a deep wound in the leg.'

The only authentic record of a gorilla killing a human occurred in 1910. The victim, a native of the Boringo tribe, had apparently gone into the forest to cut some bamboo and had surprised a family of gorillas resting in an open space. The man was immediately attacked by a large bull, and when his corpse was found later on it was minus the head and neck which were lying near by. It appeared the gorilla had grasped the unfortunate native's neck with one huge hand and his sacrum with the other and then simply *pulled* (Akeley, 1929).

Monkeys have been known to treat a bird in this way, but this method of assault is most unusual for a gorilla. As Jorge Sabater Pi has already stated, this primate normally attacks the legs, thighs and buttocks of humans with its powerful teeth and fingernails.

The smallest known primate is the rare Feather-tailed tree shrew (*Ptiolcercus lowii*) of Malaysia, Sumatra and Borneo. Adult specimens have a total length of 230–330 mm *9–13 in* (head and body 100–140 mm *3·93–5·51 in*, tail 130–190 mm *5·1–7·5 in*) and weigh 35–50 g *1·23–1·76 oz*. The Mouse lemur *Microcebus murinus* of Madagascar is approximately the same length (274–300 mm *10·8–11·8 in*), but heavier, adults weighing 45–80 g *1·59–2·82 oz* (Hill, 1953).

The greatest irrefutable age reported for a primate is 54+ years for a pair of Orang-Utans (*Pongo pygmaeus*) named 'Guas' and 'Guarina' at Philadelphia Zoological Garden, Pennsylvania, USA, who were still alive in April 1975. They were received on 1 May 1931 after having been kept for some time in Cuba.

The famous male Chimpanzee (*Pan troglodytes*) 'Heine' of Lincoln Park Zoological Gardens, Chicago, Illinois, USA, was *c.* 50 years 3 months at the time of his death on 10 September 1971. He arrived there on 10 June 1924 when aged *c.* three years. The *potential* lifespan of this species has been estimated at 60 years (Riopelle, 1963).

The rarest primate is the Hairy-eared mouse lemur (*Cheirogaleus trichotis*) of Madagascar which, until fairly recently, was known only from the type specimen (Gunther, 1875), and two skins. In 1966, however, a live example was found on the east coast near Mananara (Fisher *et al*, 1969). In 1967 it was described as a new genus, *Allocebus*.

The largest members of the monkey family are the Mandrill (*Mandrillus sphinx*) of West Africa, the Anubis baboon (*Papio anubis*) of Equatorial Africa, the Chacma baboon (*Papio ursinus*) of South Africa and the Yellow baboon (*P. cynocephalus*) of southern Africa (Dorst & Dandelot, 1970). Adult males of these four species have a head and body length of 76–101 cm *30–40 in* and weigh 29–41 kg *65–90 lb*.

There is insufficient data for one species to be cited as consistently being the largest either in poundage or length, but the only monkey which has been reliably credited with a weight in excess of 45 kg *100 lb* is the mandrill. One huge male tipped the scales at 54 kg *119 lb*, and an unconfirmed weight of 59 kg *130 lb* has been reported for another individual. Adult females are about half the size of males.

The smallest known monkey is the Pygmy marmoset (*Cebuella pygmaea*) of Ecuador, northern Peru and western Brazil. Mature specimens have a maximum total length of 304 mm *12 in*, half of which is tail, and weigh 49–70 g *1·7–2·47 oz (Walker et al, 1968)*.

The greatest reliable age recorded for a monkey is 'about 46 years' for a big male mandrill called 'George', who was deposited at London Zoo on 30 November 1906 by the Hon Walter Rothschild (later Lord Rothschild) and died on 14 March 1916. Lord Rothschild purchased this specimen in Paris for £124, and he told Major Flower (1931) that he was 'practically certain' that it was the same mandrill as one that had been imported into Europe in 1869.

*The only water-skiing chimpanzee in the world, with a dolphin acting as motorboat (*Sea World*)*

The Chacma baboon 'Jackie' with his master

Flower also gives a record of a Chacma baboon living for 45 years.

The most intelligent of sub-human primates is the chimpanzee. It is one of the very few animals that uses tools, and the stick has been described as the 'universal instrument' of this species. In its natural habitat the stick is used to extract termites or honey from nests, and in captivity to reach objects which are beyond the reach of its arms.

The orang-utan is not far behind, and in some experiments at Yerkes Primate Research Center in Atlanta, Georgia, USA, it has actually exceeded the chimpanzee in mental ability.

The most intelligent monkeys are baboons. One of the most remarkable examples was a Chacma baboon named 'Jackie', who was probably the only primate in history to reach the exalted rank of corporal in the army – and end up with a war medal!

The baboon was discovered by Albert Marr on his farm in Villieria, Pretoria, South Africa, a few years before the outbreak of the First World War and the two quickly became very attached to each other. The animal turned out to be exceptionally intelligent and took so readily to training that when Marr joined the Third South African Infantry Regiment he took along his companion as well. The friendly monkey was an instant success with the soldiers, and it wasn't long before he was made the regimental mascot. As a result he was issued with rations, a pay book and a specially made-up uniform.

Private Jackie was a perfect recruit. On the parade ground he was always smartly turned out, and every time he saw a passing officer he would stand to attention and give a very correct salute. He was also very proficient at lighting cigarettes for his comrades-in-arms. In August 1915 the two privates sailed with their regiment for the war zone, and during the next three years the inseparable pair saw front-line service against the Turks and Germans, and were also with the brigade during a campaign in Egypt.

The baboon proved to be an extremely valuable acquisition because he was a first-class guard. With his acute hearing he could detect the enemy long before his human companions, and when he picked up anything he would either give a series of short barks or tug urgently at his master's tunic. On 26 February 1916 Private Marr was hit by a bullet at the Battle of Agagia and fell to the ground. The distraught monkey refused to leave his master's side, and when a medical team arrived they found the faithful creature licking the wound.

Just over two years later, in April 1918, both Privates were injured together. It happened in the Passchendale area, W Flanders, Belgium. The brigade had come under heavy fire and, as the air filled with the sound of deafening explosions, Private Jackie could be seen desperately trying to build a fortress of stones around himself for protection. He never finished it – suddenly a shell exploded close by and a chunk of shrapnel hit him in the right leg, partly severing it. The same shell also wounded his master. Both soldiers were rushed to a British casualty clearing station, where the baboon's tattered leg was amputated by Dr R N Woodsend, who later wrote an account of the incident in *The Practitioner* of June 1959: 'It was a pathetic sight. The little fellow, carried by his keeper, a corporal, lay moaning in pain, the corporal crying his eyes out in sympathy. "You must do something for him, he saved my life in Egypt, he nursed me through dysentry."

'We decided to give the patient chloroform and dress his wounds. If he died under the anaesthetic perhaps it would be the best thing. As I had never given an anaesthetic to such a patient before, I thought it would be the most likely result. However, he lapped up the chloroform as if it had been whisky and was well under in a remarkably short time. It was a simple matter to amputate the leg with scissors and I cleaned the wounds and dressed them as well as I could. He came around as quickly as he went under; the problem was then what to do with him. This was soon settled by the corporal: "He's on the strength."'

Jackie made a full recovery and shortly before the Armistice he was promoted to corporal, and awarded a medal for valour. In 1919 he took part in the Lord Mayor's victory procession through London seated on a German howitzer which had been captured by his regiment.

In 1973 Ken Smith, a Johannesburg reporter, interviewed Albert Marr, then 84, in Pretoria and was shown Jackie's official army discharge papers, which stated that he was 'bilingual'. Jackie died in 1921 and was buried on his master's farm.

The largest rodent in the world is the Capybara (*Hydrochoerus hydrochaeris*), also called the 'Carpincho' or 'Water hog', which is found in tropical South America. Mature specimens (females are larger) have a head and body length of 101–137 cm *3 ft 3 in–4 ft 6 in* and weigh up to 79 kg *174 lb* (Zara, 1972). A smaller species (*H. isthmius*) found in Panama scales about 27 kg *60 lb* (Walker *et al*, 1968).

The only other rodent which remotely approaches this size is the Canadian beaver (*Castor canadensis*). In March 1952 a huge male weighing 39 kg *87 lb* was caught at Livingston, Wisconsin, USA (Anon, 1953).

The largest rodent found in Britain is *now* the Coypu (*Myocastor coypus*), also known as the 'Nutria', which was introduced into this country from Argentina by East Anglian fur-breeders in 1928. Soon afterwards four escaped from a nutria farm near Ipswich, Suffolk. One was killed almost immediately, but the other three founded a dynasty of wild specimens. (The coypu breeds three times a year and produces six to ten young in a litter.) When the Second World War broke out coypus from 37 of the 51 farms in East Anglia also managed to escape, and they soon became established in Norfolk and parts of Suffolk, where they began attacking crops and threatening drainage systems by burrowing in river banks. By 1960 the Ministry of Agriculture estimated there were at least 200000 coypus in the Broads area. Fortunately the very hard winter of 1963 killed about 80 per cent of the population, and by the end of 1965 the Ministry believed it had exterminated most of the others. Since then, however, there has been a succession of mild winters which have allowed the coypu to re-establish itself, and in recent years pest officers have been taking over 10000 annually.

Adult males measure 76–91 cm *30–36 in* in length (including short tail) and average 5–5·4 kg *11–12 lb* in weight, but much larger specimens have been recorded. On 5 December 1951 a coypu measuring 106 cm *3 ft 6 in* in total length and weighing 8·16 kg *18 lb* was killed in a dike by three boys at Ditchingham, Suffolk. Another individual shot at Orford, Suffolk, in September 1959 scaled 10 kg *22 lb*, and a weight of 13 kg *28 lb* was reported for a coypu caught in Suffolk in 1962. In captivity weights up to 18 kg *40 lb* have been reported for cage-fat animals.

The smallest known rodent is the Old World harvest mouse (*Microymys minutus*), of which the British form measures up to 135 mm *5·3 in* in length, including the tail, and weighs 4·2–10·2 g *0·14–0·36 oz*. It is thus about half the size of the common house mouse (*Mus musculus*).

The rarest rodent in the world is believed to be the James Island rice rat (*Oryzomys swarthi*), also called 'Swarth's rice rat'. Four specimens were collected on this island in the Galapagos group by J S Hunter for the California Academy of Sciences in 1906 (Orr, 1938), and it was not heard of again until January 1966 when the skull of a recently dead animal was found (Fitter, 1968).

The greatest reliable age recorded for a rodent is 22 years for an Indian crested porcupine (*Hystrix indica*) which died in Trivandrum Zoological Gardens, SW India, in 1942 (Simon, 1943). Flower mentions a Common porcupine (*H. cristata*) which died at London Zoo on 9 January 1886 after spending 20 years 4 months 14 days in captivity.

The largest known insectivore is the Moon rat (*Echinosorex gymnurus*), also called 'Raffles gymnure', which is found in Burma, Thailand, Malaysia, Sumatra and Borneo. Adult specimens have a head and body length of 265–445 mm *10·43–17·52 in*, a tail measuring 200–210 mm *7·87–8·26 in* and weigh up to 1400 g *3·08 lb* (Walker *et al*, 1968).

The Giant African water shrew (*Potamogale velox*) of Central Africa, which looks remarkably like a small otter, has also been credited with the title of 'largest insectivore', but it is doubtful whether this species exceeds 1000 g *2·20 lb* in weight. Mature specimens have a head and body length of 290–350 mm *11·4–13·7 in* and a tail measuring 245–290 mm *9·6–11·4 in*.

Anteaters (family Myrmecophagidae) are, of course, much larger, but although they also feed on termites and other soft-bodied insects they are not classified as insectivores but belong to the order Edentata (without teeth).

The largest insectivore found in Britain is the Common hedgehog (*Erinaceus europaeus*). Mature specimens have a head and body length of 245–310 mm *9·64–12·2 in*, a tail measuring 20–35 mm *0·7–1·37 in*, and usually weigh 450–1200 g *15·4–42·24 oz* (Van den Brink, 1967), but Burton (1969) says there is a record of a male weighing 1400 g *3 lb 8 oz*.

The smallest known insectivore is the Etruscan pygmy shrew (see page 23).

The greatest reliable age recorded for an insectivore is 10 years 6 months for a Hedgehog tenrec (*Ericulus setosus*) which died in London Zoo in 1971 (Jones, 1972).

The rarest insectivore is the tenrec *Dasogale fontoynonti*, which is known only from the type specimen collected in eastern Madagascar, and now in the Museum d'Histoire Naturelle, Paris.

The largest of all antelopes is the rare Central African Giant Derby eland (*Taurotragus derbianus derbianus*). Adult bulls average 1·65 m *5 ft 9 in* at the withers and scale about 714 kg *1575 lb*, but weights up to 898 kg *1980 lb* have been reported for old animals (Rowland Ward, 1971).

The Common eland (*T. oryx*) of East and South Africa nearly matches the Derby giant eland in height, but is not quite so massively built. (The average weight of bulls is 589–635 kg *1300–1400 lb*.) There are exceptions, however, and Meinertzhagen (1938) says he shot one in Nyasaland (Malawi) which tipped the scales at 943 kg *2078 lb*. Two other bulls collected by him in Kenya both weighed 893 kg *1969 lb*. Adult female elands rarely exceed 1·52 m *5 ft* at the withers and 453 kg *1000 lb*.

The smallest antelope is the Royal antelope (*Neotragus pygmaeus*), which lives in the dense forests of West Africa from Sierra Leone to Gabon. Mature specimens (both sexes) measure 254–304 mm *10–12 in* at the withers and weigh only 3·17–3·62 kg *7–8 lb*, which is the size of a large Brown hare (*Lepus europaeus*).

The slender Swayne's dik-dik (*Madoqua swaynei*) of Somalia, East Africa, weighs only 2·26–2·72 kg *5–6 lb* when adult, but this species stands about 330 mm *13 in* at the withers. Phillips's dik-dik (*M. phillipsi*) of the same country also stands 330 mm *13 in*, but Rowland Ward (1969) gives its weight as 3·62 kg *8 lb*. Drake-Brockman (1930) claims that some adult examples of *M. swaynei* and *M. phillipsi* tip the scales at no more than 1·81–2·5 kg *4–5½ lb*, but these figures have not been confirmed.

The rarest antelope is probably Jentink's duiker (*Cephalophus jentinki*), also known as the 'Black-headed duiker', which is thinly scattered in the deep forests of Liberia, Senegal and the Ivory Coast. The total population has been estimated at anything from a few dozen to possibly a few hundred (Fitter, 1968).

The greatest reliable age recorded for an antelope is 20 years 1 month 22 days for a female White-bearded gnu (*Connochaetes taurinus albojubatus*) which died in Philadelphia Zoological Garden, Pennsylvania, USA, on 27 July 1928 (Flower, 1931). Another specimen reportedly lived for 20 years in the Chicago Zoological Park, Illinois, USA (Robb, 1960). Crandall cites a record of a female Brindled gnu (*C. taurinus*) received at the New York Zoological Society (Bronx Zoo) on 2 September 1920 which was sold on 26 August 1940 after spending 19 years 11 months 22 days in captivity. Elands usually live 15–20 years.

In 1946 a motorised hunting party from the

Ruwullah tribe was pursuing a herd of gazelle in the Transjordan Desert when they suddenly spotted a boy running at tremendous speed in the middle of the group. After getting over their initial shock, the Arabs took off after the human racing phenomenon, and after a hard chase at speeds up to 80 km/h *50 miles/h* (*sic*) the lad collapsed in an exhausted heap.

It was believed that the child, who was aged about ten, and covered in hair, had been abandoned by his mother, probably a Bedouin, when very young and had been 'adopted' by the gazelles. The 'gazelle boy' was put in the care of Dr Musa Jalbout of the Iraki Petroleum Company, who later told a Reuter correspondent: 'He acts, eats and cries like a gazelle. He refuses the food I offer him and insists on eating grass. But I have no doubt that he is a human being who has lived all his life among gazelles.'

The largest deer in the world is the Alaskan moose (*Alces a. gigas*) which is found in the forested areas of Alaska, USA and the Yukon, Canada. Adult bulls stand *c.* 1·83 m *6 ft* at the withers and scale 499–543 kg *1100–1200 lb*. Two large bulls shot at Funny River, Alaska, measured 1·98 m *6 ft 6 in* and 2·05 m *6 ft 8¾ in* at the withers respectively, but they were not weighed. They are now on display in the American Museum of Natural History (Peterson, 1955).

Lockhard (1895), in a reference to the Alaskan moose, says: 'Those down at Peel River and the Yukon are much larger than up this way [Great Slave Lake]. There I have known two cases of extraordinary moose having been killed, the meat alone of each of them weighing about 1000 pounds.'

These weights imply a live avoirdupois of 771–816 kg *1700–1800 lb*, but Seton (1927) says the poundage of the meat was guessed. The same writer also points out that the standing height of an Alaskan moose may be 203–254 mm *8–10 in* less than the same measurement taken between pegs.

The Alaskan moose, the world's largest deer (J Boyer)

According to Whitehead (1972) the Alaskan moose reaches its maximum size on the Kenai Peninsula, specimens having been accurately measured up to 2·29 m *7 ft 6 in* at the withers, but no reliable weight data has been published for bulls taken in this region. (Estimated weights up to 1179 kg *2600 lb* have been claimed by hunters.) Cows are about three-quarters the size of bulls.

Not surprisingly, adult bull moose – four races are recognised in North America – can be formidable adversaries, and there is a record of an *Alces a. gigas* ramming a car and killing the driver (Caras, 1964).

The largest deer found in Britain is the Red Deer (*Cervus elaphus*) of Scotland, Devon, Somerset and the Lake District. A full-grown stag stands about 1·12 m *3 ft 8 in* at the withers and weighs 104–113 kg *230–250 lb*, while hinds are about two-thirds this size.

The heaviest wild red deer on record was probably a stag killed on the Isle of Islay, Inner Hebrides, Scotland, in 1940 which scaled 207 kg *456 lb* 'clean' without the liver, which would have accounted for another 2·72–3·62 kg *6–8 lb* (Whitehead, 1964). Another stag killed in Glenmore Deer Forest, Inverness-shire, Scotland, in 1877 weighed 209 kg *462 lb*.

The heaviest park red deer on record was a stag killed at Woburn, Bedfordshire, in 1836 which tipped the scales at 216 kg *476 lb* (height at withers 1·37 m *4 ft 6 in*). Another stag weighing 216 kg *472 lb* was killed in Warnham Court Park, Sussex, on 7 September 1926.

The smallest true deer (family Cervidae) is the Pudu (*Pudu mephistophiles*) of northern South America, adult males measuring 330–381 mm *13–15 in* at the withers and weighing 8·16–9·07 kg *18–20 lb*.

The Mouse deer (*Tragulus javanicus*) or Chevrotain of SE Asia, including Java, Borneo and Sumatra, is much smaller but this species is not a *true* deer because it has only three compartments to its stomach (cf. four for Cervidae). Adult specimens measure 203–254 mm *8–10 in* at the withers and scale 2·72–3·17 kg *6–7 lb*, which makes it the smallest member of the Ruminantia.

The rarest deer is Fea's muntjac (*Muntiacus feae*), which is known from only two specimens collected on the borders of Tennasserim, Lower Burma and Thailand (Tate, 1947).

The Black muntjac (*M. crinifrons*) is known only from three specimens collected in the State of Chekiang, SE China. Two were taken at Ningpo in 1885 and 1886 respectively, and the third at Tunglu in 1920 (Allen, 1938–40).

According to recent scientific observations muntjacs only sleep for 25 s at a time, which must be something of a record for a mammal. Most of their life is apparently taken up with the problems of searching for food, breeding, looking after their young and watching out for enemies like the leopard.

The greatest reliable age recorded for a deer is 26 years 6 months 2 days for a Red deer (*Cervus elaphus*) which died in the National Zoological Park, Washington, DC, USA, on 24 March 1941 (Jones, 1958). A female Malayan sambar (*Cervus unicolor equinus*) died in New York Zoological Gardens (Bronx Zoo) on 11 December 1955 aged 26 years 5 months 6 days (Crandall, 1964).

In November 1937 a hunter killed a large bull Elk (*Cervus canadensis*) near Winslow, Arizona, USA, which carried a Biological Survey ear-tag showing that it had been transported to Arizona from Wyoming in 1913 when nearly a year old (Murie, 1951). The maximum potential for this species – and the moose – is *c.* 30 years.

The largest of all marsupials is the Red kangaroo (*Macropus rufa*) of central, southern and eastern Australia.

In one series of 426 adult males or 'boomers' collected by Frith & Calaby (1969) in western New South Wales the heaviest specimen weighed 77 kg *169 lb* and measured 2·19 m *7 ft 2 in* (?) along the curve of the body nose tip to tail. Another one was 2·47 m *8 ft 1¼ in* long, but weighed only 70 kg *154 lb*, although the two researchers said it was a bigger framed animal than the other example, and would have been much heavier if it had been in better condition.

Lengths up to 3·35 m *11 ft* along the curve and weights up to 136 kg *300 lb* have been claimed by hunters in the past, but the probable size limit is *c.* 2·71 m *8 ft 11 in* and 82 kg *180 lb*.

The Eastern grey kangaroo (*Macropus giganteus*) of eastern Australia and Tasmania is also of comparable size and Lydekker (1893–6) says there is a record of a boomer measuring 2·91 m *9 ft 7 in* along the curve and weighing 91 kg *200 lb*. This report, however, appears to have been exaggerated because the skin, preserved in the Australian Museum, Sydney, NSW, only measures 2·47 m *8 ft 2 in* in length (Troughton, 1965).

The Red kangaroo – the largest living marsupial

Females of both species are much smaller and rarely exceed 27 kg *60 lb*.

Kangaroos roam in 'mobs' and the leader is known as 'the old man'. It is his task to keep the ambitious young bucks in order, and he often does so with vicious upper cuts and straights worthy of Joe Frazier – plus the odd kick or two in the abdomen with both hindlegs. Boxing thus comes naturally to large male kangaroos because their method of fighting with one another or with an enemy is to stand upright and use their strong tail, which can weigh up to 9 kg *20 lb*, to support the body, leaving them free to use their arms and legs in a combination of punching and kicking. The animal is at its most dangerous, however, when it has its back up against a tree or rock, and many an incautious hound has been disembowelled by a hindfoot kick.

In May 1972 a giant red kangaroo savagely mauled an 86-year-old man on Granite Island off the coast of South Australia. The man was rushed to hospital in Adelaide with multiple injuries and was later reported to be seriously ill, but further information is lacking.

This extraordinary attack probably took place during the mating season when adult male kangaroos battle each other for the affections of the females. In this case the boomer must have mistaken the unfortunate man – a living creature of much the same size as an adult buck – for a rival and treated him accordingly.

Both the red and grey kangaroos can leap for miles across open country at 40 km/h *25 miles/h* with seemingly boundless energy, and bursts up to 64 km/h *40 miles/h* have been recorded for young mature females ('blue flyers') over short distances, i.e. 274 m *300 yd*.

In 1963 racing-car drivers taking part in a meeting at Caversham, Western Australia, were suddenly interrupted by a red kangaroo. Doing a steady 48 km/h *30 miles/h* the marsupial bounded on to the track and leapt past cars as they dropped into low gear at a bend. Several alarmed drivers went into premature slides and others tried to frighten the animal away. But the kangaroo took no notice and pressed on, beating all the cars round the bend and scattering spectators as it bounded back into the bush.

On 21 July 1966 a circus boxing kangaroo named 'Fuji' escaped from his trainer in Tokyo, Japan, and with boxing-gloves on set off on a tour of the city with several police cars and pedestrians in hot pursuit. Police say he reached speeds of up to 64 km/h *40 miles/h* during the chase. When eventually cornered by apprehensive pedestrians Fuji put up a spirited defence and downed three of his opponents before a policeman put a Judo submission hold on him.

During the course of a chase in New South Wales in January 1951 a female red kangaroo made a series of bounds which included one of 12·80 m *42 ft*. There is also an unconfirmed record of a grey kangaroo jumping nearly 13·5 m *44 ft 8½ in* on the flat (Walker *et al*, 1968).

The high-jumping ability of the two species is equally as impressive. Usually they do not jump much higher than 1·52 m *5 ft* but 2·74 m *9 ft* fences have been cleared on occasion, and one terrified kangaroo chased by hunting dogs cleared a pile of timber 3·04 m *10 ft* high (Troughton, 1965).

Kangaroos are also excellent swimmers. One old boomer chased by dogs into the sea off the coast of Western Australia was last seen swimming strongly about 1·6 km *1 mile* offshore where it probably fell victim to sharks. In November 1974 a fisherman hooked a live 1·52 m *5 ft* long kangaroo more than a mile off the Victoria coast.

A few years ago a man was driving through Northern Territory when he accidentally knocked down a large kangaroo. Thinking the animal was dead the motorist dressed the corpse in his coat and propped it up against a tree so that he could take a photograph; the kangaroo however, had only been stunned in the crash, and at the click of the camera it recovered consciousness and bounded off – still wearing the coat which had £650 in Australian bank notes in one of the pockets! Several weeks later a kangaroo wearing an elegant sports jacket was reportedly seen in the area, but it is not known whether the well-heeled motorist ever got his money back.

It all sounds a bit far-fetched, like so many stories of the Australian bush, but it's an amusing tale nevertheless.

The smallest marsupial is the very rare Kimberley planigale (*Planigale subtilissima*), a flat-skulled marsupial 'mouse' found only in the Kimberley district of Western Australia. Adult males have a head and body length of *c.* 44 mm *1·73 in*, a 50 mm *1·96 in* long tail, and weigh about 4 g *0·141 oz*. Females are smaller.

The systematics of the genus *Planigale*, however, are still a subject of some controversy, and it may well be that *P. subtilissima* is a sub-species of *P. ingrami*, which is widely distributed across northern Australia.

The rarest marsupial is probably the little-known Thylacine (*Thylacinus cynocephalus*), also known as the 'Tasmanian tiger' or the 'Tasmanian wolf', the largest of the carnivorous marsupials, which – if not already extinct – is now confined to the remoter parts of SW Tasmania.

The last thylacine held in captivity was a male caught by a wallaby-trapper in the heavily forested Florentine Valley *c.* 96 km *60 miles* NW of Hobart in 1933. It was later exhibited at Hobart Zoo and died there the following year.

In January 1957 a thylacine was reportedly photographed and kept in sight for 2 min on Birthday Bay Beach, 56 km *35 miles* SW of Queenstown, by the pilot of a mining company helicopter, but experts who examined the pic-

The Tasmanian 'wolf', the largest of the carnivorous marsupials (Hobart Zoo)

ture said it was a large dog. A few months later three sheep were found dead in the Derwent Valley and it was claimed that their injuries corresponded with those that a thylacine might inflict. Robert Brown (1973), however, says a large dog was caught shortly afterwards in the trap-cage set up at the site.

In 1961 fishermen killed what they thought was a young Tasmanian tiger at isolated Sandy Cape on the west coast, but when they returned to their hut at the end of the day they found the body had disappeared. Hair samples taken from the site were later said to have been *positively identified* as those of a thylacine, but no scientific details have been published.

In December 1966 the traces of what was thought to have been a thylacine lair in which a female and pups had been living were found in a disused mine boiler at Whyte River, but an examination of hair samples and footprints seen near the lair revealed that the animal was in fact a Common wombat (*Phascolomis ursinus*).

On 3 November 1969 the tracks of a thylacine were said to have been *positively identified* in the Cradle Mountain National Park, and other *definite* sightings have since been made in the Cardigan River area of the north-west coast and the Tooms Lake region, but all this information is purely circumstantial.

Brown believes that many of the so-called 'sightings' in recent years have, in fact, referred to greyhounds of which there are several hundred in Tasmania. This dog is very similar in shape and size to the thylacine and, from a distance, could easily be mistaken for one of these tawny-striped marsupials.

A hundred years ago the thylacine was fairly common in Tasmania. Its prey then was the kangaroo and the wallaby, but as man invaded the primitive forests and steadily exterminated much of the indigenous wildlife, the thylacine was forced to seek alternative prey. It developed a taste for sheep and poultry, and this soon brought it into sharp conflict with the farmers, who began a relentless war of extermination.

In 1888 a Government bounty was brought in and official statistics show that between that date and 1914 2268 thylacines were killed, although the total figure was probably much higher. One Tasmanian and his brother shot 24 of these animals in one day and were paid £1 per head bounty money (Boswell, cited by Harper, 1945). On top of this, in 1910 the thylacine population was decimated by an outbreak of disease, possibly distemper (Simon & Geroudet, 1970).

The Sandhill dunnart (*Sminthopsis psammophia*), also called the 'Narrow-footed marsupial mouse', is known only from the type specimen collected by the Horn Expedition in 1894 near Lake Amadeus, Northern Territory, but this creature may eventually prove to be reasonably common in this unworked area.

In September 1973 another marsupial mouse new to science was caught in a trap in the Billiatt Conservation Park 160 km *100 miles* east of Adelaide, South Australia. This specimen, together with a mother and her recently born young, were later taken to the Institute of Medical and Veterinary Science in Adelaide for study.

The Brindled nail-tailed wallaby (*Onychogalea fraenata*), which was thought to have become extinct in 1937, was rediscovered in the Emerald District, central Queensland, in April 1974 by Dr G Gordon during a fauna survey.

The largest sirenian is the Florida manatee (*Trichechus t. manatus*), which was formerly widely distributed along the coasts of the Gulf of Mexico, the West Indies and the north-eastern parts of South America.

Adult males average 2·43–3·04 m *8–10 ft* in length and weigh 204–272 kg *450–600 lb*. Females are slightly smaller.

The largest manatee on record was one caught off the coast of Texas, USA, in *c.* 1910 which measured 4·64 m *15 ft 3 in* in length and weighed 594 kg *1310 lb* (Gunter, 1941).

The Dugong (*Dugong dugon*) of the Indo-Pacific region is a more heavily built animal than the manatee but has a proportionately shorter body. According to Mani (1960) the largest dugong on record was a female measuring 4·06 m *13 ft 4 in* in length and weighing nearly 1 tonne/*ton* landed by fishermen at Bedi Bunder off the Saurashtra coast, Indian Union, on 23 July 1959.

In October the previous year a male dugong measuring 3·09 m *10 ft 2 in* in length and weighing an estimated 453 kg *1000 lb* was captured alive off Malinda, 112 km *70 miles* north of Mombasa, Kenya. It was wanted by a film unit for a picture they were shooting and, after spending a couple of days in a hotel swimming-pool, the sirenian was returned to the sea.

Douglas Storer (1963) describes how a dugong measuring 3·65 m *12 ft* in length and weighing more than 1 tonne/*ton* hauled itself out of the sea at Punta Arenas, southern Chile, in 1942 and dragged its great bulk through the streets of the city. Eventually it tried to make its way back towards the sea, but was shot by

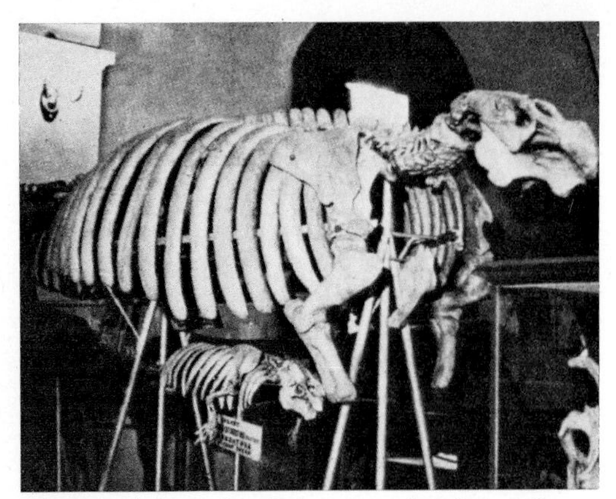

The huge skeleton of Steller's sea cow dwarfs that of an adult Dugong (E. Tratz)

policemen who thought they were dealing with a sea-monster.

This was a definite case of mistaken identity, however, because the dugong has a very weak rib cage and would soon be suffocated by its own body weight if it tried to move around on land. The 'monster' was in fact a vagrant Southern elephant seal (*Mirounga leonina*) from one of the sub-Antarctic islands.

It should be noted here that both the manatee and the dugong were dwarfed by Steller's sea cow (*Hydrodamalis gigas* = *H. stelleri*), first discovered in the waters round Bering Island and other small islands in the Komandorskie group in the Bering Sea in 1741. An adult female examined by Georg Steller, the famous German naturalist, measured 7·52 m *24 ft 8 in* in length and weighed an estimated 4000 kg *8818 lb* 3·93 tonne/*ton*, but some males may have reached 9·14 m *30 ft* and a weight of over 7 tonne/*ton*. This species reportedly became *extinct* through overhunting for food only 27 years after its discovery, but according to Russian sources the animal survived until at least 1830 in the same area.

In November 1934 the badly decomposed remains of what was believed to be a 9·14 m *30 ft* long Steller's sea cow were found on the shore of Henry Island, about 37 km *23 miles* from Prince Rupert, British Columbia, Canada. The strange creature had a 609 mm *2 ft* long head and four flipper-like appendages, two of which were still attached to the body. The skull – no teeth were evident – and vertebrae were later sent to the Government biological station at Nanaimo for examination and found to belong to a large Basking shark (*Cetorhinus maximus*) (see photograph page 141).

In July 1962 the crew of the Russian whaler *Buran* saw half a dozen unusual-looking creatures in a shallow lagoon near Cape Navarin to the south of the Gulf of Anadyr. The animals, measuring 6·09–7·92 m *20–26 ft* in length, were swimming round slowly and appeared to be browsing on seaweed, the staple diet of *H. gigas*. The description the men later gave to Russian scientists also fitted that of Steller's sea cow, but it has since been suggested that the creatures may have been female Narwhals (*Monodon monoceros*). Certainly it is difficult to believe that such a large animal could escape detection for so long.

The smallest sirenian is the freshwater Amazonian manatee (*Trichechus inunguis*). The maximum size reached by this species is 2·5 m *8 ft 3 in* and 140 kg *308 lb*. This manatee is also the most seriously threatened member of the family Sirenia, and is now confined to the Amazon Basin. It is protected by law, but there is no real enforcement and it is still regularly harpooned or shot for its meat.

Manatees reportedly live for more than 50 years in the wild, but captive specimens seldom live longer than eight years.

The longest recorded elephant tusks are a pair from the eastern Congo (Zaire). They were originally owned by King Menelek of Abyssinia (Ethiopia), who later presented them to a 'Euro-

The longest elephant tusks on record (New York Zoological Society)

pean political officer'. They were eventually put up for sale in London and were purchased by Rowland Ward Ltd, the famous Piccadilly taxidermists, who presented them in 1907 to the National Collection of Heads and Horns kept by the New York Zoological Society in Bronx Zoo. The right tusk measures 3·50 m *11 ft 5 in* along the outside curve and the left 3·35 m *11 ft*. Their combined weight is 133 kg *293 lb*.

Another huge pair of tusks measuring 3·39 m *11 ft 1¼ in* and 3·05 m *10 ft 0½ in* respectively, with a combined weight of 111 kg *244 lb 8 oz* were collected by T Christensen in Kenya in February 1959 (Rowland Ward, 1969), and another bull with tusks measuring 3·28 m *10 ft 9 in* and 3·25 m *10 ft 8 in* respectively (combined weight 115 kg *253 lb*) was shot by Mrs P Fluckiger in Kenya in March 1970 (Rowland Ward, 1971).

In 1904 W R Foran shot an old elephant bull on the slopes of Mt Kenya which carried only one complete tusk measuring 3·37 m *11 ft 1 in* in length, and weighing 77 kg *169 lb*. The other tusk, which was just a broken stump, scaled 13 kg *30 lb*.

The right tusk of a bull elephant is nearly always slightly longer and heavier than the left because the animal prefers to dig and root about with its left tusk. In the case of the Christensen elephant, however, the difference is rather pronounced.

The longest recorded Asiatic elephant tusks are a pair in the Royal Siamese Museum, Bangkok, which measure 3·01 m *9 ft 10 in* and 2·74 m *9 ft* respectively (Rowland Ward, 1928).

The heaviest recorded tusks are a pair in the British Museum (Natural History) London, which were collected from an aged bull shot by an Arab with a muzzle-loading gun at the foot of Mt Kilimanjaro, Kenya, in 1897. They were sold in Zanzibar (now part of Tanzania) in 1898 to an American company, Messrs Landesberger, Humble & Co, who in turn sent them to London for auction in 1901. The heavier of the two tusks was bought by the British Museum (Natural History) for £350, and the other one was acquired by Joseph Rodgers & Sons Ltd, a cutlery firm in Sheffield. In 1932 this tusk was sold to W B Wolstenholme Ltd, also of Sheffield, and the following year it was purchased by the British Museum (Natural History). The tusks were measured by Hill (1957) in 1955: one of them was 3·11 m *10 ft 2½ in* along the outside curve, weighing 102 kg *226 lb 8 oz*, and the other 3·19 m *10 ft 5½ in* and 97 kg *214 lb*, giving a combined weight of 199 kg *440 lb 8 oz*. Blunt (1933) claims the tusks weighed 107 kg *236 lb* and 102 kg *225 lb* respectively when they were *fresh*, i.e. 209 kg *461 lb* combined, but recent test weighings of elephant ivory indicate that the loss of weight during the drying-out period is only 453–907 g *1–2 lb* (Rowland Ward, 1971).

Another huge pair collected by Major P H G Powell-Cotton from an average-sized bull shot near Lake Albert on the Uganda–Congo border in 1905 and now preserved in the Powell-Cotton Museum at Quez Park, Kent, measure 2·74 m *9 ft* and 2·71 m *8 ft 11 in* respectively, and have a combined weight of 169 kg *372 lb* (90 kg *198 lb* and 79 kg *174 lb*).

According to Hallett (1967) the tusks of the famous Mohammed of Marsabit Mountain Reserve in northern Kenya were so long and heavy that the poor elephant couldn't raise his head properly and had to walk backwards to avoid getting them stuck in the ground. In 1950 gamewardens estimated that there was 3·04 m *10 ft* of tusk outside each gum and put their combined weight at 181 kg *400 lb*. When this elephant died of old age in 1960, however, the tusks were found to total just over 113 kg *250 lb*, the longest measuring 3·28 m *10 ft 9 in* in length and scaling 64 kg *141 lb* (Bere, 1966).

His even more celebrated successor Ahmed was said to have the longest and heaviest tusks of any living elephant, and in 1971 game circles estimated that they were both between 72 kg *160 lb* and 77 kg *170 lb* in weight. But when this legendary pachyderm died on 17 January 1974 it was discovered that the tusks, measuring 2·97 m *9 ft 9 in* and 2·84 m *9 ft 4 in* respectively, only totalled 134 kg *296 lb* (A MacKay, pers. comm. 21 April 1975).

Owing to selective shooting for sport and ivory very few large tusks are now left in Africa, and anything scaling over 32 kg *70 lb* must be considered exceptional.

The heaviest recorded Asiatic elephant tusks are a pair collected from a bull killed in the western Terai jungle, northern India, and later presented to King George V. The tusks, now in the British Museum (Natural History) have a combined weight of 146 kg *321 lb* (lengths 2·67 m *8 ft 9 in* and 2·60 m *8 ft 6½ in*).

The heaviest single tusk on record was a specimen weighing 117 kg *258 lb* collected in Dahomey, West Africa. It was put on display at the Paris Exposition of 1900, along with another tusk weighing 97 kg *214 lb* (a pair?). A H Neumann (1899) said another one weighing

The horns of the Indian buffalo have been measured up to 4·24m 13 ft 11 in *tip to tip along the outside curve*

113 kg *250 lb* was once on exhibition in the Zanzibar Custom Hall.

The longest (and heaviest) cow elephant tusks on record are a pair taken from an elephant shot by Peter Pearson, the famous Uganda game-ranger, on the NE shore of Lake Albert in 1923. The tusks, now in the British Museum (Natural History) measure 2 m *6 ft 7 in* and 1·92 m *6 ft 3½ in* respectively, and have a combined weight of 44 kg *98 lb.*

Cow tusks rarely exceed 7–9 kg *15–20 lb* in weight (some cows never grow tusks at all), but they are more valuable than those of bulls because of their closer grain.

The longest horns grown by any animal are those of the Indian buffalo (*Bubalus bubalis*). One huge bull shot in *c.* 1955 had horns measuring 4·24 m *13 ft 11 in* from tip to tip along the curve across the forehead, and two other heads measured 4·16 m *13 ft 8 in* and 4·06 m *13 ft 4 in* respectively (Powell, 1958). There is also a record of a single horn measuring 2 m *6 ft 6½ in* on the outside curve.

The Ankole cattle (*Bos taurus*) of Africa also grow exceptionally long horns and heads measuring in excess of 3·04 m *10 ft* have been reported. A measurement of 2·06 m *6 ft 9 in* has also been cited for a single horn collected near Lake Ngami, Botswana, by W C Oswell (Rowland Ward, 1907).

Texan longhorn steers have been credited with spreads up to 3·65 m *12 ft*, but any measurement over 2·13 m *7 ft* must be considered exceptional. Mr Jessie S Ivie (pers. comm. 18 October 1975) of Lancaster, Texas, USA, has four pairs of longhorn steer horns in his private collection which measure over 2·43 m *8 ft* from tip to tip in a straight line. His largest pair measure a colossal 2·97 m *9 ft 9 in.*

A pair of Texas longhorn steer horns (centre) measuring 2·97 m 9 ft 9 in *tip to tip (Jessie S Ivie)*

'Brooklyn Supreme', the 1451 kg 3200 lb *Belgian stallion (Ralph Fogleman)*

The heaviest of the 60 or so breeds of horse (*Equus caballus*) are the heavy draught breeds, all of which evolved from the Great Horse of Europe.

The heaviest horse ever recorded was a 19·2 hands 2 m *6 ft 6¾ in* pure-bred Belgian stallion named 'Brooklyn Supreme' (1928–48), owned by Ralph Fogleman of Callender, Iowa, USA. This giant weighed 1451 kg *3200 lb* in his prime and other statistics included a maximum body girth of 3·09 m *10 ft 2 in* and a 1·01 m *40 in* neck. The stallion was so powerful that one previous owner was afraid to team him up with other draught-horses on his farm in case he worked them to death (Fogleman, pers. comm. 30 April 1966).

'Wilma du Bos' (foaled 15 July 1966), a 18·2 hands 1·88 m *6 ft 2 in* Belgian mare owned by Mrs

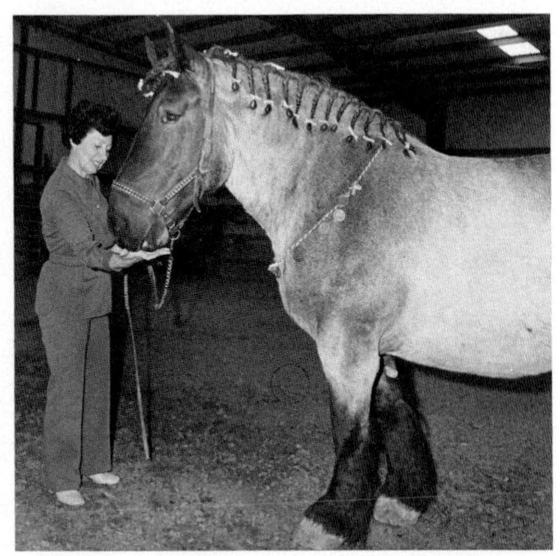

The Belgian mare 'Wilma' who, at the height of her pregnancy, may have been the heaviest horse on record (John Arden)

Virgie Arden of Reno, Nevada, USA, reportedly weighed just over 1451 kg *3200 lb* when she was shipped from Antwerp, Belgium, in December 1972, but when the horse arrived in New York just before Christmas she scaled 1400 kg *3086 lb*. These figures, however, are misleading because Wilma was heavily in foal at the time and had a girth measurement of 3·65 m *12 ft*. Her normal weight is about 1086–1133 kg *2400–2500 lb*.

The heaviest horse ever recorded in Britain was probably the 17·2 hands 1·78 m *5 ft 10 in* Shire stallion 'Honest Tom' foaled in Littleport, Cambridgeshire, in 1884, who weighed 1321 kg *2912 lb* in *c.* 1890. His 17·2 hands 1·78 m *5 ft 10 in* son 'Sandycroft Tom', foaled in Cheshire in 1892, weighed 1270 kg *2800 lb* in his prime (Keith Chivers, pers. comm. 17 November 1973).

Another Shire stallion called 'Mammoth', foaled at Toddington Mills, Bedfordshire in *c.* 1850 and later exported to America for exhibition purposes allegedly weighed nearly 1524 kg *3360 lb* 1·5 tonne/*ton*, but further details are lacking.

In 1942 a weight of 1168 kg *2576 lb* was recorded for the champion Suffolk stallion 'Monarch' at the Ipswich Stallion Show. This horse stood 17·2 hands 1·78 m *5 ft 10 in* and measured 2·53 m *8 ft 4 in* in girth.

Nowadays most breeders of heavy draught-horses aim for animals that are not unduly over-bulky otherwise they tend to lose the quality of their breeding.

The heaviest horse living in Britain today is 'Saltmarsh Silver Crest' (foaled 1955), a 18·1 hands 1·85 m *6 ft 1 in* champion Percheron stallion owned by George E Sneath of Money Bridge near Pinchbeck, Lincolnshire. He weighed 1257 kg *2772 lb* in 1967 (maximum girth 2·53 m *8 ft 4 in*).

The tallest horse ever recorded was the Clydesdale gelding 'Big Jim' (foaled 1950), bred by Lyall M Anderson of West Broomley, Montrose, Scotland, which stood 21·1 hands 2·16 m *7 ft 1 in*. He died in St Louis, Missouri, USA, in 1957 (Farrell, pers. comm. 18 March 1974). The same height was also reported for the Percheron-Shire cross 'Firpon' (foaled 1959) owned by Julio Falabella. This horse (weight 1350 kg *2976 lb*) died on the Recco de Boca Ranch, near Buenos Aires, Argentina, on 14 March 1972 (Mandel, pers. comm. 4 August 1972). In 1908 an unconfirmed height of 21·2

The 21·1 hands 2·16 m 7 ft 1 in *Percheron-Shire cross 'Fir pon' (J Falabella)*

hands 2·18 m *7 ft 2 in* was reported for a 1285 kg *2835 lb* Shire gelding called 'Morocco' in Allentown, Pennsylvania, USA.

The smallest breed of horse (*sic*) is the Falabella of Argentina which was bred down to Lilliputian size over a period of 45 years by crossing and recrossing a small group of undersized English Thoroughbreds with Shetland ponies. The first stallion bred by the Falabella family – with perfect horse conformation – was 'Napoleon', who stood 5 hands 51 cm *20 in* and weighed 32 kg *70 lb*, and the 700 Falabellas living in Argentina today are all descended from him.

Adult specimens stand 38–74 cm *15–30 in* at the shoulder and weigh 18–36 kg *40–80 lb*. The upper accepted limit for the American Miniature Horse Breeders' Association is 86 cm *34 in*, and in 1975 there were 1826 registrations.

The Falabella, incidentally, is an extremely fast runner for its size and can beat a racehorse over 91 m *100 yd*.

The smallest breed of pony (under 14 hands 1·42 m *4 ft 8 in*) is the Shetland pony which usually measures 8–10 hands 81–102 cm *32–40 in* and weighs 125–175 kg *275–385 lb*. In March 1969 a measurement of 3·2 hands 36 cm *14 in* was reported for a 'miniature' Shetland pony named 'Midnight', owned by Miss Susan Perry of Worths Circus, Melbourne, Victoria, Australia, but this animal was not fully grown.

There are several authentic records of horses living over 40 years. Edmund Crisp (1869) mentions a Suffolk mare that foaled at the age of 42, and Smyth (1937) gives a record of a 46-year-old brood mare which foaled for the 34th time at the age of 42. An Orkney mare died at Fintray, Aberdeenshire, Scotland, in 1936 aged 45 years.

The greatest reliable age recorded for a horse is 62 years in the case of 'Old Billy' (foaled 1760) believed to be a cross between a Cleveland and Eastern blood, who was bred by Mr Edward Robinson of Wild Grave Farm in Woolston, Lancashire. In 1762 or 1763 he was sold to the Mersey and Irwell Navigation Company and remained with them in a working capacity (i.e. marshalling and towing barges) until 1819 when he was retired to a farm at Latchford, near Warrington, where he died on 27 November 1822. The skull of this horse is preserved in the Manchester Museum, and his stuffed head is now on display in the Bedford Museum (Seyd, 1973).

A more recent and better documented case was that of a 17 hands 1·73 m *5 ft 8 in* light

Botswana's domestic Ankole cattle, which sometimes grow horns measuring more than 2·74 m 9 ft *tip to tip (G L Wood)*

The Falabella, the world's smallest breed of horse (J Falabella)

draught-horse named 'Monty', owned by Mrs Marjorie Cooper, of Albury, New South Wales, Australia, who was euthanased on 25 January 1970 aged 52 years because of progressing debility. This extreme age was later confirmed by Dr J W Watson, Reader in Veterinary Anatomy at the University of Melbourne (pers. comm. 17 June 1974), who said that the gelding was foaled in Laverton, NSW, in *c.* October 1917. The head (skull and mandible) and heart of this horse are preserved in the Museum of the School of Veterinary Science at the University of Melbourne.

In February 1960 an ex-Italian Army horse named 'Topolino', foaled in Libya on 24 February 1909, died in Brescia, Italy, aged *c.* 51 years.

Despite its diminutive size, the Falabella is also very long-lived. The stallion Napoleon, for instance was 43 years old at the time of his death, and the maximum life potential for this breed has been put at 50–55 years.

Ponies are generally longer-lived than horses on an average. Lord Rothschild told Flower (21 October 1919) that he had an authentic record of a pony living for 54 years in France, and there is an unconfirmed claim of 58 years for a Shetland pony (*The Times*, 3 May 1944).

In July 1970 a 66-year-old Welsh pony cob with 'a greying face and the arched back of extreme age' was reported to be living on a farm near Pebbles Bay, Gower Peninsula, South Wales (Cooper, pers. comm. 24 July 1970), but this longevity record was exaggerated. The actual age of this animal at the time of its death in April 1973 (killed in a car accident) was 38 years.

All heavy draught breeds are renowned for their great strength and pulling power, and in the

'Old Billy', the world's oldest horse (Manchester Museum)

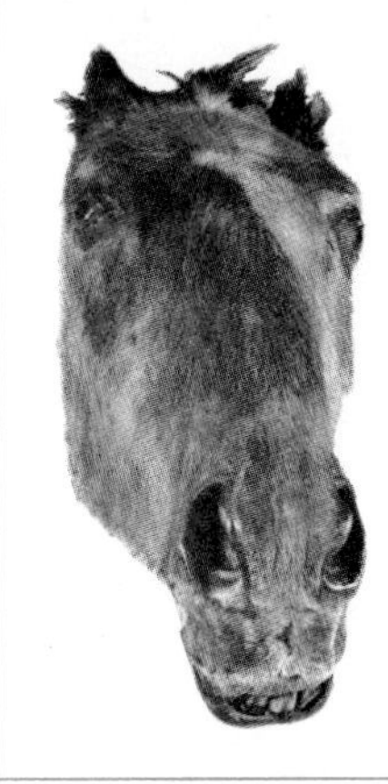
The preserved head of 'Old Billy' (Bedford Museum)

18th century hauling tests for big wagers were regularly held in England. Horses would often have to pull carts full of sand with the wheels partly sunk in the ground and wood blocks placed in front of them to make the test more difficult. According to Lady Wentworth (1946) 'these fearful weights could not be moved till the horses went down on their knees, a cruel test which must have ruptured many'.

In a carefully conducted test carried out at the Royal Agricultural Hall, London, on 24 February 1924 two eight year-old Shire geldings named 'Vesuvius' and 'Umber', yoked tandem-fashion and on slippery granite setts, moved off with 18·77 tonne *18 ton 9 cwt 2 qr* behind them, the shaft horse moving the load before the trace horse got into his collar. Six tonne/*ton* was moved on the tan and 16·5 tonne/*ton* on the wooden blocks. In an earlier test in Liverpool the same pair (combined weight 1590 kg *3498 lb*) hauled a 20 tonne/*ton* load of cotton, plus the 2 tonne/*ton* of the wagon, for 46 m *50 yd* over

The greatest load ever pulled by a pair of draught horses

granite setts, but this was an easier load to pull than the deadweight of metal in the later demonstration. They also pulled against a dynamometer at the Wembley Exhibition and registered a pull equal to a starting load of 51 tonne *50 ton*.

On 24 September 1935 two Percheron stallions named 'Rock' and 'Tom' owned by George G Statler Farms, Ohio, USA, pulled a load of 1769 kg *3900 lb* a distance of 11·4 m *37 ft 6 in* at Hillsdale, Michigan. The load was placed on a specially built wagon with four dual pneumatic-type wheels, chains over the tyres and all wheels locked. The pull exerted by these two horses was stated to have been equivalent to hauling 22·4 tonne/*ton* on a wagon for 20 consecutive starts on granite block pavement (Imhof, pers. comm. 27 October 1966).

The heaviest load ever hauled by a pair of draught-horses was an incredible 130·6 tonne *128·5 ton* by two Clydesdales (combined weight 1587 kg *3500 lb*) at the Nester Estate, Michigan, USA, on 26 February 1893. The load, consisting of 50 logs of white pine 36055 board ft of timber was placed on a special sledge litter and pulled across snow for a distance of 259 m *50 rods 284 yd*. After the test the logs were transported to the World Fair in Chicago and put on display. When the fair ended a man bought the wood and used it to build himself a home (Kebler, pers. comm. 29 November 1966).

The largest breed of heavyweight cattle is the Chianini which was brought to Italy from the Middle East by the Etruscans. Mature bulls usually stand 1·83 m *6 ft* at the forequarters and scale about 1360 kg *3000 lb*, but they can grow to 1·98 m *6 ft 6 in* and 1814 kg *4000 lb*.

The highest poundage recorded for a specimen of heavyweight cattle is 2140 kg *4720 lb* for a Hereford-Shorthorn named 'Old Ben', owned by Mike and John Murphy of Miami, Indiana, USA. When this animal died at the age of eight in February 1910 he had attained a length of 4·92 m *16 ft 2 in* from nose to tip of tail, a maximum girth of 4·16 m *13 ft 8 in* and a height of 1·93 m *6 ft 4 in* at the forequarters. The stuffed and mounted steer is now on display in Highland Park Museum, Kokomo, Indiana, as proof to all who would otherwise have said 'there ain't no such animal'.

A Holstein-Friesian bull named 'Mac' owned by Mr J D Avery of Massachusetts, USA, weighed 2099 kg *4628 lb* at slaughter and measured 4·85 m *15 ft 11 in* in total length.

The British record is held by the famous 'Bradwell Ox', owned by William Spurgin of Orpland Farm, Bradwell on Sea, Essex, which weighed 2036 kg *4480 lb* in 1830. It had a nose to tail length of 4·57 m *15 ft* and a maximum girth of 3·35 m *11 ft*.

The Hereford-Friesian bull 'Big Bill Campbell' owned by Major C H Still of Hall Farm,

The stuffed body of the enormous Hereford-Shorthorn 'Old Ben' (Highland Park Museum)

The famous 'Bradwell Ox', which weighed 2036 kg 4480 lb *in 1830 (University of Reading Museum)*

Northamptonshire, weighed 1778 kg *3920 lb* at his peak in 1954, but by 1956 his weight had dropped to 1651 kg *3640 lb*. This animal stood over 1·83 m *6 ft* at the forequarters and measured 3·81 m *12 ft 6 in* from nose to tail. He was destroyed on 13 June 1958.

The largest pig ever recorded was 'Big Boy', a hog bred by B Liles and H A Sanders of Black Mountain, North Carolina, USA, which recorded a weight of 863 kg *1904 lb* on 5 January 1939.

The British record is held by a pig bred by B Rowley of Doncaster which scaled 609 kg *1344 lb* at the age of four years (*c.* 1955). In September 1957 a live weight of 549 kg *1211 lb* (465 kg *1025 lb* dead) was reported for a three-year-old Large White owned by G Osborn of Thorndon Park, Brentwood, Essex.

In 1773 a weight of 639 kg *1410 lb* was claimed for a pig owned by Joseph Lawton of Astbury, Cheshire, but this poundage was wrongly recorded. According to the Annual Register for that year the pig, which measured 2·94 m *9 ft 8 in* in total length and had a girth of 2·51 m *8 ft 3 in*, had a live weight of 492 kg *1085 lb* (421 kg *928 lb* dead).

The smallest breed of pig in the world is the Miniature Japan, a long-haired black species bred by Hiroshi Ohmi of the Tochigi Prefecture. Mature specimens average 51 cm *20 in* in length and weigh *c.* 30 kg *66 lb*.

The largest litter of piglets on record was one of 34 thrown on 25–6 June 1961 by a sow owned by Aksel Egedee of Denmark. In February 1955 a Wessex sow owned by Mrs E C Goodwin of Paul's Farm, Leigh, Kent, also had a litter of 34, but 30 of these were stillborn.

The huge Old English mastiff 'Monty', who attained a weight of 120 kg 264 lb *shortly before his death (G L Wood)*

The heaviest breed of Domestic dog (*Canis familiaris*) is the St Bernard. The heaviest one on record was 'Schwarzwald Hof Duke', also known as 'Duke', owned by Dr A M Bruner of Oconomowoc, Wisconsin, USA. He was whelped on 8 October 1964 and weighed 134 kg *295 lb* on 2 May 1969, dying three months later aged 4 years 10 months. This dog measured 82·5 cm *32½ in* at the shoulder, and 139 cm *55 in* round the chest (Bruner, pers. comm. 8 August 1969).

The heaviest St Bernard ever recorded in Britain is 'Burton Black Magician', also called 'Shane' (whelped 11 January 1969) owned by Mrs Sheila Bangs of Mitcham, Surrey. In October 1973 he scaled a peak 120·8 kg *266 lb 4¾ oz*. His weight now fluctuates between 111 kg *245 lb* and 114 kg *252 lb*.

'Montmorency of Hollesley', also known as 'Monty', an Old English mastiff owned by Mr Randolph Simon of Wilmington, Sussex, was also of comparable size. He was whelped on 1 May 1962 and weighed 117 kg *259 lb* in July 1969. In April 1970 his weight was estimated to be 120·6 kg *266 lb*, at which time he measured 142 cm *56 in* round the chest and 84 cm *33 in* round the neck. He was put to sleep in April 1971 aged 8 years 11 months.

'Dominic', the tallest Great Dane on record

The only other breeds of dog which have been known to exceed 91 kg *200 lb* are the Great Dane and the Irish wolfhound. In January 1959 a weight of 102 kg *225 lb* was reported for a Great Dane named 'Zazon of Clarendon', owned by Mr Ernest Booth of Romiley, Cheshire. Another specimen named 'Simon', owned by Mr Terry Hoggarth of North Wingfield, near Chesterfield, Derbyshire, scaled 105 kg *231 lb* in September 1972. Miss Sheelagh Seale, owner of the Ballykelly Irish Wolfounds in Arklow, Co. Wicklow, Ireland, said she once had two dogs which both weighed over 101 kg *224 lb* (pers. comm. 27 October 1971). In September 1971 a weight of *308 lb* 140 kg was reported for a champion Irish wolfhound (shoulder height 91 cm *36 in*) living in Plymouth, Devon, but this extreme poundage was believed to have been a misprint for *208 lb* 94 kg.

The tallest breeds of dog are the Irish wolfhound and the Great Dane, both of which can exceed 99 cm *39 in* at the shoulder. In the case of the Irish wolfhound the extreme recorded example was 'Broadbridge Michael' (whelped in 1920) owned by Mrs Mary Beynon of Sutton-at-Hone, Kent. He stood 100·3 cm *39½ in* at the age of two years. The Great Dane 'Brynbank Apollo', also called 'Dominic' (whelped in 1970), owned by Mrs Iris Bates of Harlow, Essex, recorded a measurement of 99·7 cm *39¼ in* on 21 April 1973.

The smallest breed of dog is the Chihuahua which came originally from Mexico. New-born pups average 99–127 g *3·5–4·5 oz* and weigh 0·9–1·8 kg *2·4 lb* when fully grown, but some 'miniature' specimens can weigh less than 453 g *16 oz*.

The world's smallest fully grown Chihuahua is a bitch of 283 g *10 oz* owned by Rodney M Sprott of Clemson, South Carolina, USA.

The smallest British breed is the Yorkshire terrier. In April 1971 a weight of 283 g *10 oz* was reported for an adult bitch named 'Sylvia' owned by Mrs Connie Hutchins of Walthamstow, Greater London.

Authentic records of dogs living over 20 years are extremely rare, but even 34 years has been accepted by one authority (Lankester, 1870).

The greatest reliable age recorded for a dog is *c.* 27 years 11 months for a Welsh sheepdog cross called 'Toots', owned by Mr Barry Tuckey of Glebe Farm, Stockton, Warwickshire. The dog arrived at the farm as a twelve-week-old puppy in *c.* April 1948 and died on 4 November 1975 (Tuckey, pers. comm. 9 April 1976).

On 28 November 1963 the death was reported of a black Labrador gun-dog named 'Adjutant', aged 27 years 3 months. He was whelped on 14 August 1936 and died in the care of his lifetime owner James Hawkes, a gamekeeper at the Reversby Estate, near Boston, Lincolnshire.

In 1974 the death was reported of a Cocker spaniel bitch aged 29 years in Greytown, Ohio USA (Foster, pers. comm. 4 July 1974), but this record has not yet been substantiated. There is also an unconfirmed claim of 29 years 5 months for a Queensland 'heeler' named 'Bluey' who died in Melbourne, Victoria, Australia, in February 1940.

In July 1975 a Pekingese bitch named 'Trixie' owned by Miss Elizabeth Reeves of Bentley, Staffordshire, celebrated her 25th birthday.

The rarest breed of dog recognised by the Kennel Club is the Portuguese water dog (Caes de Agua). In February 1974 Mrs Herbert H Miller, Jr of New Canaan, Connecticut, USA, Corresponding Secretary of the Portuguese Water Club of America, told the compiler that the world population was no more than 50, of which total 32 were in the USA and 18 in Canada, Portugal and Spain.

In January 1975 only 14 specimens were known to exist of the Shar-pei or Chinese Fighting Dog – four in Hong Kong and the rest in the USA – but this breed is only recognised by

The Portuguese water dog – the world's rarest breed of dog

The Shar-pei or Chinese Fighting Dog (R. Albright)

the Hong Kong Kowloon Kennel Association. The dog, which has been likened to a wild boar, became extinct in China because of the restrictions on trading and ownership (Albright, pers. comm. 28 January 1974).

The Broholmer, a breed only recognised in Denmark, reportedly became extinct in the late 1960s, but in December 1974 a dog turned up at the home of a pharmacist in Helsinge. The Danish Royal Veterinary College in Copenhagen tried to set up a frozen sperm bank for 'Bjoern', as he was called, in the hope that a bitch would eventually be found to continue the line, but the experiment was unsuccessful. The dog died on 4 January 1975 (Boultwood, pers. comm. 16 January 1975).

The fastest breed of dog (excluding the greyhound) is the Saluki, also called the 'Arabian gazelle hound'. Speeds up to 69 km/h *43 miles/h* have been claimed for this breed (Ash, 1927), but tests in the Netherlands have shown that it is not as fast as the racing greyhound which has attained a measured speed of 66·7 km/h *41·7 miles/h* on a track. It is generally conceded, however, that the saluki has more stamina than the greyhound and can easily outstrip it over 914 m *1000 yd*.

The Ibizan hound is also very swift, but an American claim that a well-trained male 'can run at 65 mph over a great distance' must be considered ludicrous. However, the compiler was recently told by the owner of one of these dogs that he had once seen it *out-sprint* a saluki on reasonably level ground.

The Afghan hound, which is often confused with the saluki, has a top speed of *c*. 48 km/h *30 miles/h*.

The canine high jump record is held by an Alsatian named 'Crumstone Danko', owned by the De Beers mining company, who scaled a 3·43 m *11 ft 3 in* high wall in a leap and scramble at a demonstration in Pretoria, Cape Province, South Africa, in May 1942. The same dog was also credited with a 5·03 m *16 ft 6 in* jump off a springboard. The British record is held by another Alsatian named 'Lancon Sultan IV', handled by PC John Evans, who scaled a 3·20 m *10 ft 6 in* wall at Hutton, Lancashire, on 8 August 1973.

The Alsatian dog 'Crumstone Danko' scaling a 3·43 m 11 ft 3 in *high wall in Pretoria, South Africa*

The St Bernard dog 'Worldtop's Kashwitna V Thor' with his owner after pulling 2721 kg 6000 lb *for a new world record (Barbara E Wolman)*

The canine long jump record is held by a greyhound named 'Bang' owned by the Hon Grantley Berkeley, who cleared a distance of 9·14 m *30 ft* over a gate while coursing a hare at Brecon Lodge, Gloucestershire, in 1849. The dog damaged its pastern bone on landing but still managed to kill his quarry (Ash, 1927). Crumstone Danko (see p. 75) once jumped 7.31 m *24 ft*, and the same distance has also been reported for a saluki.

The greatest load ever shifted by a dog was 2721 kg *6000 lb* of lead ingots pulled by a 81 kg *177 lb 8 oz* St Bernard named 'Worldtop's Kashwitna V Thor' at Bothell, Washington, USA, on 9 August 1974. The dog, owned by Francis and Arloa B Good of Palmer, Alaska, pulled the weight on a four-wheeled carrier across a cement surface for a distance of 4·57 m *15 ft* in less than 20 s. (Under International Sled Dog Association Rules the weight must be pulled the required distance in under 90 s.) The same dog has pulled over 4535 kg *10 000 lb* in exhibitions in Alaska, but Mrs Barbara E Wolman, President of the Northwest Newfoundland Club, which organises the annual world championship dog weight hauling contests, said that the feats involved the use of mining carts on rails. She added: 'One of our members is a well known engineer for the railroad, and he reports that by using steel to steel contact, with fine ball bearings and superior lubrication, it is possible to move an almost infinite amount of weight' (pers. comm. 3 September 1975).

On 28 July 1975, in *unfavourable* conditions, a 40 kg *89 lb* Newfoundland yearling bitch named 'Barbara-Allen's Sea Duchess', owned by Mr and Mrs David Clutter of San Ramon, California, USA, pulled 1415 kg *3120 lb* the required 4·57 m *15 ft* in 40 s at Bothell. The animal thus pulled 15·7 kg *34·8 lb* per pound of body-weight, making her proportionately the strongest canine in the world (cf. 15·3 kg *33·87 lb* for Worldtop's Kashwitna).

A 73 kg *160 lb* Alaskan malamute named 'Lobo' owned by Mr Howard Baron of Big Bear City, California, USA, once pulled a 4535 kg *10 000 lb* truck and trailer for a distance of 6·09 m *20 ft* on a level road, but the weight was moved on highly inflated pneumatic tyres.

The greatest ratter of all time was Mr James Seale's bull terrier bitch 'Jenny Lind', who killed 500 rats in 1 h 30 min at The Beehive, Old Crosshall St, Liverpool, on 12 July 1853. Another specimen named 'Jacko', owned by Mr Jemmy Shaw, was credited with killing 1000 rats in 1 h 40 min, but the feat was accomplished over a period of ten weeks in batches of 100 at a time. The last 100 were accounted for in 5 min 28 s in London on 1 May 1862.

The record litter of 23 puppies thrown by the foxhound bitch 'Lena' in 1944 (F Curran)

The St Bernard bitch 'Careless Ann' with the 15 survivors of her 23-pup litter (R. Rodden)

The largest recorded litter of puppies was one of 23 thrown on 19 June 1944 by 'Lena', a foxhound bitch owned by Commander W N Ely of Ambler, Pennsylvania, USA. On 6/7 February 1975 'Careless Ann', a St Bernard bitch owned by Robert and Alice Rodden of Lebanon, Missouri, USA, also produced a litter of 23 puppies, 15 of which survived (Rodden, pers. comm. 19 February 1975). The British record is held by 'Settrina Baroness Medina', a Red setter bitch, owned by Mgr M J Buckley, Director of the Wood Hall Centre, Wetherby, West Yorkshire, who gave birth to 22 puppies, 15 of which survived, on 10 January 1974 (Buckley, pers. comm. 24 February 1975). On 4/5 June 1972 a five-year-old Alsatian bitch named 'Sheba', owned by Mr Michael Hawes of Manchester, reportedly gave birth to 24 puppies (20 survived), but this claim has not been authenticated.

The most improbable litter of puppies on record was one produced by a Great Dane in December 1972 after an illicit love-affair with . . . a Dachshund! No one quite knows for certain how the sausage dog managed this seemingly impossible feat, but Mrs Paddy Homer of Llandovery, South Wales, owner of the two animals, thinks the canine Romeo sneaked up on the love of his life when she was sleeping on the lawn. The 13 'Little Danes', or 'Great Dachshunds' as they have also been called, are all short in the leg like their father, but some have the Great Dane head and ears.

The greatest sire of all time was the champion greyhound 'Low Pressure', nicknamed 'Timmy', whelped September 1957 and owned by Mrs Bruna Amhurst of Regent's Park, London. From December 1961 until his death on 27 November 1969 he fathered 2414 registered puppies, with at least 600 others unregistered. A Basset hound called 'Hamlet' (whelped in 1967), owned by Mr Keith Keen of Kensal Rise, London, sired 1973 puppies before he was withdrawn from stud in 1973 after a groin injury.

The most valuable dog on record is the champion greyhound 'Super Rory' (whelped October 1970), owned by Mrs Judith Thurlow of Great Ashfield, Suffolk. In June 1972 she turned down an offer of £14000 for him. In January 1956 Miss Mary de Pledge of Bracknell,

Berkshire, refused an offer of £10500 from an American breeder for her champion Pekingese 'Caversham Ku-Ku of Yam'.

The greatest tracking feat on record was performed by a Doberman named 'Sauer' trained by Detective-Sergeant Herbert Kruger. In 1925 he tracked a store-thief 160 km *100 miles* across the Great Karroo, a dry tableland region of Cape Province, South Africa, by scent alone.

On 13 June 1974 an eight-year-old miniature fox terrier named 'Whisky' wobbled into Mambrey Creek Station, 240 km *150 miles* north of Adelaide, South Australia, after travelling approximately 2720 km *1700 miles* across the country's vast central wilderness. The dog had been lost by his owner, Adelaide truck-driver Geoff Hancock, at Hayes Creek 192 km *120 miles* south of Darwin, Northern Territory, in October 1973 (Hancock, pers. comm: 7 October 1974).

The greatest canine mountaineer of all time was a Beagle-type bitch named 'Tschingel' who, between 1868 and 1876 made 53 heavy ascents, eleven of them virgin summits, and hundreds of less important peaks or partial ascents in the Alps.

This incredible dog was born in a small village in the Bernese Oberland, Switzerland, in 1865. Like all good Alpinists she started her climbing career when still very young, and her first important expedition was the crossing of the Tschingel Pass after which she was subsequently named. In 1868 her owner Christian Almer, one of the greatest of all Swiss guides, presented the four-legged mountaineer to the Rev William Coolidge, one of his best clients, and during the next eight years the dog followed her new master up nearly every peak worth climbing in the Alps, including the Jungfrau (4165 m *13668 ft*) and the Eiger (3974 m *13040 ft*). In between times she also managed to give birth to 34 puppies! Tschingel had a marvellous instinct for guessing where a hidden crevasse lay, and also for finding the best way across an open one. Eventually she became so experienced in snow conditions that she could tell by merely looking at a snow bridge whether it was safe to cross, and when she drew back from a certain spot the two-legged climbers with her knew at once that it would be too dangerous to venture along that route. On one occasion she actually showed a local guide the way down some difficult rocks. On dangerous slopes Tschingel was roped through the collar to the human members of the party. In 1875, Tschingel, who had been made an honorary member of the famous Alpine Club, accompanied Coolidge up Mont Blanc (4810 m *15668 ft*), Europe's highest mountain, and when she returned to Chamonix she was given a cannon salute. Tschingel died in 1881 and was buried at Dorking, Surrey.

The 9-year-old heavyweight ginger and white cat 'Spice' after reducing to 15·9 kg 35 lb *(Loren Caddell)*

The heaviest domestic cat on record (*Felis catus*) was a nine-year-old ginger and white tom named 'Spice', owned by Mrs Loren Caddell of Ridgefield, Connecticut, USA, who tipped the scales at 19·54 kg *43 lb* on 26 June 1974 (Pandolfi, pers. comm. 1 September 1974). This obesity was due to an acute hypothyroid condition, and prescribed medication and dieting have since reduced the cat's weight to a more reasonable 15·87 kg *35 lb*.

The British record is held by a female tabby named 'Gigi' (1959–72), owned by Miss Anne Clark of Carlisle, Cumbria. This cat *normally* fluctuated between 16·78 kg *37 lb* and 18·14 kg *40 lb*, but in April 1970 she recorded a peak weight of 19·05 kg *42 lb* (total length 91 cm *36 in*, girth 94 cm *37 in*). A ginger tom called 'Dinkie' living in Minchinhampton, Gloucestershire, was also credited with the same poundage in April 1955 (total length 101 cm *40 in*, girth 81 cm *32 in*, neck 46 cm *18 in*), but further information is lacking. The average adult weight is 5 kg *11 lb*.

The smallest adult cat on record was probably a dwarf tabby named 'Jasper' (b 1946), owned by Mrs G Avery of West Ham, London, who never got above 1·58 kg *3 lb 8 oz* during his 15 years of life.

Cats are normally longer-lived animals than dogs, and there are a number of authentic records over 20 years. Mellen (1940), for instance, found 16 in North America, including two cats which

The 20-year-old twin cats 'Ginger' and 'Sandy' (Pat Hillman)

were both 31 years old. Information on the subject, however, is often obscured by two or more cats bearing the same nickname in succession.

The oldest cat ever recorded was probably the tabby 'Puss', owned by Mrs T Holway of Clayhidon, Devon, who died on 29 November 1939 aged 36 years 1 day. A more recent and better-documented case was that of the female tabby 'Ma', owned by Mrs Alice St George Moore of Drewsteignton, Devon. She was put to sleep on 5 November 1957 aged 34 years. Another cat named 'Bobby', owned by Miss B Fenlon of Enniscorthy, Co. Wexford, Ireland, died on 5 July 1973 aged 32 years 3 weeks 1 day.

In November 1958 the American Feline Society received a report that a cat living in Hazleton, Pennsylvania, had just celebrated its 37th birthday, but when an official of the Society made further inquiries he discovered that the claim was based on the combined life-spans of at least two cats! (Kendell, pers. comm. 19 March 1966.)

On 2 January 1972 a man living in Dumfriesshire, Scotland, announced that his cat (unnamed) had just passed the 43-year mark, but when Mr W Ferguson, a member of the Scottish Cat Club, wrote to him for further information he received the following answer: 'In reference to the cat, I am sorry to say it was killed this morning by a train. My brother brought it from Milligonbush Farm where he worked to Horsoholm in the summer of 1939. I have lived with the cat all my life. We were on the farm for 36 years. My brother who fetched the cat was killed by a train in the same place a month ago. I am the only one left but I assure you everything I say is true. I am 64 years old and I gave up farming and bought a house and have lived here ever since.'

On 18 May 1976 'Ginger' and 'Sandy', twin cats owned by Miss Pat Hillman of Birmingham, celebrated their 20th birthdays.

The largest live litter ever recorded was one of 14 kittens born in December 1974 to the Persian cat 'Bluebell', owned by Mrs Elenore Dawson of Wellington, Cape Province, South Africa. A litter of 14 kittens was also born in the NAAFI club at the RAF station Elsham Wolds, Lincolnshire, in 1946 or early 1947, but two of them were stillborn (Rouse, pers. comm. 16

The 11 survivors of the one-year-old Siamese cat 'Seeley's' 13-kitten litter.

December 1971). Another litter of 14 kittens was also reported for a calico cat named 'Kelly', owned by Miss Kim J Bean of Seneca, Missouri, USA, in May 1972, but only nine were born alive. On 27 April 1975 a 26-month-old Siamese called 'Tikatoo', owned by Mr Laurie Robert of Havelock, Ontario, Canada, reportedly gave birth to 15 kittens, but this record had not yet been fully substantiated.

The British record is held by a seal point Siamese named 'Seeley', owned by Mrs Harriet Browne of Southsea, Hampshire, who on 23 April 1972 gave birth to a litter of 13 kittens, eleven of which survived. On 30 April 1971 a litter of 13 (eleven stillborn) was also reported for a nine-month-old cat named 'Spur', owned by Mrs Grace Sutherland of Walthamstow, London. Shortly afterwards the mother went permanently blind.

In July 1970 a litter of 19 kittens (four incompletely formed) was allegedly born by Caesarean section to 'Tarawood Antigone', a brown Burmese, owned by Mrs Valerie Gane of Church Westcote, Kingham, Oxfordshire, but this may have been an abortion from more than one mating.

A cat named 'Dusty', aged 17, living in Bonham, Texas, USA, gave birth to her 420th kitten on 12 June 1952. A 21-year-old cat called 'Tippy' living in Kingston-upon-Hull, Humberside, gave birth to her 343rd kitten in June 1933, but died shortly afterwards.

Domestic cats are surprisingly hardy creatures and can survive for long periods without food and water. The endurance record is probably held by a cat of unknown parentage which was discovered, emaciated and starving, in a crate containing unassembled motor-car components in Durban, Natal, South Africa, on 1 November 1955. The packing-case had been sealed at the Morris Works at Cowley, Oxfordshire, on 1 August, which means the unfortunate moggy had been trapped for three months. Sadly the feline prisoner died early the next morning despite the best medical attention and a post-mortem revealed that the animal had survived the voyage by living on grease-smeared pieces of paper.

There was a happier ending for 'Peter', a three-year-old cat who survived eight days under water in a cabin of the Dutch MV *Tjoba* when it capsized and sank in the River Rhine near St Goar on 14 December 1964. The cat, which somehow managed to keep his head in an air-pocket while the rest of his body lay under water, was rescued when the vessel was raised and made a full recovery.

In July 1974 a kitten survived 53 hours of entombment in a concrete wall in Skopje, Yugoslavia. The cat, which had been adopted by the men working on a new cultural hall on the banks of the Vadar River, disappeared one day when the labourers were setting up the planks to mould a concrete wall. When the timber was removed 48 hours later the kitten was found jammed between the bottom plank and the cement. It had managed to survive in its prison by breathing through a small crack in the wood. The impression made by the cat's body in the concrete was later preserved for future generations to see.

The greatest height from which a cat has survived a fall is *c.* 55 m *180 ft* for a female named 'Quincy' who slipped from a 19th storey balcony on Broadway Avenue, Toronto, Ontario, Canada, on 21 April 1973. According to her owner, Mr Peter Thompson (pers. comm. 2 August 1973) who witnessed the accident the cat plummeted into shrubs directly beneath his balcony and suffered a broken leg and slight lung bruises which necessitated a nine-day stay in a veterinary hospital. On 25 April the previous year an 18-month-old part-Persian cat named 'Paul', left behind by his owner in a locked apartment on the 26th floor (*c.* 75 m *246 ft*) in the same city, was found an hour later huddled on the lawn below. This fall, however, was not witnessed by human eyes, and the fact that the cat suffered no broken bones suggests that the animal may have made his way to the ground by an easier route.

The British record is held by a tom cat named 'Pussycat', of Maida Vale, London, which slipped and fell 36·5 m *120 ft* from the balcony of his mistress's 11th-storey flat on 7 March 1965. The cat landed unhurt on all paws, but the shock of the impact blunted his appetite for several days afterwards.

Although dogs have a much better sense of direction than cats, there are a number of authentic cases of cats walking more than 160 km *100 miles* to their old homes when their owners had moved.

The greatest distance covered by a cat under these circumstances is 1528 m *950 miles* for a ginger tom named 'Rusty' who followed his owner from Boston, Massachusetts, to Chicago, Illinois, USA, in 1949. It took him 83 days to complete the journey.

The British record is held by a three-year-old tabby named 'McCavity' who in 1960 walked 800 km *500 miles* to his old home at Kea, near Truro, Cornwall, from Cumbernauld near Glasgow, Scotland, in *three weeks*! The cat was so exhausted after his marathon journey that he was unable even to lap milk and died the following morning.

The most travelled cat on record is a female Siamese called 'Princess Truman Tao-Tai', who has notched up more than 2 400 000 km *1 500 000 sea miles*. She joined the crew of the British iron-ore carrier *Sagamire* as a kitten in 1959 and has never been allowed ashore because of quarantine laws. In February 1975 the owners of the ship, Furness-Withy, sold the vessel to an Italian firm, but only on the condition that the buyers would do their best to keep the Siamese happy in her old age.

The greatest mouser on record was a tabby named 'Mickey' owned by Shepherd & Sons Ltd of Burscough, Lancashire, who killed more than 22 000 mice during his 23 years with the firm. He died in November 1968.

The greatest ratter on record was a female tabby called 'Minnie' who, during the six-year period 1927–33, accounted for 12 480 rats at the White City Stadium, London.

The most eccentric cat on record was probably a black female named 'Mincha' who ran up a 12 m *40 ft* high tree in Buenos Aires, Argentina, and never came down again. The local people used to push food up to her on a pole and a milkman delivered daily. In 1954 the cat was still going strong after spending six years aloft, and had even managed to have three lots of kittens during her marathon tree squat!

Section II

BIRDS

(class Aves)

A bird is a warm-blooded, air-breathing bipedal vertebrate covered with feathers and having the forelimbs modified into wings which are sometimes rudimentary and useless for flight. The brain is well developed, and the jaws are covered with horny sheaths forming a beak. Young are produced from eggs. Other features include a four-chambered heart and an essentially high constant body temperature because flight requires a tremendous output of energy over a long period.

The earliest known birds were the magpie-sized *Archaeopteryx* and *Archaeornis*, which lived about 140000000 years ago. Unlike modern birds, however, they had well-developed teeth in their jaws and a long jointed bony tail like a lizard's, but their wings bore characteristic feathers. Their remains were first discovered in Bavaria, West Germany, in 1861.

The class *Aves* was originally divided into two subclasses, the *Ratite* (running birds) and the *Carinatae* (flying birds); but now it is split up into 28 orders of equal status comprising 157 families and about 8650 species. One of these orders, *Passerines* (perching birds), contains about 5100 species or nearly 60 per cent of the total number.

The largest living bird is the Ostrich (*Struthio camelus*) of Africa. There are now five geographical races – the last Syrian ostrich (*S. c. syriacus*) was killed and eaten by Arabs during the Second World War – and they differ slightly in size, shape and colour. The largest sub-species is the Northern ostrich (*S. c. camelus*) which is now found in reduced numbers south of the Atlas Mts from Upper Senegal and Niger across to the Sudan and central Ethiopia. Adult cock examples of this ratite or flightless bird stand about 2·43 m *8 ft* tall (height at back 1·37 m *4 ft 6 in*) on the average and weigh 120–127 kg *265–280 lb* (Duerden, 1919), but heights up to 2·74 m *9 ft* and weights up to 157 kg *345 lb* have been reliably reported. (In 1883 an unconfirmed weight of 190 kg *420 lb* was reported for an outsized specimen killed in the Sudan.) Adult hens are smaller, standing about 2·13 m *7 ft* tall and weighing 90–100 kg *200–220 lb*.

The Southern ostrich (*S. c. australis*), which is now found only in the wilder parts of the NE Transvaal, South Africa, southern Rhodesia and the southern part of what was Portuguese East Africa, as well as Botswana and South West Africa, is not as heavy as its northern cousin and has a shorter neck and legs. Adult cocks are usually 2·13–2·43 m *7–8 ft* tall and scale 100–114 kg *220–250 lb*. Larger domesticated birds have been recorded but these were usually crosses between *S. c. camelus* and *S. c. australis*. According to William A Hooper (pers. comm. 18 August 1971), owner of the famous Highgate

Ostrich Farm at Oudtshoorn, Cape Province, South Africa, abattoir birds measuring a few inches over 2·43 m *8 ft* and weighing up to 148 kg *326 lb* have been reported. (A 145 kg *320 lb* ostrich yielded 6·4 m^2 *21* ft^2 of skin.)

During the mating season the cock ostrich is one of the most dangerous animals on earth and its war-dance has to be seen to be believed. It stamps its feet, waves its neck in erratic circles and generally works itself up into a towering rage before charging at the object of its wrath, which it then proceeds to kick with its huge powerfully clawed legs. (According to one writer the kicking power of an angry ostrich is equivalent to being hit by the knock-out punches of five heavyweight boxers simultaneously!)

At times like this the ostrich is apparently completely devoid of fear. Cronwright Schreiner (1897) relates the story of a vicious old cock which caught sight of a goods train travelling at speed down a steep gradient. The bird immediately rushed on to the track and advanced fearlessly towards the screeching monster. A sudden sprint forward . . . one mighty kick . . . and the next moment the ostrich was cut to pieces.

Not surprisingly there are a number of records of people being disembowelled or fatally kicked in the head by an irate ostrich. In 1965 a 21-year-old man died at Oudtshoorn after one had fractured his skull and Clarke (1969) says there are two or three serious accidents in that area every year. Even a man on horseback is not immune from attack, and a number of riders have received serious injuries from the high-kicking bird.

Although the ostrich is not rated very highly in the intelligence stakes – the brain of a 123 kg *270 lb* cock bird weighed only 42·11 g *1·48 oz* or 0·03 per cent of the total body-weight (Quiring, 1950) – it is nowhere near as stupid as tradition would have us believe. In South West Africa they have been trained to herd sheep, and a farmer near Durban has taught an ostrich to chase off birds that attack his crops. Probably the most unusual role ever given to ostriches was that by a South African sportsman living near Pietermaritzburg who trained four birds to act as goalposts when he and his friends played football!

The maximum size attained by flying birds is limited by surface-volume ratio and the speed of flight.

The deadly scimitar-like claw of the cock Ostrich, one of the most dangerous animals on earth (G L Wood)

Ostriches – the world's fastest running birds – being raced at Oudtshoorn, South Africa (William A Hooper)

The heaviest flying bird or carinate is the Kori bustard or Paauw (*Otis kori*) of East and South Africa. Very little weight data has been published for this species, but a 14·54 kg *30 lb* cock bird would be considered a large specimen. Hens are smaller and much lighter in weight. One of the heaviest birds on record was a specimen shot by H T Glynn, a well-known sportsman, in South Africa which weighed exactly 18·1 kg *40 lb*. The head and neck of this bird were presented to the British Museum (Natural History), London (Bryden, 1936).

Another huge bird with a 2·54 m *8 ft 4 in* wing span shot by William Baldwin (1894) in the western Transvaal was 'the fattest and largest I have ever seen', and he estimated its weight at 28·5 kg *54 lb*. Baldwin, however, was not always a reliable observer – he once shot a thoroughbred horse on a farm in the Orange Free State in the mistaken belief that it was a new species of wild animal! – so this figure should be treated with extreme caution.

Mr Carl Schneritz told Harting (1906) that he shot two big Kori bustards on the high veld between Scheenspruit and Rustenberg in the Transvaal in 1899. The larger of the two measured 1·37 m *4 ft 6 in* in length, 2·74 m *9 ft* in expanse of wing, and weighed 16·8 kg *37 lb*.

In 1959 a scientific expedition from the Durban Museum shot a Kori bustard weighing 16·1 kg *35 lb* in East Africa which kept them in meat for nearly a week (Clancy, pers. comm. 8 February 1965).

Because the male Kori bustard is very close to the upper size limit above which flight is impossible it only flies reluctantly and then only for short distances when threatened. Also, like other high wing-loading birds, it requires a very long run before it can get airborne, and a 18·18 kg *40 lb* example would probably be exhausted after a 91 m *100 yd* low-level flight.

The Great bustard (*Otis tarda*) of western Europe, North Africa and Central Asia has also been credited with the title of the world's heaviest flying bird (Hvass, 1963), but this statement is based on an isolated record of a cock bird shot in the USSR which tipped the scales at a colossal 21 kg *46 lb* and probably could not fly (Baturin, 1935). The normal weight range for adult cock birds is 10·9–15·9 kg *24–35 lb*. This species bred in England until 1832 and one bird with a 1·98 m *6 ft 6 in* wing span shot in Norfolk weighed 10·9 kg *24 lb* (Harting, 1906).

Weights in excess of 13·63 kg *30 lb* have also been reliably reported for the Arabian bustard (*Choriotis arabis*). Col Richard Meinertzhagen (1954) says the average weight of a series of 19 males he shot in Irak was 10·9 kg *24 lb*, the heaviest scaling 16·81 kg *37 lb*. The average weight of 18 females was 7·72 kg *17 lb*, the heaviest weighing 9·54 kg *21 lb*.

Blandofrd (1892) asserts that weights up to 19·09 kg *40 lb* have been recorded for the now rare Great Indian bustard (*Choriotis nigriceps*) and he gives the normal weight range of cock birds as 11·37–13·63 kg *25–30 lb*. Jerdon, on the other hand, quotes a maximum weight of 12·72 kg *28 lb*, and Stuart Baker (1921) says the heaviest weight he could trace for a cock bird was 12·04 kg *26 lb 8 oz*. Earlier a 'sportsman' writing in the *Oriental Sporting Magazine* of August 1830 says that between 1809 and 1830 he bagged 961 Great Indian bustards near Ahmednagar, Bombay Province, India. According to him the cock birds weighed 8·12–14·5 kg *18–32 lb* and the hens 3·63–6·81 kg *8–15 lb*.

Gilliard (1958) claims the Australian bustard (*Choriotis australis*) 'is probably the heaviest of all flying birds', but Gould, who killed a great number of them on the plains of the Lower Namoi, New South Wales, and also in South Australia, gives the weight of the cock bird as between 5·9 and 7·2 kg *13* and *16 lb*. The maximum weight recorded for this species is 14·4 kg *32 lb* for a cock bird killed in Victoria (Serventy & Whittell, 1948).

The Mute swan (*Cygnus olor*), which is resident in Britain, the Trumpeter swan (*Cygnus c. buccinator*) of North America and the Whooper swan (*Cygnus c. cygnus*) of northern Europe and Asia also exceed 13·6 kg *30 lb* on occasion, the average weight of adult cobs being 12·2 kg *26 lb 14 oz*, 11·9 kg *26 lb 4 oz* and 10·8 kg *23 lb 13 oz* respectively (Boyd, 1972). Audubon (1840) mentions a trumpeter swan which scaled 17·2 kg *37 lb 13 oz* and there is a record from Poland of a cob mute swan weighing 22·5 kg *49 lb 8 oz* (Sanden, 1935).

A Wandering albatross (*Diomedea exulans*) nestling weighed by Tickell (1968) on Bird Island, South Georgia, Antarctica, shortly before its departure scaled 16·13 kg *35 lb 7½ oz*, but a third of this weight consisted of heavy fat deposition which would be shed in the first few weeks of flight. In one series of 108 weighings of adult *D. exulans* the average weight was found to be 8·18 kg *18 lb*. The heaviest bird scaled 11·3 kg *25 lb* and the lightest 5·9 kg *13 lb*. Gibson & Sefton (1960) point out, however, that 'being voracious feeders, the considerable stomach contents of well-fed birds influence the weight accordingly'.

The heaviest bird of prey is the Andean condor (*Vultur gryphus*), which ranges from western Venezuela to Tierra del Fuego and Patagonia. Adult males scale 9·09–11·3 kg *20–25 lb*, which is the weight of a good-sized domestic turkey, but one old male shot on San Gallan Island off the coast of Peru in 1919–20 weighed 12 kg *26 lb 8 oz* (Murphy, 1925). In October 1965 a weight of 11 kg *24 lb 3¼ oz* was reported for an Andean condor named 'Friedrich' at Frankfurt Zoo, West Germany, but the bird (hatched 1959) weighed only 8·3 kg *18 lb 5¾ oz* at the time of its death on 24 February 1971 (Grzimek, pers. comm. 25 July 1972).

The now very rare Californian condor (*Gymnogyps californianus*) is slightly smaller than its South American cousin, the average adult bird scaling 9·09 kg *20 lb* – but higher weights have been reported. The heaviest bird collected by Stephens (1895) scaled 9·77 kg *21 lb 8 oz*, and Henshaw (1920) quotes a weight of 10·8 kg *23 lb* for another specimen. Three other examples formerly in the private collection of E B Towne, Jr, and now in the California Academy of Sciences, Los Angeles, reportedly weighed 12·2 kg *27 lb*, 13·1 kg *29 lb* and 14·09 kg *31 lb* respectively, but Carl B Koford (1953), the leading authority on this species, thinks that Towne may have obtained this weight data second-hand from the men who collected the birds for him.

The only other birds of prey which *regularly* exceed 9·09 kg *20 lb* are the Cinereous vulture (*Aegypius monachus*) of southern Eurasia and the Himalayan griffon (*Gyps himalayensis*) of Central Asia. In one series of 20 male examples of *A. monachus* the weights ranged from 7 to 11·5 kg *15·4* to *25·3 lb* (average 9·22 kg *20·3 lb*), and in another series of 21 females the range was 7·5–12·5 kg *16 lb 8 oz–27 lb 8 oz* (average 10 kg *22 lb*). In *G. himalayensis* the weight ranged from 8 to 12 kg *17·6* to *26·4 lb* giving an average weight of 10 kg *22 lb* for both sexes (Brown & Amadon, 1968).

The heaviest eagle in the world is the Harpy eagle (*Harpia harpyja*) which ranges from southern Mexico to eastern Bolivia, southern Brazil and northern Argentina. Adult females average 7·27–7·72 kg *16–17 lb*, but they sometimes exceed 9·09 kg *20 lb*. Stanley E Brock, former Manager of the vast Dadanawa cattle ranch in Guayana and now General Manager of Wild Kingdom Park in Orlando, Florida, USA, once owned an enormous female named 'Jezebel' who tipped the scales at 12·27 kg *27 lb* (Brock, pers. comm. 10 August 1972), but this was exceptional. The male is much smaller than the female, the weight ranging from 4 to 4·6 kg *9·8 to 10·1 lb*.

In 1972 Mr Sam Barnes of Pwllheli, North Wales, supplied the compiler with details of 'Atlanta', his enormous pet female Berkut/Burkut golden eagle (*Aquila chrysaetos daphanea*), which he collected in the Tien Shan Mts in the Kirghiz Republic of Soviet Central Asia some years ago. The Kirghiz use these eagles to hunt deer and wolves – the bird is trained to grip the head of the animal it is pursuing and slow it to a halt – and never trade them to Westerners, but because the eagle was sick and believed to be dying, Mr Barnes was able to purchase her for two bottles of brandy and a few hundred cigarettes. Adult females of this race normally weigh about 6·5 kg *14·3 lb*, but Mr Barnes flies his bird at an extraordinary 12·1 kg *26 lb 10 oz* and says she probably weighs 13·63 kg *30 lb* after eating a 2·1 kg *4 lb 8 oz* rabbit!

Steller's sea eagle (*Haliaeetus pelagicus*) of NE Siberia is another huge bird, females weighing 6·8–8·97 kg *14·96–19·73 lb*, and there is an old record of a White-tailed sea eagle (*Haliaeetus albicilla*) killed at Stornaway, Lewis, Outer Hebrides, in the 19th century which scaled 7·5 kg *16 lb 8 oz* (Harting, 1906). This latter poundage probably constitutes a weight record for a British eagle.

The greatest expanse of wing attained by any living bird is that of the Wandering albatross (*Diomedea exulans*) of the southern

This Wandering albatross caught at South Georgia had a wing expanse of 3·45 m 11 ft 4 in

The 12·3 kg 27 lb *Harpy eagle 'Jezebel' with her owner (Stanley Brock)*

oceans. In one short series of birds examined by Tickell (1960) on Bird Island, South Georgia, Antarctica, adult males averaged 3·09 m *10 ft 1·8 in* with wings tightly stretched, and adult females 3·1 m *10 ft 2·2 in*, but females are normally smaller than males.

In an earlier study during June–August 1959 at Malabar (Sydney) and Bellambi, 56 km *35 miles* to the south, New South Wales, Australia, the average wing span of the 119 birds netted while on the water was found to be 2·99 m *9 ft 10 in*, with a maximum of 3·23 m *10 ft* $7\frac{1}{4}$ *in* and a minimum of 2·71 m *8 ft 11 in*. The expanse of wing was measured by means of a brass scale attached to the gunwhale of the boat along which the wings were spread and *gently* stretched to their greatest extent. 'To what extent, if any, the moult condition of the primary feathers affects the wing-span measurement at this time of the year has not been determined', write Gibson & Sefton (1960), 'but it is thought to be not great'.

The largest accurately measured specimen on record was a male caught by banders in Western Australia in *c.* 1957 which spanned 3·60 m *11 ft 10 in* (Disney, pers. comm. 11 April 1967). Another male found stranded on the beach at Bunbury, 176 km *110 miles* south of Fremantle, Western Australia, on 17 July 1930 had a wing span of 3·50 m *11 ft 6 in* (Whitlock, 1931).

Some unmeasured examples of this dynamic soarer probably exceed 3·65 m *12 ft*, and the maximum expansion may be as much as 3·96 m *13 ft*. Scouler (1826) claims he caught one bird with a 3·65 m *12 ft* wing span and the same measurement is quoted by Hutton for his largest specimen. According to Spry (1876) all the birds collected by him on Marion Island in the southern Indian Ocean measured 3·35–3·65 m *11–12 ft* and Lord Campbell (1877) says he measured several birds of 3·65–3·96 m *12–13 ft* in the Indian Ocean during the cruise of the *Challenger*.

Dr Robert Cushman Murphy (1936), one of the world's foremost authorities on the sea-birds of the southern oceans, is more critical, however, and considers a measurement of 3·96 m *13 ft* to be excessive. The largest bird examined by him in South Georgia measured 3·45 m *11 ft 4 in* and he put the maximum wing expanse of this species at 'about $11\frac{1}{2}$ ft, with the wings of the dead bird stretched out as tightly as possible'.

As already noted, this measurement has since been surpassed by 101 mm *4 in*, and some new evidence has recently come to light which would suggest that the maximum expanse of wing attained by this bird may be in excess of 3·96 m *13 ft*. Mr J Strand Jones (pers. comm. 30 May 1971) of Llanbedr, Ceredigion, Wales, described how, as an apprentice, he helped measure a wandering albatross with a wing span of 4·21 m *13 ft 10 in* aboard the SS *Devon* between the Cape of Good Hope, South Africa, and Fremantle, Western Australia, in April/May 1929: 'We caught it by throwing a heavy nut attached to a cord across its back as it flew alongside the ship, then hauling it aboard. The cord was thrown from the boat deck and the bird hauled in to the after well deck. What struck me about the bird apart from the incredible expanse of wing was the size of the beak, but that was too dangerous to measure. Afterwards we took out the stanchions, as when loading cargo, and it flopped overboard.'

Unfortunately the capture of the bird was not mentioned in the ship's log (the SS *Devon* was owned by the New Zealand Shipping Co, which is now part of the P & O Line), but the measurement – albeit extreme – *could* be genuine.

Figuier (1894) mentions a wandering albatross in the Leverian Museum, London, which measured 'thirteen ft from the tip of one wing to the tip of the other', but further information is lacking. (The museum collection was auctioned in 1806 and went into a number of private hands.)

The enormous size of *D. exulans* (maximum weight 12·11 kg *26 lb 12 oz*) is matched by its voracious appetite and the bird has to seek food almost incessantly in order to stay alive. Its main diet is fish, squid and crustaceans found on the surface, but it will also feed on refuse dumped overboard from ships and has even been known to attack a man in the water with its vicious 203 mm *8 in* long beak.

Such an incident occurred after a naval battle off the Falkland Islands in the South Atlantic during the First World War when members of the crew of the British cruiser HMS *Kent* went to pick up German survivors of the stricken cruiser *Nürnberg*. As their boats approached the sailors struggling to stay afloat in the icy water the British ratings saw several albatrosses attacking the unfortunate men.

The wandering albatross's closest rival in terms of wing expanse is the Andean condor (*Vultur gryphus*), which has the largest 'sail area' of any living bird. Very little accurate data has been published on size, but it is believed that the adult male has an average expanse of 2·81 m *9 ft 3 in* and the female 2·74 m *9 ft*. Fisher quotes a figure of 2·99 m *9 ft 10 in* for a specimen preserved in alcohol and the same measurement is given by Koford (1953) for another condor collected by J R Pemberton which is now in the Museum of Comparative Zoology, Harvard University, Cambridge, Massachusetts, USA. The old male shot on San Gallan Island (see page 85) measured a fraction of an inch over 3·05 m *10 ft* and Friedrich at the Frankfurt Zoo (see page 85) had a wing expanse of 3·07 m *10 ft 1¼ in*.

According to Brown & Amadon (1968) the Andean condor has been reliably measured up to 3·23 m *10 ft 6 in* and this figure probably refers to the huge bird seen by Mr Edward Whymper at Antisana, north central Ecuador. Humboldt

The magnificent head of the Berkut golden eagle, one of the heaviest birds of prey (Keystone Press)

The Andean condor has the largest 'sail area' of any living bird (Associated Press)

says he did not see a single condor in Ecuador over 2·74 m *9 ft*, and he was assured by the inhabitants of Quito that they had never shot any that exceeded 3·35 m *11 ft*.

The now very rare California condor (*Gymnogyps californianus*) is slightly smaller than its South American cousin, the adult male and female both averaging about 2·74 m *9 ft* in wing expanse. Frank Stephens told Koford (pers. comm. 24 February 1907) that of five adults collected by him four measured 2·84 m *9 ft 4 in* or fractionally more. Wingspreads up to 3·35 m *11 ft* have been claimed in the literature for this species, but Koford gives the greatest substantiated measurement as 2·92 m *9 ft 7 in* for a male collected by E B Towne which is now in the Museum of Vertebrate Zoology at the University of California, Berkeley, California, USA. Three other males preserved in the California Academy of Sciences, Los Angeles, measure 2·91 m *9 ft 6½ in*, 2·89 m *9 ft 6 in* (this is the bird which allegedly weighed 14·09 kg *31 lb*) and 2·87 m *9 ft 5 in* respectively, and a female in the Museum of Comparative Zoology, Harvard University, Cambridge, Massachusetts, has a span of 2·89 m *9 ft 6 in*.

The Cinerous vulture (*Aegypius monachus*) and the Himalayan griffon (*Gyps himalayensis*) are also of comparable size, and Brown & Amadon say large specimens are as big as most California condors. In one series of *A. monachus* the spans ranged from 2·53 to 2·70 m *8 ft 3 in to 8 ft 10 in*, but a measurement of 2·99 m *9 ft 10 in* has been reported elsewhere for a female (Harting, 1906). There is also an authentic record of a female *G. himalayensis* measuring 3·06 m *10 ft 0½ in*.

Some ornithologists believe that the vulture-like Marabou stork (*Leptoptilus crumeniferus*) of Africa south of the Sahara has a larger expanse of wing than both species of condor, and Fisher & Peterson (1964) even go so far as to state that the marabou has the largest wingspread of *any living bird* with a maximum expanse of *c.* 3·65 m *12 ft*

Dr Roger Tory Peterson, the doyen of American ornithologists, told the compiler (pers. comm. 10 July 1975): 'James Fisher and I based our report on personal correspondence with Col Meinertzhagen (now deceased) in which he indicated that he had taken a specimen with a wing span of 13 ft 4 in. This was an unpublished record. When we pressed Meinertzhagen as to the accuracy of this size, he said, "Well, I may have stretched the wings just a bit." James Fisher attempted to confirm Marabou wing spans by measuring specimens from the British Museum. However, as there is considerable variance between a dried skin and that of a freshly taken specimen, we decided to stay on the conservative side of Meinertzhagen's largest specimen and give it as *c.* 12 ft.'

Col Richard Meinertzhagen, who shot this marabou in Central Africa, in the 1930s, was a very reliable observer so, even if he had stretched the wings as tightly as possible, it would hardly have added more than 101–127 mm *4–5 in* to the natural expanse; thus, the actual wingspread must have been nearer 3·96 m *13 ft* than 3·65 m *12 ft*. This measurement, however, is misleading because there are very few records of marabou storks exceeding 2·43–2·74 m *8–9 ft*.

Captain (later Major) Stanley Flower shot what he considered to be a *large* marabou at Kaka, in the Sudan, in March 1900 and found it had a wing span of 2·53 m *8 ft 4 in*. Two other storks killed in East Africa measured 2·64 m *8 ft 8 in* and 2·68 m *8 ft 9½ in* respectively, and another specimen shot in Zambia in 1966 reportedly had a wing expanse of 2·77 m *9 ft 1½ in*. In 1913 a wing expanse of 3·27 m *10 ft 9½ in* was reliably reported for an Adjutant stork (*Leptoptilus dubius*) shot on the banks of the Godavari River, Bombay Province, India.

A Mute swan (*Cygnus olor*) named 'Guardsman', who died at the famous swannery at Abbotsbury on the Fleet in Dorset in 1945 was credited with a wing span of 3·65 m *12 ft*, but photographic evidence suggests the expanse was nearer 2·89 m *9 ft 6 in* There is, however, an authentic record of a Trumpeter swan (*Cygnus c. buccinator*) measuring 3·09 m *10 ft 2 in* (Breland, 1948). Both swans (cobs) have an average expanse of approximately 2·43 m *8 ft*.

Having large wings can sometimes be a handicap as witness the sad story of an Old World white pelican (*Pelecanus onocrotalus*) with an 2·43 m *8 ft* span which escaped from its cage in a small British zoo in 1951 during the cleaning process. It tried out its wings for the first time, but unfortunately one of its aerial manoeuvres went wrong and it landed upside down in the near-by lion grotto – where it ended its life a few seconds later.

The greatest expanse of wing recorded for an eagle (there are 59 recognised species) is 2·84 m *9 ft 4 in* for a female Wedge-tailed eagle (*Aquila audax fleayi*) killed in Tasmania (Fleay, 1952), but this was exceptional. In one series of 126 specimens measured by research workers near Adelaide, South Australia, in 1932 the average expanse was found to be 2·31 m *7 ft 5 in* and the average weight 3·63 kg *8 lb*, although Chisholm (1948) says some individual birds were much larger. Earlier, in December 1931, a span of 2·84 m *9 ft 2 in* and a weight of 5·8 kg *12 lb 12 oz* had been reported for a wedge-tailed eagle found drowned in a dam near Molong, New South Wales (Wood, 1972). Less reliable is a claim that a specimen caught in a dingo trap on the Nullambar Plains, South Australia, had a span of 3·12 m *10 ft 3 in*, and that another one shot by a boundary rider near the Werribee Gorge, Victoria, in 1914 measured a full 3·35 m *11 ft* from tip to tip (Roche, 1914).

According to Mr P Murrell, Director of the National Parks and Wildlife Service at Sandy Bay, Tasmania (pers. comm. 21 November 1973) on 21 May 1968 his department received a complaint that two enormous wedge-tailed eagles had carried off two well-grown lambs on Bruny Island, off the south-east coast. 'The irate farmer estimated their wing span at 11 ft', Mr Murrell said, 'but an Inspector who investigated the matter could not find any eagles in the area so the measurement remains unconfirmed.'

Mr Sam Barne's enormous Berkut golden Atalanta (see page 85) has a wing expanse of approximately 2·76 m *9 ft 1 in*. A measurement was taken as the bird was held aloft in the face of a gale-force wind. As the eagle extended her wings to maintain her balance a bobbin of cotton was stretched across from the fourth scutchulated primary in each wing (the longest feather in the Berkut). On the basis of this evidence Atalanta would probably make 2·84 m *9 ft 4 in* if the wings were stretched out tight after death. Adult females of this race normally have a wing span of about 2·13 m *7 ft*.

The greatest expanse of wing claimed for any race of Golden eagle is 2·85 m *9 ft 4½ in* for a female *Aquila c. chrysaetos* captured alive at Stockfield Park near Wetherby, Yorkshire, on 29 November 1804 after being wounded by gunshot (*Annual Register*, 1805), but this figure must be considered excessive. Earlier, on 4 December 1796, another outsized bird was captured in a fox trap near Wareham in Dorset. This specimen had a more acceptable wing-span measurement of 2·48 m *8 ft 2 in* and was said to be the largest flying bird ever seen in England.

In one series of seven adult males the wing expanses ranged from 1·89–2.12 m *6 ft 2½ in–6 ft 11½ in*, and in a series of five females the range was 2·15–2·27 m *7 ft 0⅔ in–7 ft 5⅓ in* (Brown & Amadon, 1968), but some of these measurements may have been obtained in Europe.

In January 1973 a golden eagle, one of Ireland's rarest birds and fully protected, was shot and killed by an angry farmer after the bird had attacked a goose on his farm at Scibbagh, Co. Fermanagh. The eagle had a wing span of 2·13 m *7 ft* and weighed 4·54 kg *10 lb*.

The Harpy eagle (*Harpia harpyja*) of South America is forced to hunt below the tree-line because of its great weight and is compensated with wings that are short but very broad. No one has yet recorded a series of wing measurements for this species, but Leslie Brown (1970) thinks an adult female harpy might span about 2·26 m *7 ft 6 in* and a male more than 1·83 m *6 ft*.

Incidentally, when a young eagle leaves the nest it is larger than its parents by as much as 0·45 kg *1 lb* and up to 228–304 mm *9–12 in* in wing span. Contraction of the bones in maturity and strenuous exercise, however, bring the raptor down to size.

Although eagles are extremely powerful birds and can kill prey three or four times their own size, the general consensus of opinion is that they cannot carry a load much in excess of their own body-weight. Usually they kill animals half their own weight or less that can be lifted easily, but some fast-moving species have been known to carry off much heavier prey.

One of the strongest species is Pallas's sea eagle (*Haliaeetus leucoryphus*). A O Hume (quoted by Harting, 1906), writing of this bird, says: 'A grey goose will weigh on the average 7 lb (much heavier are recorded) but I have repeatedly seen good-sized grey geese carried off in the claws of one of these eagles, the bird flying slowly and low over the surface of the water, but still quite steadily.' Harting also says the same writer once saw another specimen capture a fish on the River

Jumna, north central India, that was so large that the bird only just succeeded in reaching a low sandbank in the river with its prey: 'As it made for this bank it flew so low, and with such difficulty, that the writhing fish in its claws struck the water every few yards, and twice seemed likely to pull its persecutor under water. On reaching the sandbank some 250 yd distant from the observer, a shot from his rifle caused it to quit the fish, which was then recovered and found to be a Carp (*Cyprinus rohita*) weighing over 13 lb, that is considerably heavier than its captor.'

The weight of an adult *H. leucoryphus* ranges from 2·62 to 3·27 kg *5·76* to *7·19 lb*.

Another formidable performer in the lifting stakes is Steller's sea eagle (*Haliaeetus pelagicus*) which, apart from large fish, also includes young seals and Arctic foxes in its diet. According to Brown & Amadon (1968) the food of this bird is taken from the ground or water 'often with very little effort', which suggests that a very large female, i.e. 8·63 kg *19 lb* is probably capable of carrying a 9·09 kg *20 lb* carcass low over the ground (or sea) for a distance of several hundred yards.

Dan McCowan (quoted by Lane, 1955) mentions two hunters in the Canadian Rockies who once saw an American bald eagle (*Haliaeetus leucocephalus*) descend from a considerable height with a 6·81 kg *15 lb* mule deer fawn in its talons which it dropped when the men shouted at it. There is also a reliable account of another bird carrying a lamb weighing at least 3·63 kg *8 lb* over a distance of 8 km *5 miles* to its eyrie, but the eagle must have taken advantage of upcurrents of air or thermals to accomplish this feat. Females of this species have been weighed up to 6·3 kg *13·86 lb*.

A Natal mountaineer named Arthur Bowland once persuaded a Verreaux's eagle (*Aguila verreauxi*) to snatch a 9·09 kg *20 lb* pack while in flight, but the bird only managed to carry it a few yards before it was forced to let go (Clarke, 1969). In another study it was found that Verreaux's eagle could carry a full-grown Hyrax weighing 2·72–3·63 kg *6–8 lb* in one talon, but had to land for a breather now and again. This bird is slightly smaller than *H. leucocephalus*, the weight ranging from 3·68 to 5·8 kg *8·98* to *12·76 lb* in adult females.

Although many lurid stories of eagles carrying off human babies have appeared in print, most ornithologists tend to look upon such accounts with suspicion because eagles normally kill their prey before bearing it away. Leslie Brown (1963) goes even further and says 3·63 kg *8 lb* ('the weight of a baby before it takes the air') is beyond the power of any eagle to lift. 'If you attach 5 lb to the feet of a golden eagle and cast it off a crag', he writes, 'it will be brought ignominiously to the ground a short distance away. Attach 8 lb to its feet and it cannot get off the ground at all, except perhaps down a slope and against very strong wind. It is therefore physically impossible for an eagle to lift an average baby, and agitated mothers have lived in fear far too long.'

Brown's information was based on experiments carried out with a tame golden eagle in the USA in 1940 to determine its lifting ability. The bird was placed on the top of a 9·14 m *30 ft* platform and various weights were attached to its feet. It was then launched into the air. These tests, however, were not a true guide, because the muscularity and power of flight of a captive eagle are considerably less than those of a wild specimen; also, the eagle may have been an adult male which averages only some 3·86 kg *8 lb 8 oz*.

Reliable evidence that the golden eagle *can* lift weights far in excess of 2·27 kg *5 lb* (albeit in an emergency!), however, is given by Gordon (1915) who describes a fight he saw in the Cairngorms, Scotland, between a golden eagle and an adult fox. (The dog fox averages 6·81 kg *15 lb* and the vixen 5·45 kg *12 lb*): 'The eagle was devouring the carcase of a blue hare when a fox sprang from the surrounding heather and seized the great bird by the wing. A well-contested struggle ensued in which the eagle made a desperate attempt to defend itself with its claws and succeeded in extricating itself from its enemy's grasp, but before it had time to escape Reynard seized it by the breast and seemed more determined than ever. The eagle made another attempt to overpower its antagonist by striking with its wings, but that would not compel the aggressor to quit its hold. At last, the eagle succeeded in raising the fox from the ground, and after a few minutes Reynard was suspended by his own jaws between heaven and earth. Although now placed in an unfavourable position for fighting his courage did not forsake him, as he firmly kept his hold and seemed to make several attempts to bring the eagle down, but he soon found the strong wings of the eagle were capable of raising him, and that there was no way of escape unless the bird should alight somewhere. The eagle made a straight ascent and rose to a considerable height in the air. After struggling for a time Reynard was obliged to quit his grip, and descended much quicker than he had gone up. He was dashed to the earth, where he lay struggling in the agonies of death. The eagle made his escape, but appeared weak from exhaustion and loss of blood.'

Little or nothing is known about the lifting ability of the Harpy eagle (*Harpia harpyja*), the world's heaviest eagle (see page 85), but the combination of enormous talons – the hind talons of the female harpy are thicker but not quite so long as those of the Kodiak bear, the world's largest land carnivore – and short, broad wings means it can lift prey like Capuchin monkeys, Opossums and Coatis almost vertically.

Another interesting point about the harpy is its remarkable tenacity for life. D'Orbigny discovered this to his cost when he encountered a large specimen on the banks of the Rio Securia in Bolivia. His two Indian guides immediately set off in pursuit and after a hectic chase they managed to bring the bird down with two arrows through the body. Despite its terrible wounds, however, the harpy was still full of fight, and the men were forced to hit the unfortunate creature repeatedly over the head with their machetes before it finally went limp. The Indians then plucked off most of the eagle's feathers and threw the carcase in the bottom of their dug-out canoe. An hour or so later, as the party continued their journey down-river, the harpy suddenly came to life and hurled itself at the astonished explorer. A desperate battle followed, this time the bird giving as good as it got, and at one point the canoe nearly capsized. Eventually one of the Indians managed to despatch the courageous harpy, but D'Orbigny was very badly clawed on the arms and bore the scars to his grave.

The smallest bird in the world is the female Bee hummingbird *Mellisuga helenae*, also known as 'Helena's hummingbird' or 'the fairy hummer' of Cuba and the Isle of Pines (James Bond, pers. comm. 23 December 1968). Very little data has been published on this species, but an average sized specimen measures 58 mm *2·28 in* from beak-tip to tail-tip, the head and body accounting for 15 mm *0·59 in*, the tail 28 mm *1·10 in* and the bill 15 mm *0·59 in*. The weight is about 2 g *0·070 oz*, which means *M. helenae* is lighter than a large Sphinx moth (2·2 g *0·077 oz*), and only one-quarter the weight of the European wren (*Troglodytes troglodytes*).

The Bee hummingbird *Acestrura bombus* of Ecuador is almost equally as diminutive, the average adult weighing *c.* 2·2 g *0·077 oz*.

Hummingbirds also have the highest energy

Helena's hummingbird, the smallest member of the class Aves (Cuban Academy of Science)

output per unit of weight of any living warm-blooded animal and Terres (1968) has calculated that a hummingbird uses about 155 000 calories a day in energy, compared with 3 500 calories for an average adult man.

The smallest bird of prey (order Falconiformes) is the sparrow-sized Bornean falconet (*Microhierax latifrons*) of the forest country of NW Borneo. Adult birds – the sexes are similar in size – measure 140–152 mm *5½–6 in* in length and weigh about 35 g *1¼ oz*.

The smallest ratite bird, and probably the smallest flightless bird known to have existed, is the Flightless rail (*Atlantisea rogersi*) of Inaccessible Island in the Tristan da Cunha group, South Atlantic. This species, which has degenerate, hair-like plumage, is only the size of a newly hatched domestic chick. It lives in burrows and is reported to be a very fast runner.

The smallest regularly breeding British bird is the Goldcrest (*Regulus regulus*), also known as the 'Golden-crested wren' or 'Kinglet'. Adult specimens measure 90 mm *3·52 in* in length and weigh between 3·8 and 4·5 g *0·108* and *0·127 oz*. The Firecrest (*Regulus ignicapillus*), a passage migrant which has bred in Hampshire in small numbers since 1961 and sometimes winters in SW England, also measures about 90 mm *3·54 in* in length but may be slightly heavier. The Wren (*Troglodytes troglodytes*), Britain's second smallest regularly breeding bird, measures only 95 mm *3¾ in* in length, but is twice as heavy as *Regulus*.

The most abundant species of bird – and the most destructive – is the Red-billed quelea (*Quelea quelea*) which is distributed throughout the drier parts of Africa south of the Sahara. Although it is impossible to calculate their numbers with any degree of accuracy, Dr Peter Ward of the Quelea Investigations Project in Maiduguri, Nigeria (pers. comm. 2 March 1974) who has been studying this bird for 15 years, believes that the total population 'must be reckoned in thousands of millions – perhaps as many as 10 000 000 000'.

Today the 'feathered locust' as it is popularly known, poses a serious threat to the growers of cereal crops in 25 developing African countries and its concentrations stretch from Senegal and Mauritania in the west across the continent to Ethiopia and Somalia, then south through Uganda, Kenya, Tanzania and Zambia to South Africa and up into Namibia and Angola. Breeding colonies and night roosts – which may contain up to 10 000 000 individuals – are attacked with flame-throwers, dynamite bombs and poisonous aerial sprays, but even though the birds are being slaughtered at an estimated rate of 1 000 000 000 a year it is not having any obvious permanent impact on the population size. The trouble is that, contrary to the habits of most species of birds, the red-billed quelea (adult birds weigh from 17 to 24 g *0·607* to *0·857 oz*) can breed three or four times a year – often in areas completely inaccessible to control units – and the eggs, usually three in number, hatch in eleven to twelve days.

The only other species of wild land bird credited with such prodigious numbers was the now-extinct Passenger pigeon (*Ectopistes migratorius*) of North America. It has been estimated that there were between 5 000 000 000 and 9 000 000 000 of these birds before 1840, forming 24–40 per cent of the total bird population of the USA. Thereafter the pigeons were killed in vast numbers by commercial hunters ('pigeon pie' being a popular table dish of the day) and the last recorded specimen, a female named 'Martha', died in Cincinnati Zoological Gardens, Ohio, USA, at 1 pm Eastern Standard Time on 1 September 1914 aged *c.* twelve years. She had been collected with several others in Wisconsin in 1902 (Deane, 1908). After death the carcass of this bird was frozen in a 136·3 kg *300 lb* block of ice and shipped to Washington, DC, where Shufeldt (1951) made a detailed examination. The mounted specimen is now on exhibition in the US National Museum.

The most abundant wild land bird after the red-billed quelea is the Starling (*Sturnus vulgaris*). In North America alone there are more than 500 000 000, and the world population may be as high as 2 000 000 000.

According to Fisher (1950) the most numerous wild breeding birds in England and Wales in the early spring are the Blackbird (*Turdus merula*) and the Chaffinch (*Fringilla coelebs*), each numbering some 10 000 000 individuals, followed by the Starling (*Sturnus vulgaris*) and the Robin (*Erithacus rubecula*) with 7 000 000 each. Whether these figures still stand up today is open to question, but the starling must now be a more numerous bird than the chaffinch.

In 1963 the total population of the House sparrow (*Passer domesticus*) in Britain was computed at 9 568 000 (Summers–Smith, 1963), or one bird for approximately every six people. It has also been calculated that there are about 8 000 000 Wood pigeons (*Columba palumbus*) in Britain – mainly in the Home Counties and East Anglia – at the end of each breeding season. Some 3 500 000 – 4 000 000, however, die from natural causes or are shot each year (the bird is classified as a pest by the Ministry of Agriculture because of the damage it does to farm seedling crops)

Mention should also be made of the Pheasant (*Phasianus colchicus*), which has made a remarkable come-back since its numbers were decimated by fowl pest and poor breeding conditions in 1970 and 1971. In 1973 about 4 000 000 birds were shot, and the following year 6 000 000, which suggests that the British population now probably exceeds 10 000 000 at the end of the breeding season. Of this total, however, some 2 000 000 birds are reared in woods in preparation for the opening of the shooting season, which means it is not strictly a wild breeding species.

During the three-year period 1970–73 the British Trust for Ornithology carried out a survey of the birds most commonly found in gardens and discovered that the most abundant species was still the Blackbird (*Turdus merula*), followed by the Starling (*Sturnus vulgaris*), House sparrow (*Passer domesticus*), Blue tit (*Parus caeruleus*), Robin (*Erithacus rubecula*), Greenfinch (*Carduelis chloris*), Hedge sparrow (*Prunella modularis*), Great tit (*Parus major*), Song thrush (*Turdus philomelos*) and Chaffinch (*Fringilla coelebs*).

The most abundant sea-bird is Wilson's petrel (*Oceanites oceanicus*), which flies to the North Atlantic every summer from its breeding-grounds at the edge of the Antarctic. No population studies have been published for this species world-wide, but its numbers must run into hundreds – possibly thousands – of millions.

Britain's most abundant sea-bird is probably the Fulmar (*Fulmarus glacialis*) with something like 100000 pairs nesting along the coastlines. The Herring gull (*Larus argentatus*), another cliff-nester, may be even more populous now that it breeds inland as well, but further information is lacking.

The most abundant species of domesticated bird is the Chicken (*Gallus gallus domestica*). There are believed to be about 3500000000 in the world, or nearly one chicken for every member of the human race. In 1965 the fowl stock in Britain was estimated at 90000000, producing nearly 200000000 chicks each year.

In 1967 it was estimated that 250000 pigeon-fanciers owned an average of 40 racing pigeons per loft, making a population of about 10000000 in Britain.

Because of the practical difficulties involved in assessing bird populations in the wild, which vary according to species and habitat and whose numbers are constantly changing, it is impossible to establish the identity of the world's rarest bird. A better criterion would be to mention those traditional world rarities which are *believed* to have a total population of less than 100. They include the Short-tailed albatross (*Diomedea albatrus*), the Nippon ibis (*Nipponia nippon*), the California condor (*Gymnogyps californianus*), the Monkey-eating eagle (*Pithecophaga jefferyi*), the Siberian white crane (*Grus leucogeranus*), the Seychelles magpie robin (*Copsychus seychellarum*), the Giant pied-billed grebe (*Podilymbus gigas*), the Grenada dove (*Leptotila wellsi*), the Puerto Rican parrot (*Amazona dufresniana*), the Noisy scrub bird (*Atrichornis clamosus*) and the Hawaiian crow (*Corvus tropicus*).

The famous Greater sulphur-crested cockatoo 'Cocky Bennett', who was reputedly 120 years old when he died in 1916

In 1975 the total population of the Mauritius kestrel (*Falco punctatus*) was down to 7, and the Hawaiian O-o (*Moho braccatus*) to just one pair.

According to the British Ornithologists' Union there are 28 species of birds which have been recorded only once in the British Isles – 18 of them since the end of the Second World War. That which has not recurred for the longest period is the Black-capped petrel (*Pterodroma hasitata*), also known as the 'Diablotin'. A specimen was caught alive on a heath at Southacre, near Swaffham, Norfolk, in March/April 1850. A Red-necked nightjar (*Caprimulgus ruficollis*) was shot at Killingworth, near Newcastle, Northumberland, on 5 October 1856.

The most tenuously established British bird is now the Snowy owl (*Nyctea scandiaca*), with only one pair breeding regularly on the island of Fetlar in the Shetland Islands. In 1975 this pair successfully raised four healthy chicks which were given a 24-hour guard by members of the Royal Society for the Protection of Birds. At the same time another pair of snowy owls successfully reared two chicks in the Highland Wildlife Park near Aviemore. Four other eggs were infertile.

In June 1975 the Nature Conservancy Council announced that an attempt was being made to reintroduce into Scotland the White-tailed sea eagle (*Haliaeetus albicilla*) which became extinct at the turn of the century. (The last occupied nest in Scotland was recorded on Skye in 1916.) Four specimens imported under licence from Norway are now living on the island of Rhum in cages and, as they develop, they will be allowed to forage for their own food and eventually freed to fend for themselves. A similar exercise was carried out in 1968 when four eaglets were

The Red-billed quelea, the world's most abundant species of wild bird (Peter Ward)

introduced by the RSPB into Fair Isle between the Shetlands and the Orkneys. Three of the birds reached maturity but have not yet returned to the island.

Britain's rarest breeding raptor (bird of prey) is the Marsh harrier (*Circus aeruginosus*), with less than five pairs in Suffolk and one other English county.

Although birds are generally longer-lived than mammals of comparable size and activity, very few species reach or exceed 40 years in the captive or wild state and reports that they sometimes live for 100 years or more should be treated with considerable suspicion. A classic example was the Egyptian vulture (*Neophron percnopterus*) which died in the menagerie at Schönbrunn, Vienna, Austria, in 1824 allegedly aged 118 years. Major Stanley Flower (1925) discovered the menagerie was not founded until 1752 'so even if it were proved that it was one of the original inmates of that famous collection, there is still a previous 36 years to be accounted for'. Other dubious records in the same category include a Greater sulphur-crested cockatoo (*Cacatua galerita*) of 140 years, an African grey parrot (*Pstittacus erythacus*) of 120 years, a Griffin vulture (*Gyps fulvus*) of 117 years, a Golden eagle (*Aquila chrysaetos*) of 104 years and a Mute swan (*Cygnus olor*) of 102 years.

In July 1840 the death was reported of a mute swan named 'Old Jack' in St James's Park, London, reputedly aged 70 years. He was said to have been hatched 'on the piece of water attached to Buckingham Palace' in 1770 (*Morning Post*, 16 July 1840).

A Greater sulphur-crested cockatoo (*Cacatua galerita*) named 'Cocky Bennett', owned by Mrs Sarah Bennett, licensee of the Sea Breeze Hotel at Tom Ugly's Point, near Sydney, New South Wales, Australia, was said to be over 120 years old when he died in 1916, but Kinghorn (1930) could not find any authentic information regarding the true age of this bird. The cockatoo was in the possession of Mrs Bennett for 26 years, and had previously been owned by Captain George Ellis, skipper of a South Sea sailing-ship, who claimed the bird was alive when he was only a nine-year-old ship's apprentice. During the last 25 years of his life 'Cocky Bennett' was practically featherless, and he was often heard to scream: 'One more *******! feather and I'll fly!'

On 8 March 1968 the death was reported of

another specimen named 'Cocky' in the Nottingham Park Aviary aged 114 years, but this record is not considered reliable because several cockatoos of that name had been kept in the aviary over the years.

The greatest authenticated age recorded for a captive bird is 72+ years for a male Andean condor (*Vultur gryphus*) named 'Kuzya' which died in Moscow Zoo, USSR, in 1964. The bird arrived at the zoo already full-grown in 1892 and lived out of doors all the year round (Sosnovski, pers. comm. 21 February 1975).

In April 1973 the death was reported of another ancient Andean condor in the Menagerie du Jardin des Plantes, Paris, France, aged 71+ years. In a letter (*vide* Andrew Palmer, Information Secretary at the British Embassy in Paris) Dr Guy Chauvier (pers. comm. 3 January 1974) Deputy Director of the zoo, said however, that the Menagerie did not wish to make any official or published claim that this bird was aged 71 or more, because although it had been established that a condor, probably aged about one year, was received at the Menagerie in 1902 their records were seriously damaged during the Second World War. 'Thus, it cannot be ruled out that the 1902 bird died during World War II and that its place as oldest inhabitant was taken – unrecorded – by another bird.' Clinically, however, Dr Chauvier was pretty confident that the condor was indeed the age claimed for it, although scientifically an element of doubt remained.

The Ostrich (*Struthio camelus*) is another potentially long-lived bird under domestic conditions but it is very accident prone and seldom dies of natural causes. (Fay Goldie was told by one ostrich-farmer with over 60 years experience that 90 per cent of ostriches die after breaking their legs.)

Barring such accidents and bearing in mind that ostriches are rarely kept longer than 15 years for feather production because the quality and quantity of plumage decreases as the birds get older, there is nothing to prevent an ostrich from living 50 years or more. During a visit to the Highgate Ostrich Farm at Oudtshoorn in April 1973 the compiler was told by Alex Hooper, son of the owner, that the previous year a cock bird aged 62 years 7 months had been killed in the unique ostrich abbatoir nearby. Another ostrich reportedly lived for 68 years (Clay, 1962).

Duerden (1919) mentions several domesticated ostriches in South Africa aged between 32 and 42 years which were still breeding, adding: 'It is known an ostrich can breed until well over 40 years and probably live to 100 years and more, though his productive powers fail much sooner.'

On 22 March 1967 the death was reported of an Asian white crane (*Grus leucogeranus*) named 'Pops' at the National Zoological Park, Washington, DC, USA, aged 62+ years. Three days earlier the bird had sustained a compound fracture of the left leg which was set and a cast applied, but the combination of shock and old age proved too much for the bird. 'Pops' was received as a young adult on 26 June 1906 and spent a total of 61 years 8 months and 25 days in captivity (Hamlet, 1968).

The greatest authentic age recorded for a sea-bird in captivity is 44 years for a Herring gull (*Larus argentatus*) which lived in the Menagerie du Jardin des Plantes, Paris, France, from 1830 to 1874 (Gurney, 1899). Another specimen which died at Musselburgh, near Edinburgh, Scotland, on 10 July 1937 was '41 years at least' (Prof James Ritchie, quoted by Flower, 1937). Dorst (1971), on the other hand, quotes 49 years as the greatest age recorded for a herring gull in captivity, but further details are lacking. Less reliable is a claim of 55+ years for another bird which was wing-clipped in the USA, in May 1882, when it was already in full plumage and so

The Asian white crane 'Pops', who spent nearly 62 years in captivity (National Zoological Park, Washington, D.C.)

presumably not less than three years old, which reputedly lived until July 1935. Its mate, allegedly at least 45 years old, and three descendants of more than 30 years, were said to be still living in June 1938.

Until recently it was generally believed that sea-birds in aviaries lived longer on the average than those living under natural conditions because they had a regular food supply and were protected from enemies, but since the introduction of bird-banding it has been discovered – albeit surprisingly – that wild sea-birds live practically as long as captive ones, although fewer of them achieve real old age. Among the records given by Prof W Rydzewski (1962), editor of the international ornithological bulletin *The Ring* are a Herring gull (*Larus argentatus*) of 31 years 11 months 10 days (25 July 1929 to 5 June 1961, Goteborg, Sweden) and a Black-headed gull (*Larus ridibundus*) of 32 years 1 month 0 days (14 May 1922 to 14 June 1954, Heligoland, Germany).

The same authority (pers. comm. 10 February 1975) also has details of a Guillemot (*Uria aalge*) and a Curlew (*Numenius arquata*) living for 32 years 1 month 3 days and 31 years 6 months 21 days respectively, and in the USA a Brown pelican (*Pelecanus occidentalis*) for 31 years 1 month 26 days (2 September 1933 to 15 November 1964) and an Arctic tern (*Sterna paradisae*) for 33 years 1 month 26 days (24 July 1936 to 19 June 1970).

The greatest authenticated age recorded for a wild sea-bird is 35 years 11 months 16 days for an Oystercatcher (*Haematopus ostralegus*) which was ringed at Heligoland, Germany, on 18 June 1927 and was found dead there on 4 June 1963. Three years later, on 20 June 1966, a Herring gull (*Larus argentatus*), aged exactly 36 years, was reportedly found on the shore of Little Traverse Bay, Lake Michigan, Michigan, USA. The bird had been ringed on 29 June 1930 when a ten-day-old chick at Duck Rock, an islet just off Monhegan Island, Maine, USA (Pettingill, 1967). According to Prof Rydzewski, however, this record has since been found to be incorrect; he says the bird was actually banded on 8 July 1948, making it only 17 years 11 months 12 days. Another claim of 41 years 7 months 18 days (10 September 1911 to 28 April 1953, Leiden, Holland) for a Mallard (*Anas platyrhynhos*) has also been knocked down by Rydzewski, who thinks an error was made in reading the number on the band (cf. 23 years for the Domestic duck, see page 107).

Albatrosses (family Diomedeidae) are believed to live *c.* 30–40 years in the wild – Rydzewski has supplied the compiler with details of a Laysan albatross (*D. immutabilis*) which lived 34 years 0 months 6 days (24 November 1938 to 3 November 1972, USA), but Westerskov (1936) thinks the potential life-span may be as much as 80 years.

In January 1887 a wandering albatross was captured by the crew of the British frigate *Duchess of Argyll* near Cape Horn. Attached to its neck was a compass-case containing the information that the bird had previously been caught in the North Atlantic by an American ship, the *Columbus*, on 8 May 1840. The albatross was later released after a new case with details of its second capture was attached to the dynamic soarer's neck. The story should have ended there . . . but eight months later, on 18 Septemer 1887, what was believed to be the very same bird was caught near Triggs Island, Western Australia. This time, however, the compass-case had a new message which read in French '13 shipwrecked sailors have taken refuge on the Crozet Islands, August 4, 1887'. This vital information was immediately telegraphed to the French authorities, and after consultations at the highest Government level the warship *La Meurthe* was ordered to sail from Madagascar to the group of small islands in the southern Indian Ocean. There it was established that the message had been attached to the albatross by the crew of the French sailing-ship *Tamaris* which was wrecked in the Crozets on 9 March 1887. Tragically the French seamen did not live to see the successful result of their avian experiment, because they perished in a desperate attempt to reach near-by Possession Island two months before the arrival, on 2 December 1887, of the French warship.

It has been calculated that some 60–75 per cent of all birds die between the ages of three and six months through disease, predation or starvation.

Two of the shortest-lived birds are the Robin (*Erithacus rubecula*) and the Swallow (*Hirundo rustica*), both of which have an average life-expectancy of 13 months in the wild state, and the life expectations of the Mallard (*Anas platyrhynchos*) and the Starling (*Sturnus vulgaris*) have been assessed at 14½ months and 17 months respectively.

A great deal of nonsense has been written about the maximum flying speeds attained by birds, the tendency being to exaggerate rather than underestimate the velocity. It is extremely difficult to time a bird accurately over a measured distance, even using elaborate tracking and recording devices, because too many other factors are involved like wind velocity, gravity, angle of flight, etc., and other methods used to determine the speed like a car running on a parallel course or an aircraft are not really reliable because the angles of ascent or descent are not measured. The question is also complicated by the fact that most published estimates of bird velocities are for ground speed – which is very different from air speed.

The air speed of a bird is defined as the velocity with which it flies in relation to the air, and ground speed as the velocity with which it flies in relation to the ground. Thus, a bird flying at 64 km/h *40 miles/h* with a tail-wind of 48 km/h *30 miles/h* has an air speed of 64 km/h *40 miles/h*, but the ground speed is 112 km/h *70 miles/h*.

According to Meinertzhagen (1955) birds have two speeds – 'a normal rate which is used for everyday purposes and also in migration and an acceleration speed, which in some cases nearly doubles the rate of their normal speed'. This latter velocity, however, cannot be maintained for any length of time.

Probably 50 per cent of the world's flying birds cannot exceed an air speed of 64 km/h *40 miles/h* in level flight, and Fisher and Peterson (1964) believe only a very small number of the rest – most of them ducks or geese – can reach or exceed 96 km/h *60 miles/h*.

The fastest-flying bird is the Spine-tailed swift (*Chaetura caudacuta*) of Asia. Air speeds up to 170 km/h *106¼ miles/h* in level flight have been reliably measured in the USSR (Gladkov, 1942).

In 1934 ground speeds ranging from 275 to 351 km/h *171·8 to 219·5 miles/h* were recorded by E C Stuart Baker (1942) with a stopwatch for spine-tailed swifts passing over his bungalow in the Cachar Hills, NE India, to a ridge exactly 3·2 km *2 miles* away behind which they *seemed* to dip into another valley. This claim, however, was shot down in flames by Wing (1956) who said the spine-tailed swift cannot be seen at a distance of 1·6 km *1 mile* even with standard binoculars.

Gatke (1895), who spent 50 years in Heligoland, Germany, studying bird migratory movements, was firmly of the opinion that much greater velocities were attained by birds flying at high altitudes than at lower ones because of the more rarefied air, but his assertion that birds like the Golden plover (*Pluvialis apricarius*), the Curlew (*Numenius arquata*) and the Black-tailed godwit (*Limosa limosa*) reach speeds up to 384 km/h *240 miles/h* at altitudes of 12 192 m *40 000 ft* must be discounted because no birds can reach this height (see page 100).

Probably more exaggerated statements have been made about the velocity attained by the Peregrine falcon (*Falco peregrinus*) than any other species of bird. McLean says he once timed a specimen hunting over a 366 m *400 yd* field in California and found its *average* speed was over 264 km/h *165 miles/h* and its maximum rate of flight 288 km/h *180 miles/h*. Other writers have credited the bird with velocities as high as 440 km/h *275 miles/h* in a stoop and 272–320 km/h *170–200 miles/h* in level flight, and Dorst (1971) even goes so far as to state that the velocity reached by a peregrine falcon in a stoop is 'the highest speed recorded for a bird and no doubt for any animal'.

Unfortunately the truth falls a long way short of any of these declarations. In recent experiments by British falconer Phillip Glasier in which miniature air speedometers were fitted to a peregrine the maximum diving speed recorded was 131 km/h *82 miles/h*, and the top speed in level flight 96 km/h *60 miles/h* (Lane, 1968). Col Richard Meinertzhagen (1955) says no bird of prey, including the peregrine, can catch swift birds like the Racing pigeon (*Columba livia*) or the Pintailed sandgrouse (*Pterocles alchata*) in level flight, and Portal puts the maximum speed of this bird through still air in level flight at 99 km/h *62 miles/h*.

Ground speeds in excess of 160 km/h *100 miles/h* have also been reported for the Golden eagle (*Aquila chrysaetos*), but the pointed wings of this raptor are designed for slow, powerful, sustained flight.

Leslie Brown (1953) once timed a golden eagle over a distance of 22·5 km *14 miles* and found that its ground speed was 134 km/h *84 miles/h*, but ground speeds in excess of 320 km/h *200 miles/h* have been reported.

The fastest-flying sea-bird is the Magnificent frigate-bird (*Fregata magnificens*), also known as the 'Man-o'-War bird'. Stolpe & Zimmer (quoted by Lane, 1955), timed one specimen at an air speed of 153 km/h *95·69 miles/h*, but ground speeds in excess of 320 km/h *200 miles/h* have been quoted.

In June 1928 a ground velocity of 395 km/h *247 miles/h* was attributed to a small group of magnificent frigate-birds in level flight over a 40·4 km *25¼ miles* course off the Cardagos Garajos Islands, 400 km *250 miles* NE of Mauritius in the Indian Ocean (Ricks, 1934), but this reading must be considered excessive. It is interesting to note, however, that *F. magnificens* has the greatest wing area in proportion to weight (1247 g *43½ oz*) of any living bird and – excluding hummingbirds – is probably the finest flying-machine in nature. Love (1911) says he once saw one of these birds pursue, overtake and seize a flying fish in the air while half a gale was blowing.

The diminutive Hummingbird (family Trochilidae) has also been credited with amazing bursts of speed. In 1945 Wager timed the courtship flight of a *Colibri thalassinus*, between two trees, and found the average velocity to be 88 km/h *55 miles/h*. He believed that two birds chasing each other in courtship flight reached speeds up to 144 km/h *90 miles/h*. Greenewalt (1960), on the other hand, using a wind tunnel equipped with a feeding-bottle, discovered that a Ruby-throated hummingbird (*Archilochus colubris*) could only manage 46·4 km/h *29 miles/h*, although Scheithauer (1967) doubts whether this figure represented its maximum speed – 'there are far stronger driving factors than the stimulus of reaching a food site, for example, aggressive pursuit and the mad manœuvres of the courtship flight'. Scheithauer timed the daily chases of two Blue-throated sylphs (*Aglaiocercus kingi*) over a figure-of-eight course 67·3 m *73·68 yd* long during the courtship period and found that the average speed was 61 km/h *38·4 miles/h* and the maximum 75·5 km/h *47·2 miles/h*. He felt that on a straight course the birds might have attained a flight velocity of 88 km/h *55 miles/h*.

The winter of the 297 km *186 miles* Ulster race for pigeons in 1961 averaged 156 km/h *97½ miles/h*, but this was wind-assisted and the average air speed was about 72 km/h *45 miles/h*. In level flight in windless conditions it is very doubtful whether any pigeon can exceed 96 km/h *60 miles/h*, although Schorger (1955) reckons the Passenger pigeon (*Ectopistes migratorius*) had a potential speed of 112 km/h *70 miles/h* when hard-pressed.

The fastest standard game bird is the Red-breasted merganser (*Mergus serrator*). On 29 May 1960 a flock of six birds was flushed from the river ahead of an aircraft on a low aerial reconnaissance flight for the US Atomic Energy Commission along the Kukpuk River, Cape Thompson, northern Alaska. When the ducks took flight all the birds turned aside, except one male which flew slightly below and ahead of the aircraft. This loner, with a burst of speed, managed to keep his position in relation to the aircraft for about 457 m *1500 ft* before finally losing ground and turning aside. The air speed of the plane during the chase was 148 km/h *80 miles/h*.

The fastest ratite or running bird is the Ostrich (*Struthio camelus*) which can travel at 45–48 km/h *28–30 miles/h* for 15–20 min without showing undue signs of fatigue; Stevenson-Hamilton (1947), however, says their habit of running in wide circles 'deprives them of much of the advantage derived from their speed'. (Hunting dogs turn this weakness to good account by running the diameter of the circle described by their quarry.) The maximum speed attained by a *frightened* ostrich has not yet been established with any degree of certainty, and more information is needed on the subject. Guggisberg (1964) says he once paced an ostrich running parallel with his Land Rover at 72 km/h *45 miles/h* for 805 m *½ mile* before the giant bird sprinted and crossed the track in front of his vehicle; this means the ostrich must have been travelling at approximately 88 km/h *55 miles/h* at the peak of its acceleration if the speedometer was reliable.

The fastest-running carinate or flying bird is the American roadrunner (*Geococcyx californianus*) of the SW USA, which has been clocked at 42 km/h *26 miles/h* when hard-pressed by a car. On another occasion a horseman came upon one of these birds standing about 91 m *100 yd* ahead of him on a level road and decided to give chase. The roadrunner took up the challenge and, with straightened neck and slightly extended wings acting as stabilisers, tore furiously along the road for a 402 m *¼ mile* before seeking refuge in a thicket. Even then, the rider was still at least 46 m *50 yd* behind.

The fastest-swimming bird is the Gentoo penguin (*Pygoscelis papua*) of Antarctica. In January 1913 Dr Robert Cushman Murphy of the American Museum of Natural History, New York, USA, was able to time a few of these birds under water in a transparent pool at the summit of a coastal hill south of the Bay of Isles, South Georgia. 'They dashed straight away under water the length of the pond and back again, with a velocity which I then had an opportunity to compute as about ten metres a second. They chased each other round and round, flashing into the air twice or thrice during their bursts of speed, every action plainly revealed through the clear, quiet water.'

Ten metres a second is equivalent to a speed of 35·6 km/h *22·3 miles/h*, which is a respectable flying speed for some birds but, in an emergency, i.e. fleeing from the very swift Leopard seal (*Hydrurga leptonyx*) this penguin probably reaches speeds of 40–43 km/h *25–27 miles/h*.

The fastest recorded wing-beat of any bird is that of the hummingbird *Heliactin cornuta* of tropical South America with a frequency of 90 beats a second, followed by the Amethystine (*Calliphlox amethystina*) with 80 beats and the *Acestrura mulsanti* with 79 beats (Stolpe & Zimmer, 1939; Scheitthauer, 1967). These rates are probably exceeded by the tiny Bee hummingbirds *Mellisuga helenae* and *Acestrura bombus* – the frequency *normally* increases in proportion to the decrease in wing length – but further information is needed on the subject.

In 1951 Edgerton *et al* reported up to 200 wing-beats per second for the Ruby-throated hummingbird (*Archulochus colubris*) and the Rufous hummingbird (*Selasphorus rufus*) of the eastern USA during courtship flights, but these frequencies were for the narrow tips of the primaries only and not the complete wing. By comparison the wandering albatross can remain airborne for up to 6 hours without moving its wings once.

The deepest-diving bird in the world is the Emperor penguin (*Aptenodytes forsteri*) of Antarctica. In 1969 a team of American scientists carried out a series of experiments at Cape Crozier to determine the diving ability of this species. All the birds used for the experimental dives were collected from groups gathered at the edge of the ice and then taken to the diving station – an isolated hole in the ice to which the birds were forced to return after each feeding dive. Depth measurements were obtained with small (4·5 g *0·152 oz*) capillary depth-recorders sutured to the backs of the birds' necks; these offered very little resistance to swimming and consequently did not have much of an inhibiting effect during the dives. Two–three hours after attaching the depth-tubes the birds were recaptured and the instruments removed. According to Kooyman *et al* (1971) a total of 238 dives were measured during the experimental studies and the greatest depth recorded was 265 m *869 ft* – by a small group of ten penguins. The duration of most of the dives was less than 1 min, but one bird which did not return to the dive station was seen swimming near the observation chamber some 10 m *32 ft* from the hole after 18 min submersion. The birds were never seen to exhale under water, and diving was usually preceded by a few rapid breaths and then a deep inhalation.

The deepest-diving flying birds are the Common loon (*Gavia immer*), also known as the 'Great Northern Diver', and the Old-squaw or Long-tailed duck (*Clangula hyemalis*), both of which can reach a depth of 45·7 m *150 ft*. Jourdain (1913) quotes a record of a common loon caught in a trammel-net at 54·8 m *30 fathoms 180 ft*, and Roberts (1932) was told by a fisherman living at the mouth of the Cascade River, Minnesota, USA, that he had netted loons at a depth of 61 m *200 ft* in Lake Superior. There is also an unconfirmed record of a loon being caught at a depth of 73·15 m *240 ft*. Similarly, a fisherman at St Joseph, Michigan, USA, told Barrows (1912) that he had taken old-squaw ducks frequently at a depth of 54·8 m *30 fathoms 180 ft*, and Butler (1898) says they were often caught at Michigan City, Indiana, USA, at the same depth. The greatest depth is recorded by Tarrant (1883) who was told by Captain Nathan Saunders that he had taken old-squaw ducks on lines set in Green Bay, Wisconsin, USA, at a depth of 6 m *200 ft*.

It has been suggested that birds caught in nets at depths greater than 10 fathoms 18·28 m *60 ft* become entangled in the net while it is being raised, but Schorger (1947) thinks it is 'wholly improbable that they were all caught during the raising of the nets'. He concluded: 'There are apparently no physical or physiological reasons why some exceptionally skilful individuals among diving birds cannot descend to a depth of 200 ft. There is ample evidence that this depth is actually reached.'

Another fine diver is the King eider (*Somateria spectabilis*), which is able to reach mussels in 25 fathoms 45·7 m *150 ft* of water (Eaton, 1910), but a statement by Horring (1919) that this bird can descend to 73 fathoms 131·6 m *432 ft* in search of food can be discounted.

Elsewhere Faber (1826) reports that Common and Black guillemots (*Uria aalge* and *Cepphus grylle*) have been recovered from the stomachs of Spiny dog-fish (*Squalus* sp.) which are rarely found in water less than 91·4 m *300 ft* deep, and Forbush (1922) states that Crested auklets (*Aethia cristatella*) have been recovered from the stomachs of Cod (*Gadus callarias*) caught at a depth of 60·9 m *200 ft*. These claims, however, are not accepted by Dewar (1924) who says there is no proof that these birds were actually swallowed by the fish on the sea-bed. Clay (1911) describes how he once saw several Baird's cormorants (*Phalacrocorax bairdii*) rise to the surface with kelp in their beaks from water with a reputed depth of over 146·3 m *80 fathoms 480 ft*, but there is no evidence that this depth (if reliable) was reached in the course of the dive.

The greatest distance covered by a ringed bird during migration is 22 400 km *14 000 miles* for an Arctic tern (*Sterna paradisaea*) which was banded as a nestling on 5 July 1955 in the Kandalaksha Sanctuary on the White Sea coast about 200 km *125 miles* from Murmansk, European Russia, and was captured alive by a fisherman 13 km *8 miles* south of Fremantle, Western Australia, on 16 May 1956. The bird had flown south via the Atlantic Ocean and then headed east past the Cape of Good Hope (Dorst, 1962).

The Arctic tern also holds the record for the greatest distance covered in a year between breeding seasons – *c.* 38 400 km *24 000 miles*. It has been known to nest within 720 km *450 miles* of the North Pole and flies every year to the South Pole and back again.

The American golden plover (*Pluvialis dominica*) also covers incredible distances during migration. Breeding in Alaska and the Arctic, it flies to South America in the autumn and some of them get as far south as New Zealand before returning to their breeding-grounds in the spring. So, in little more than six months, they cover between 24000–27200 km *15 000–17 000 miles*.

Between breeding seasons the Sooty albatross (*Phoenetria fusca*) flies round the world at 40 degrees south latitude, covering a distance of 30400 km *19 000 miles* in some 80 days, which must also be something of a record.

Most hummingbirds do not migrate very far, but there are exceptions. One of the most remarkable flights in the avian world is made by the diminutive Rufous hummingbird (*Selasphorus rufus*) which every autumn leaves Alaska and flies along the entire west coast of North America to its winter quarters in Mexico. The following spring it makes the return journey, covering 6400 km *4000 miles* in all.

The greatest recorded homing flight by a pigeon was one made by a bird owned by Mr Peter Robertson of Northumberland. Released at Beauvais, northern France, on 16 June 1972 to fly back to England, the pigeon turned up exhausted in Durban, South Africa, two months later, having flown an airline route of 9600 km *6000 miles*, but an actual distance of possibly 12 160 km *7600 miles* to avoid the Sahara Desert.

The shortest bird migrations are made by species like the Mountain quail (*Oreotyx pictus*), the Black-capped chickadee (*Parus atricapillus*) and Clark's nutcracker (*Nucifraga columbiana*) of North America which merely descend from exposed mountain ridges to the sheltered valleys below during the autumn months. The mountain quail, which nests at altitudes up to 8686 m *9500 ft* in the central California mountains, leaves the region of deep snow in September and, in groups of 10–30 *walk* in single file down below the 1524 m *5000 ft* mark. In the spring they make the return trek, again on foot, to the higher altitudes (Tyne & Berger, 1971).

At one time it was believed that most birds flew at heights of 6096 m *20 000 ft* or more during migration because of the physical advantages of low pressure at high altitudes (see Gatke, 1895), but this theory no longer holds. Most migrating birds, in fact, fly at relatively low altitudes (i.e. below 91·44 m *300 ft*) and only a few dozen species fly higher than 914 m *3000 ft*.

The celebrated record of a skein of 17 Egyptian geese (*Alopochen aegyptiacus*) photographed by an astronomer at Dehra Dun, northern India, on 17 September 1919 at an estimated height of between 17 708 and 19 293 m *58 080 and 63 360 ft 11 and 12 miles* has been discredited by experts because the picture is not clearly defined. (The calculation was based on the known diameter of the sun and the wing span of a goose in flight (Meinertzhagen, 1955).) A more reasonable height of 8046 m *26 400 ft 5 miles* has also been quoted for this skein (Dorst, 1971), and this is probably a more accurate reading.

The highest acceptable altitude recorded for a bird is 8200 m *26 902 ft* for a small number of Alpine choughs (*Pyrrhocorax graculus*) which followed the British Everest expedition of May 1924 to Camp V, and Hingston (1936) believes that if a camp could have been established on the summit of Everest 'I have little doubt that it would be visited by choughs.' The same writer also saw 55 other different kinds of birds above the 4267 m *14 000 ft* mark, and at Camp III (6400 m *21 000 ft*) he spotted a Jungle crow (*Corvus* sp.) and a Rose-breasted finch (*Carpodacus puniceus*) migrating across the range.

On 23 May 1960 the Indian Everest Expedition found three species of raptor lying dead on the South Col at a height of nearly 7925 m *26 000 ft*. One of these birds was brought down and later identified as the Steppe eagle (*Aquila nipalensis nipalensis*) (Gyan Singh, 1961). In 1921 a Lammergeier (*Gypaetus barbatus*) was seen at 7620 m *25 000 ft* on the same mountain (Wollaston, 1922), and this magnificent bird probably sails over the summit on occasion. (A dynamic soarer like the lammergeier does not need a great deal of oxygen for flight like other birds.)

A Wall creeper (*Tichodroma muraria*) has been seen at 6400 m *21 000 ft* in the Karakorum Range, north India (Ingram, 1919), a Crane (*Grus.* sp.) at 6096 m *20 000 ft* in the Himalayas, an Andean condor (*Vultur gryphus*) at 6035 m *19 800 ft* in the Andes (Jacks, 1953) and Godwits (*Limosa* sp.), Curlews (*Numenius* sp.) and even Jackdaws (*Corvus monedula*) at 6004 m *19 700 ft* on Mt Everest. On 9 July 1962 a Western Airlines L-188 Electra was flying over Lander County, central Nevada, USA, at an altitude of 6400 m *21 000 ft* when the pilot noticed a light thud. When the plane landed one of the crew found a large blood-stained dent on the leading edge of one of the plane's horizontal stabilisers with a feather sticking to it, which was forwarded to the US Fish & Wildlife Service for identification. The unfortunate bird was a Mallard (*Anas platyrhynchos*) (Terres, 1968).

On three separate occasions in 1959 a radar station in Norfolk picked up flocks of small passerine night migrants flying in from Scandinavia at altitudes up to 6400 m *21 000 ft*. According to Dr David Lack (1960) of the Edward Grey Institute, Oxford, they were probably warblers (Sylviidae), chats (Turdidae) and flycatchers (Muscicapidae).

Two years later, in September 1961, I Nisbet, an American ornithologist-physicist studying the height of bird migration with radar at Cape Cod, Massachusetts, picked up some echoes at an altitude of 6400 m *20 000 ft*. He thought they were Black-bellied plovers (*Pluvialis squatarola*), Semi-palmated sandpipers (family Scolopacidae) and certain other waders known to migrate over Cape Cod at that time of the year.

The greatest depth underground recorded for a bird is 183 m *600 ft* for a European robin (*Erithacus rubecula*) found at the bottom of a mine-shaft in Wigan, Lancashire, in 1935.

Birds have very large eyes for their size – they often weigh more than the brain – because they are more dependent on sight than other vertebrates, and the vision of eagles, hawks and falcons is extremely acute. Leslie Brown (1970) once watched a Verreaux's eagle (*Aquila verreauxi*) launch a stoop at a Martial eagle (*Polemaetus bellicosus*) carrying a Rock hyrax from a distance of 2414 m *2640 yd 1½ miles* and says this predator has such keen eyesight that it could probably see a small green grasshopper 25 mm *1 in* long 274 m *300 yd* away. The same authority also states that a Golden eagle (*Aquila chrysaetos*) can detect a 450 mm *18 in* long hare at a range of 1965 m *2150 yd* (possibly even 3·2 km *2 miles*) in good light and against a contrasting background, and a falcon a pigeon at a range of over 1066 m *3500 ft*.

The visual acuity of the large bird of prey is at least eight to ten times stronger than that of human vision.

Owls (Strigiformes) have a sensitivity to low light intensities that is 50–100 times greater than that of human night vision. Tests carried out at the University of Michigan, USA, between 1938 and 1943 showed that under favourable conditions the Barred owl (*Strix varia*), the long-eared owl (*Asio wilsonianus*) and the Barn owl (*Tyto alba pratincola*) could swoop on a dead mouse from a distance of 1·83 m *6 ft* or more in an illumination of only *0·000 000 73 ft-candle* (the light from a 6 watt inside-blue bulb, reduced by the rheostat, and passed through nine sheets of paper) equivalent to the light from a standard candle at a distance of 356·6 m *1170 ft*, and that the same birds could see dead prey (with difficulty) in an illumination of only *0·000 000 8 ft-candle*, which is equivalent to the light from a standard candle at a distance of 1076 m *3536 ft* (Dice, 1945).

The 'intellectual giants' of the bird world are generally considered to be the Crows (Corvidae) and the Parrots (Psittaciformes), but the subject is so complex that it would need a separate study that cannot be undertaken here.

The question of the 'dimmest' bird, however, is much easier to resolve. The general consensus of opinion is that **the least intelligent living bird** is the domesticated Turkey (*Meleagris gallopavo*) and this is not difficult to believe. For instance, there is one authentic case of a farmer who left an empty barrel in his yard. Six of his best turkeys promptly scrambled into it, piled up one on top of the other and died of suffocation because none had the sense to get out. Other breeders have known turkeys to stand out in the open during a downpour and be drowned because they weren't intelligent enough to walk a few yards to their hutches. Each year thousands of turkeys freeze to death on cold nights because they stubbornly refuse to seek refuge in their warm sleeping quarters, and some turkeys are so dim that they even have to be persuaded to eat. Some breeders rear chickens alongside turkeys in the hope that the turkeys will copy the actions of their more intelligent companions but this is generally wishful thinking. Not surprisingly turkeys are great panickers and will run head-long into walls or entangle themselves on wire

The night vision of the Barn Owl is 100 times more acute than that of a human being (Aquila Photographics)

fences if they hear an unfamiliar noise. There are even reports of turkeys being frightened to death by pieces of paper fluttering in front of their path.

The largest egg produced by any living bird – and the largest single cell in the animal world today – is that of the North African ostrich (*Struthio c. camelus*). The average example measures 156 × 136 mm *6·15 × 5·35 in* and weighs 1·65 kg *3 lb 11 oz*, compared to 152 × 127 mm *6 × 5 in* and 1·56 kg *3 lb 8 oz* for the Southern ostrich (*S. c. australis*) (Duerden, 1919). The shell is 1·6 mm *0·0625 in* thick and can support the weight of an 114·5 kg *18 stone 252 lb* man.

The largest egg laid by any bird on the British list is that of the Mute swan (*Cygnus olor*). In one series of 88 samples the measurements ranged from 100 to 122 mm × 70 to 88 mm *3·93 to 4·80 in × 2·75 to 3·14 in*, the average being 112·5 × 73·5 mm *4·42 × 2·89 in* (Schonwetter, 1960–61). The weight is 336–364 g *12–13 oz*.

The largest egg produced by a sea-bird is that of the Wandering albatross (*Diomedea exulans*), which is highly variable in shape and size. In one series of 87 from Gough Island, in the Tristan de Cunha Group, South Atlantic, the average was 127 × 76·9 mm *5 × 3·02 in* (Verrill, 1895). The weight of six freshly laid South

The enormous egg laid by the hen Kiwi (Schram Photos)

Georgia eggs ranged between 429 and 487 g *15·3 and 17·3 oz* (Murphy, 1936), but weights up to 588 g *21 oz* have been claimed for specimens collected on Auckland Island, 320 km *200 miles* south of New Zealand (Ross, 1847).

The largest egg laid by any bird in proportion to its own size is that of the Kiwi (Apterygiformes) of New Zealand. There is an old record (1889) of a 1·68 kg *3 lb 12 oz* Mantell's kiwi (*Apteryx australis mantelli*) laying an egg weighing 406 g *14·5 oz*, or nearly one-quarter the body-weight of the hen, but weights up to 504 g *18 oz* have been reliably reported for other specimens. The problems of producing such an enormous cell, however – the egg of the chicken-sized kiwi is more than ten times as large as a hen's egg – can be disastrous, and not infrequently female kiwis containing fully developed eggs have been found dead in their nesting burrows. The bee hummingbirds (see this page) are also strong contenders.

Most parasitic cuckoos lay relatively small eggs, but Tyne & Berger (1971) say there is a record of a single egg of a non-parasitic cuckoo (*Crotopha major*) which was one-third of the female's body-weight.

The smallest mature egg laid by any bird is that of the Bee hummingbird *Mellisuga helenae*, the world's smallest bird. On 8 May 1906 two specimens were collected at Boyate, Santiago de Cuba by O Tollin, who presented them to the US National Museum of Natural History, Washington, DC, USA, in 1909. One of these eggs has since been mislaid, but the other one – which has a broken shell – measures 11·4 × 8 mm *0·45 × 0·31 in* and weighs 0·5 g *0·176 oz* (Watson, pers. comm. 20 August 1975). This egg is closely matched in diminutiveness by that of the Vervain hummingbird (*M. minima*) of Jamaica and Haiti. Two examples collected by James Bond (pers. comm. 22 August 1967) on Ile Tortue, Haiti, measure 11·6 × 8·4 mm *0·45 × 0·33 in* and 11·5 × 8·15 mm *0·45 × 0·32 in* respectively.

Details of even smaller eggs have been published for other species of birds, but these were all 'sports', which means they were emitted from the oviduct *before reaching maturity*. A typical example was an egg laid by a Zebra finch (*Taniopygis castanotis*) which measured only

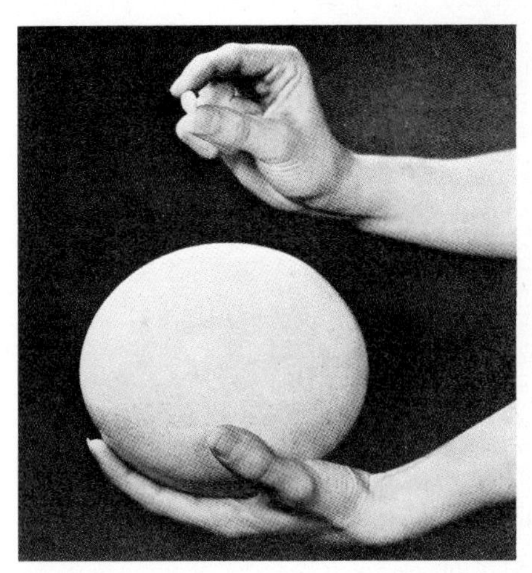

The eggs of the hummingbird and the ostrich – extremes in reproduction

The largest and smallest eggs of British birds, the mute swan and goldcrest

9·5 × 7 mm *0·37 × 0·27 in* (C Taylor, pers. comm.).

The smallest egg laid by any bird on the British list is that of the Goldcrest (*Regulus regulus*), which measures 12·2–14·5 mm *0·48–0·57 in* × 9·4–9·9 mm *0·37–0·39 in*. The weight is 0·6 g *0·211 oz*, which means the egg is almost as small as the one deposited by the Bee hummingbird *Mellisuga helenae*.

The smallest egg laid by any bird in relation to body-weight is that of the Emperor penguin (*Aptenodytes forsteri*), which constitutes only 1·4 per cent of the total body-weight.

The longest incubation period of any bird is that of the Wandering albatross (*Diomedea exulans*) with a normal range of 75–72 days, but Tickell (1968) says he observed two eggs on Bird Island, South Georgia, Antarctica, which took 85 days to hatch. The Royal albatross (*D. epomophora*) has a normal range of 75–81 days, and the Common kiwi (*Apteryx australis*) 75–80 days. The incubation period of the Mallee fowl (*Leipoa ocellata*) of Australia is normally 62 days, but Petersen (1963) cites a record of an egg taking 90 days to hatch.

The longest incubation period of any bird on the British list is that of the Fulmar (*Fulmarus glacialis*) the single egg taking 52–53 days to hatch out.

The shortest incubation periods are to be found among the passerines or perching birds. A small number of species have been credited with a ten-day incubation period (the time that has elapsed between the laying of the last egg in a clutch and the hatching of that egg when all the eggs hatch), but most of these birds are now known to require from 11 to 14 days. They include the Siskin (*Carduelis spinus*) 11–12 days, House sparrow (*Passer domesticus*) 12–14 days, Wryneck (*Jynx torquilla*) 12 days and the Wren (*Troglodytes troglodytes*) 14 days.

Hummingbirds have been credited with incubation periods as short as eight days (Brehm, 1861; Gentry, 1876), but the actual duration is 14–21 days depending on the species.

The North American cowbird (*Molothrus ater*) has traditionally been credited with a ten-day incubation period, but Nice (1953) found the time in one series of 62 hatchings was eleven to twelve days. Sometimes, however, an egg may be retained for 12–24 hours in the oviduct, and the same authority mentions another observer (Hoffman, 1929) who caught a cowbird that appeared to be egg-bound and gave it a home for the night. The following morning he found two eggs in the cage. 'One of these eggs', writes Nice, 'would have had an extra day of development before it was laid, and if immediately incubated, might conceivably have hatched in ten days.'

The only species of birds with well-authenticated records of ten-day incubation periods are the Great spotted woodpecker (*Dendrocopus major*) and the Black-billed cuckoo (*Coccyzus erythropthalmus*). In the first case Bussmann (1946) gave details of six eggs which were laid between 26 April and 1 May. Four young hatched out on 10 May, and the last two were out by 10 am on 11 May. In the other case reported by Spencer (1943) one egg was found on 2 July, and additional eggs were laid on 16 July. Nice, however, thinks the last egg in the cuckoo clutch may have been retained in the oviduct for an extra 24 hours and been laid in a more advanced state than the others.

The shortest incubation period of any bird on the British list is that of the Siskin and the Hawfinch (*Coccothraustes coccothraustes*), both averaging about eleven to twelve days.

The number of eggs laid by a bird in a 'single clutch' varies considerably among species and is largely related to food supplies and the physical condition and age of the hen.

The Bobwite (*Colunus virginianus*) of North America is generally credited with laying the largest clutches, with a normal range of 12–24 eggs, followed by the Blue tit (*Parus caeruleus*) with 7–24 eggs, the Pheasant (*Phasianus colchicus*) 6–22 eggs, the Scaup (*Aytha marilla*) 6–22 eggs, the Red-crested pochard (*Netta rufina*) 6–21 eggs and the Partridge (*Perdix perdix*) 12–20 eggs, but Campbell and Ferguson (1973) believe the larger numbers were probably the produce of two or even three hens laying in the same nest.

By continually removing eggs from a clutch, however, experimenters have succeeded in tricking a Wryneck (*Jynx torquilla*) into laying 48 eggs, a House sparrow (*Passer domesticus*) 51, a Cape gilded flicker (*Colaptes chrysoides*) 71 and a Bobwhite (*Colinus virginianus*) 128 eggs before stopping (Phillips, 1887; Wing, 1956). The Mallard duck (*Anas platyrhynchos*) may lay 80–100 eggs and there is a record of one hen producing an incredible 146 eggs in an effort to achieve a nest complement (Mouquet, 1924).

At the other end of the scale, most true seabirds lay only one egg in a clutch, and the Emperor penguin (*Aptenodytes forsteri*), which produces only a single egg annually, has the lowest egg output of any living bird.

The heaviest clutch of eggs on record was probably one consisting of 94 ostrich eggs which the explorer Major R Bagnold saw and photographed in the Sudan. The eggs lay in a circle measuring 3·65 m *12 ft* across and their combined weight was probably in excess of 136·4 kg *300 lb*. This total, however, must have been made up of a number of separate broods, because a single female normally lays 15–20 eggs. (In 1962 a game-warden counted 285 chicks with a single cock and hen ostrich near Isiolo, Kenya.)

The largest nests (i.e. constructed of sticks or twigs) are built by eagles (Accipitridae). One nest built by a pair of Bald eagles (*Haliaeetus leucocephalus*) in Vermilion, Ohio, USA, and possibly their successors over a period of 35 years, measured 2·59 m *8 ft 6 in* in width and 3·65 m *12 ft* in depth. When it finally crashed to the ground during a storm – killing the eaglets inside – its weight was estimated at 1818 kg *2 tons 4000 lb*. Another nest built by a pair of bald eagles near St Petersburg, Florida, USA, had a width of 2·89 m *9 ft 6 in* and a depth of 6·09 m *20 ft* (Petersen, 1963).

The incubation mounds of dry sandy earth with a core of vegetable matter built by the Brush turkeys (Megapodidae) of Australia, New Guinea and neighbouring islands are much larger but several pairs of birds help in the construction. According to Frith (1962) the mounds of most Mallee fowl (*Leipoa ocellata*) measure about 4·57 m *15 ft* in diameter and 60–91 cm *2–3 ft* high, but he did see one in southern Australia which measured 5·48 m *18 ft* in diameter. When dug out it was found to be 1·37 m *4 ft 6 in* deep, and when fully mounted in the summer it was 1·21 m *4 ft* high. Other mounds have reportedly been discovered in Australia measuring up to 4·57 m *15 ft* in height and 10·66 m *35 ft* in diameter. It has been calculated that the nest site of a Malee fowl may involve the mounding 299 m³ *300 yd³* of matter weighing 304 tonne *300 ton*.

The smallest nests are built by Hummingbirds (Trochilidae). That of the Bee hummingbird *Mellisuga helenae* is about the size of a thimble, and the nest of the Crested tree swift (*Hemiprocne mystacea*) of Malaysia and Australia is almost as minute.

The most valuable nests are those built by the little Cave swiftlets (*Collocalia esculenta*) of SE Asia and Indonesia which produce the raw material for bird's-nest soup. The most sought-after nests are those made entirely of salivary secretion which do not require extensive cleaning. (Nests made of secretion mixed with feathers or vegetable matter fetch lower prices.)

There are also a number of birds which use very strange and sometimes uncomfortable material to build their nests. Madoc mentions a crow's nest which was made entirely of barbed wire, and there is a record of a pigeon's nest in Sheffield which consisted of 152 mm *6 in* nails plus a few feathers. In August 1940 a Chaffinch's nest built almost entirely of confetti was found in a garden at Reedham, Norfolk. It was later exhibited at Norwich Castle Museum.

The highest price ever paid for a stuffed bird is £9000 for a specimen of the Great auk (*Pinguinus impennis*) in summer plumage collected by the naturalist Count F C Raben in Iceland in *c.* 1821. The bird was purchased at a Messrs Sotheby & Co, London, auction on 4 March 1971 by the Director of the Iceland Natural History Museum, who later said that his museum would have gone as high as £23 000 for this magnificent example. It had spent the previous 150 years in Aalhom Castle, Denmark. Only about 80 stuffed specimens are preserved. The last great auk was killed on Eldey, Iceland, in 1844. The last British sightings were at Co. Waterford, Eire, in 1834 and St Kilda in *c.* 1840.

In December 1970 a Passenger pigeon (*Ectopistes migratoria*) fetched a mere £65 at another Sotheby auction. The last wild example had been killed at Babcock, Wisconsin, USA, in September 1899.

The highest price ever paid for a live bird is £6000 for a male racing pigeon named 'Motta', purchased by Mr Louis Masarella of Kirby Muxloe, Leicestershire, England, on 29 March 1975. Earlier, on 7 December 1973, he had paid £5000 for another male racing pigeon named 'Champion Workman'.

The longest feathers grown by any bird are those of the Onagadori ('long-tailed fowl'), a strain of *Gallus gallus*, which have been bred in SW Japan for over 300 years. Only the roosters grow the very long tail coverts, and an extreme measurement of 10·6 m *34 ft 9½ in* was reported in 1972 for a specimen owned by Mr Masasha Kubota of Kochi in Shikoka.

The normal body temperature of passerine birds varies between 41·1 °C *106 °F* and 39·4–40·0 °C *103–104 °F* when asleep, but it rises during intense activity.

The highest body temperature ever recorded for a bird is 44·8 °C *112·7 °F* for a Western pewee (*Myiochanes richardsoni*). This is closely matched by the Spine-tailed swift (*Chaetura caudacuta*), the fastest-moving living creature, which has a body temperature of 44·7 °C *112·5 °F*.

No bird has a body temperature regularly as low as that of an adult human being (36·9 °C *98·4 °F*), but the species that comes nearest is the King penguin (*Aptenodytes patagonica*) with a reading of 37·7 °C *100·00 °F*. The Western grebe (*Aechmophorus occidentalis*) has a body temperature of 38·5 °C *101·3 °F*, the Ruby-throated hummingbird (*Archilochus colubris*) 38·9 °C *102 °F*, the Brown kiwi (*Apteryx australis*) 39 °C *102·2 °F*, the Ostrich (*Struthio camelus*) 40 °C *104 °F*, the Wandering albatross (*Diomedea exulans*) 40·7 °C *105·3 °F* and the Mute swan (*Cygnus olor*) 41 °C *105·8 °F* (Wetmore, 1921; Portmann, 1950; Dorst, 1971).

The lowest body temperatures are found in those species of birds which enter a torpid or hibernating state during which the body temperature is lowered to a level near that of the surrounding air, and in the case of hummingbirds (Trochilidae) readings as low as 13·3 °C *56 °F* have been recorded (Chaplin, 1933). In December 1946 Jaeger (1949) found a torpid Nuttall's poor-will (*Phalaenoptilus nuttallii*) in a rocky crevice on the side of a deep canyon in the Chuckawalla Mts, California, USA. He took the body temperature of the bird on five different days and found that it ranged from 18 to 19·2 °C *64·4 to 67·7 °F*. The surrounding air temperatures on the same days ranged from 17·5 to 22·9 °C *63·5 to 73·3 °F*.

The most aerial of all birds is the Sooty tern (*Sterna fuscata*). After leaving the nesting-grounds the young birds do not come to land or alight on water until they return to their breeding-grounds on tropical and sub-tropical islands in the Atlantic, Pacific and Indian Oceans for the first time at the age of three or four years.

The most aerial of all land birds is the Common swift (*Apus apus*) which remains aloft for at least nine months at a stretch and possibly much longer. There is no truth, however, in the claim that it sleeps on the wing, although Locke (1970) says it 'somehow manages to get enough rest during the night-flight to enable it to maintain a steady level without exhaustion'.

The largest domesticated bird (excluding the Ostrich) is the Turkey (*Meleagris gallopavo*), which was introduced into Britain via Spain from Mexico in 1549.

The greatest live weight recorded for a turkey is 34 kg *75 lb*, reported in December 1973 for a 'holiday' turkey reared by Signe Olsen of Salt Lake City, Utah, USA.

The world record for a dressed bird is 32·53 kg *71 lb 12 oz* for a stag named 'Mr Chukie', reared by Dale Turkeys Ltd of Ludlow, Shropshire, which won the National Heaviest Turkey Competition held in London on 3 December 1975. Its live weight was probably slightly in excess of 36·28 kg *80 lb*.

Today male turkeys are bred so big that they are no longer physically capable of mating. Hens are therefore artificially inseminated. Just after the Second World War the average weight of a big male turkey in Britain was 9·09–12·7 kg *20–28 lb*, but it was narrow-breasted and long-legged. Now the housewife demands a bigger-breasted bird.

The largest breed of domestic goose is the Toulouse, ganders weighing between 9·09 and 13·6 kg *20* and *30 lb*.

The largest breed of domestic chicken is one developed by Mr Grant Sullens of West Point, California, USA, over a period of seven years by crossing and recrossing the bigger members of different varieties, including the Rhode Island Red. One of his super-chickens, a monstrous rooster named 'Weirdo', reportedly weighed 10 kg *22 lb* in January 1973 and was so ferocious that he had already killed two cats and crippled a dog which came too close. The largest breed of chicken in Britain is the Dorking, with roosters weighing up to 6·36 kg *14 lb*. This size is closely matched by the Buff Cochin, roosters weighing 4·09–5·45 kg *9–12 lb* (Clausen & Ipsen, 1970).

The largest breed of domestic duck is the Muscovy (*Cairina moschata*), adult drakes weighing 4·54–6·36 kg *10–14 lb*. This species came originally from South America.

In August 1974 Cherry Valley Farms Ltd of Rothwell, Lincolnshire, England, announced that they had developed a new strain of giant duck which was about twice the weight of a normal bird bred for the table. In one growing trial 80 drakes and 80 ducks achieved an average weight of 4·04 kg *8·9 lb* in eight weeks. The heaviest drake 'Jumbo' scaled 5·04 kg *11·1 lb*, which is believed to be a world record for an eight-week-old duck. The new breed, a Pekin strain called 'Line 151', was the result of a 15-year

Super-duck 'Donald Jumbo', who weighed 5·03 kg 11·1 lb *at the age of 8 weeks (Fearnley Photograph)*

project involving new genetic, nutritional, veterinary and management techniques and the heavyweight ducks are now being used for breeding with ordinary ducks (H Nott, pers. comm. 17 March 1975).

The longest-lived domesticated bird (excluding the ostrich) is the Domestic goose (*Anser anser domesticus*), which normally lives about 25 years. In April 1976 a gander named 'George', owned by Mrs Florence Hull of Pilling, Lancashire, celebrated his 49th birthday, but unconfirmed claims up to 80 years have been reported (Gurney, 1899).

The Domestic chicken (*Gallus gallus domestica*) rarely lives longer than 13 or 14 years. In October 1952 the death was reported of a Black Minorca hen named 'Blackie' in Cahir, Co. Tipperary, Eire, aged 16 years 7 months, and in July 1974 the compiler received a letter from Mr Neville Coop of Whangerei, New Zealand, who said he owned a 21-year-old Bantam hen which was 'still very hale and hearty'. According to Flower (1938) there are reliable records of birds living 22, 23, 24, 25 and even 30 years, but the latter figure takes some believing.

The Domestic duck (*Anas plalyrhynchos domestica*) has a normal life-span of ten years, but ages in excess of 18 years have been reported. A duck named 'Bibbler', hatched by an adopted mother hen in 1942 at Barnham, Sussex, died in April 1965 aged 23 years. The specimen is now on display in the Bognor Museum, Sussex (Cutten, pers. comm. 14 August 1970).

The longest-lived small cagebird is the Canary (*Serinus canaria*), which was imported into Europe from the Canary Islands in the 16th century. The average life-span is 12–15 years, but ages in excess of 20 years have been reliably reported. In June 1972 it was reported that Mrs Kathleen Leck (b 1940) of Hull, Yorkshire, still had a 31-year-old canary which was bartered by her father in Calabar, Nigeria, when she was one year old.

On 8 April 1975 a cross-bred canary named 'Joey' celebrated his 28th birthday in his cage at Mrs Edna Porter's home in St Anns, Nottingham.

The average life-span of the Budgerigar (*Melopsittacus undulatus*), which was introduced into this country from Australia in 1840, is six to eight years, but there are several authentic records of birds living beyond ten years. On 13 March 1975 the Gateshead Budgerigar Society, Durham, presented a virtually featherless budgie named 'George', owned by Mrs Elsie Ramshaw of Denaby Main, Yorkshire, with a special black and white rosette to mark his 25th birthday.

The highest authentic rate of egg-laying by a domestic bird is 361 eggs in 364 days by a Black Orpington in an official test at Taranki, New Zealand, in 1930 (Breland, pers. comm. 17 June 1967).

The UK record is 353 eggs in 365 days in a National Laying Test at Milford, Surrey, in 1957 by a Rhode Island Red owned by W Lawson of Welham Grange, Retford, Notts.

The largest hen's egg reported is one of 454 g *16 oz* (average weight 58 g *2·07 oz*) with double yolk and double shell laid by a White leghorn at Vineland, New Jersey, USA, on 25 February 1956.

The largest egg reported in the UK was one of 'nearly 12 oz' 339 g for a five-yolked egg measuring 31 cm *12¼ in* round the long axis and 22·8 cm *9 in* round the shorter axis laid by a Black Minorca at Mr Stafford's Damsteads Farm, Mellor, Lancashire, in 1896.

The ancient gander 'George', who will be 50 next year (Gazette & Herald, Blackpool)

The smallest hen's egg on record was a 'sport' weighing only 1·29 g *0·044 oz* or 98 per cent less than the average weight (Szuman, 1926).

Although there are a number of records of chickens flapping wildly round the farmyard for several minutes after being decapitated, **there is only one authentic case of a chicken surviving 17 days without its head!**

The incredible story started on 12 November 1904 when Mr Herbert V Hughes, proprietor of the Belvidore Hotel in Sault Sainte Marie, Michigan, USA, was beheading chickens for the Sunday dinner. Some of them were given to a member of the kitchen staff to pluck and clean. Suddenly the girl let out a high-pitched scream and rushed terror-stricken from the kitchen, and when Mr Hughes investigated he found a Black Minorca hen . . . headless . . . walking slowly round the room.

The news spread like wildfire and for more than two weeks the Belvidore Hotel was invaded by crowds eager to see the bizarre sight of a living headless hen.

Mr Hughes fed 'Biddy', as she was called, by means of a syringe injected into the raw end of the food pipe and the bird flourished. Sometimes she would stretch up and flap her wings and then attempt to preen her ruffled feathers with her headless neck; other times she made strange croaking noises as if she was trying to sing.

The hen lived until 30 November. She would probably have lived even longer if it hadn't been for the carelessness of an attendant. Each day the end of the neck had been gradually healing up, and by the 17th day the skin had grown so closely over the end of the windpipe that poor Biddy suffocated.

The most talkative bird of all time was probably a budgerigar (*Melopsittacus undulatus*) called 'Sparkie Williams' (1955–62), owned by Mrs M Williams of Bera Cross, near Bournemouth, Hampshire, England, who had a vocabulary of 531 words, including the words 'budgerigar' and 'chatterbox'. His last words were reportedly 'I love Mama.'

Very few parrots acquire a vocabulary of more than 20 words. In 1974 it was reported that a French lawyer had taught two pet parrots called 'Ito' and 'Jocotte' to utter 500 French words and a few in English. Mention should also be made of 'Prudle', the world-famous male African grey parrot (*Psittacus erythacus*), owned by Mrs Lyn Logue of Golders Green, London, who has won the annual 'Best talking parrot-like bird' title at the National Cage and Aviary Bird Show held in London for the past eleven years (1965–75). Prudle was taken from a nest in a tree about to be felled at Jinja, Uganda, in 1958.

Section III

REPTILES

(class Reptilia)

A reptile is a cold-blooded, air-breathing vertebrate covered with protective scales or bony plates. Unlike mammals and birds it has no effective mechanism for the regulation of body heat, and its temperature rises and falls according to the temperature of the surrounding air or water. The brain is small and poorly developed. Paired limbs, when present, are generally short and project so awkwardly from the sides of the body that the animal is compelled to crawl. Young are usually produced from eggs deposited on land, but in some species of lizard the eggs are retained within the oviduct of the mother and the young are born alive.

The earliest known reptiles were *Hylonomus*, *Archerpeton*, *Protoclepsybrops* and *Romericus* which lived about 290 000 000 years ago. Their remains have been discovered in Nova Scotia.

There are about 5175 living species of reptile and the class is divided into four orders. These are the *Rhynchocephalia* (the tuartara); the *Squamata* (lizards and snakes); the *Crocodilia* (crocodiles, alligators and gharials); and the *Chelonia* (turtles, tortoises and terrapins). The largest order is *Squamata*, which contains about 4900 species or 94 per cent of the total number.

The largest reptile in the world is the Estuarine or Saltwater crocodile (*Crocodylus porosus*) which ranges from India, Sri Lanka, southern China and the Malay Archipelago to northern Australia, New Guinea, the Philippines and the Solomon Islands.

Fully grown bulls average 4·26–4·87 m *14–16 ft* and scale 408–521 kg *900–1150 lb*, but old individuals may be half as heavy again. Adult females average 3·04 m *10 ft* in length.

As with many other large animals the size attained by the estuarine crocodile has often been exaggerated in literature. Sir Samuel Baker (1874), for instance, says those found in Sri Lanka – *Crocodylus p. porosus* – were usually larger than those found on the Indian coast, and lengths of 6·70 m *22 ft* were quite common, but this measurement has never been substantiated. According to Deraniyagala (1939) the official record (between pegs) for the whole of Sri Lanka is 6 m *19 ft 7⅛ in* for a notorious man-eater shot in Eastern Province, and he also gives details of another one killed at Dikvalla, Southern Province, which measured 5·4 m *17 ft 8 in*. In 1924 a length of 6·4 m *21 ft* was reported for a crocodile shot by game-wardens at Kumana, Northern Province, who said it was so huge that a man could only leap over the carcass with difficulty (*sic*). This measurement was probably reliable, because the preserved skull has a 'dry' length of 724 mm *28·54 in* = *c.* 762 mm *30 in* 'green'

The Estuarine crocodile, the world's largest reptile (National Zoological Park, Washington, D.C.)

length. (The head of a large *C. porosus* – in excess of 4·87 m *16 ft* – accounts for one-eighth of the total length.)

A 698 mm *27½ in* long crocodilian skull preserved in the Raffles Museum, Singapore, is said to have belonged to a 6·70 m *22 ft* saurian, but the over-all length was more probably in the region of 5·94 m *19 ft 6 in*. (A 5·03 m *16 ft 6 in* bull weighing 589 kg *1300 lb* shot in Sumatra had a 635 mm *25 in* skull.)

In former times, before heavy persecution, this species reached a greater size than it does today because it was allowed full opportunity for uninterrupted growth, and crocodiles measuring 9·14 m *30 ft* or more in length were allegedly killed. None of these records, however, have been verified although, as we shall see later, such a measurement is not beyond the realms of possibility.

The most famous saurian in this category was an alleged 33-footer *10·05 m* killed in the Bay of Bengal in 1840 which had a girth round the belly of 4·16 m *13 ft 8 in* – the body was obviously distended by internal gases – and weighed an estimated 3048 kg *3 ton*. Fortunately the skull of this giant reptile was preserved and later presented to the British Museum (Natural History), London, by Mr Gilson Row, and the measurements of this exhibit were given as 927 × 475 × 312 mm *36½ × 18¾ × 12⅓ in* (Boulenger, 1889). This was 'dry' bone, however, and the owner of this skull must have measured just over 7·72 m *25 ft 6 in* in life.

Earlier, in 1823, another huge estuarine crocodile measuring 8·22 m *27 ft* was killed by Paul de la Gironière (1854) and George R Russell at Jala Jala, near Lake Taal, on the island of Luzon in the Philippines, after a struggle lasting more than six hours. The crocodile, a notorious man-eater, had apparently attacked and killed one of the Frenchman's shepherds while he was crossing a river, and Paul de la Gironière decided to avenge his death. In the monstrous saurian's stomach were found a horse, bitten into seven or eight pieces, plus about 68 kg *150 lb* of pebbles, varying from the size of a fist to that of a walnut, and the head weighed 204 kg *450 lb* before the ligaments were detached.

The skull of the Jala Jala crocodile, now preserved in the Museum of Comparative Zoology at Harvard University, Massachusetts, USA, measures 990 × 482 × 330 mm *39 × 19 × 13 in*, which makes it the largest crocodilian skull on record, if we exclude fossil remains.

This skull is closely matched in size by another one in the Indian Museum, Calcutta, which measures 965 × 457 × 342 mm *38 × 18 × 13½ in* and weighs 24·49 kg *54 lb*. It came from a 25-footer *7·62 m* killed in the Hooghly River, in the Alipore district of Calcutta, in 1931. Another huge example in the United Services Club, Calcutta, measures 850 × 406 × 292 mm *33½ × 16 × 11½ in* (Prashad, 1931).

Dunbar Brander (1931) says he examined the skull of an estuarine crocodile in the Elgin Museum, Morayshire, Scotland, which measured 937 × 469 × 317 mm *37 × 18½ × 12½ in* and weighed 24·49 kg *54 lb* (without teeth), but this exhibit disappeared during the last war.

The greatest authentic measurement recorded for an estuarine crocodile is 8·63 m *28 ft 4 in* for a bull shot by Mrs Kris Pawlowski on MacArthur Bank, in the Norman River, south-eastern Gulf of Carpentaria, Northern Territory, Australia, in July 1957. In normal circumstances this measurement would be rejected because nothing of this monstrous saurian was preserved, although a photographic record existed until 1968. However, as Mrs Pawlowski's husband Ron is one of the world's leading authorities on this species, and farmed the estuarine crocodile at Karuma, Mount Isa, Queensland until 1969, this record must be regarded as one with a high probability of accuracy.

Mr Pawlowski (pers. comm. 14 July 1974), who has examined 10287 *C. porosus* (his biggest catch was an 18-footer *5·48 m*), said the only means of dragging such a bulky weight (*c.* 2032 kg *2 ton*) on to dry land away from tidal reach would have been by tractor but, as one of these vehicles wasn't conveniently to hand, this was completely out of the question! He also added that the freshly severed head was so heavy that he could not move it.

The maximum length reported for a female estuarine crocodile in Australia is 3·96 m *13 ft*.

Some very large estuarine crocodiles have also been killed in the Segama River, North Borneo (now Sabah).

According to James R Montgomery (pers. comm. 27 May 1974), who ran a small rubber estate about half-way between Hilia and the mouth of the river during the period 1926–32, the Segama was a veritable crocodile haven, both in numbers and size. 'My interest in this reptile', he says, 'arose firstly from the protection of my native labour force who bathed and laundered in the river, and my main difficulty at first was to learn how to kill and not lose the body, which always managed to get back into the water even if dead. I eventually found that a soft-nosed .375 bullet in the spinal column, just behind the head, was the answer.

'In most cases it was not possible to do a peg measurement as there were little means of moving the body. What we did was to straighten it out, drawing a line from the snout and tip of tail and measuring that. In the whole of this period, I shot and measured as best I could some 20 reptiles between 20 and 26 ft.

'My last kill was of particular interest to me. The day before he had turned over my dugout and got my best hunting dog. I went after him the next day and it was not difficult finding him. I got him first shot at 50 yd kneeling in a large dugout. He was on a slope, the snout at the water's edge. He was unable to move his body but thrashed with his jaws for a bit. We managed to peg measure him at 20 ft 3 in.'

Mr Montgomery also mentioned a fabulous brute which lived some 16 km *10 miles* down-river from his quarters. This reptile had been a legend for years and the River People (Seluke) estimated that he was more than 200 years old.

'On one memorable occasion I came across him asleep on a sand bank in the middle of the river. His snout was in the water at one end and the tip of his tail in the water at the other. We got him to move off and measured the sand bank very carefully, with a 50 ft surveyor's tape. The sand bank was 32 ft 10 in long, making him in the region of 33 ft plus.

'We never shot at him as the Seluke looked on him as the Father of the Devil. Silver money was always thrown into the river whenever he was seen to ward off harm. On one occasion he broke the back of my kerosene launch (seven tons) in trying to overturn it. I suppose I must have seen him six or seven times in all, and he made my twenty-footer look like a baby.'

Major Moulton, Chief Secretary to the Government of Sarawak, told Barbour (1926) that he had a reliable Bornean record of an estuarine crocodile measuring 7·01 m *23 ft*, and as recently as April 1966 one measuring 6·32 m *20 ft*

9 in was shot at Liaga on the south-east coast of Papua.

On 26 June 1960 Mr Keith Adams, of Perth, Western Australia, harpooned a 6·14 m *20 ft 2 in* long esturine crocodile in the McArthur River, near Borroloola, Northern Territory. The preserved skull measures 762 mm *30 in* in length.

In January 1964 'Big Gator', a huge semi-legendary *C. porosus* of the River Adelaide, Northern Territory, was shot by Fred Bennett, a Darwin crocodile-hunter. This saurian measured exactly 6·09 m *20 ft* in length and weighed 1097 kg *2418 lb*.

As recently as April 1975 a very large estuarine crocodile attacked and ate a man while he was swimming in Mission River, 32 km *20 miles* east of Weipa, Queensland. Two hunting friends of the victim, Mr Peter Reimers, 32, found his clothing, rifle and ammunition on the bank. They also discovered the man's footprints leading to the edge of the water and nearby the tracks of a big crocodile. The police at Weipa were alerted and the crocodile was blown to the surface by a dynamite charge and shot. (The police found the dead man's legs still in the jaws of the saurian.) According to newspaper reports this man-eater measured 5·79 m *19 ft* in length, but Mr Tom Spence, Director of Perth Zoological Gardens, Western Australia, told Mr Ken Sims, a reptile-collector/dealer based in Penang, Malaysia, who has been corresponding with the compiler for about three years, that *c.* 1 m *39·37 in* of the crocodiles tail was missing – giving a theoretical length of 7·70 m *22 ft*.

The largest estuarine crocodiles ever held in captivity are two bulls owned by George Craig of Daru Island, southern New Guinea. One of them called 'Oscar' measures 5·48 m *18 ft* in length and weighs an estimated 1 tonne/*ton*, and the other, 'Gomik', is 5·33 m *17 ft 6 in* long. Both these giants were caught in the Fly River. Another huge bull caught by Craig on a baited hook in the same river measured 5·94 m *19 ft 6 in* in length but died shortly after being dragged on to a shelving mud-bank (Pinney, 1976).

Although the estuarine crocodile *looks* very sluggish in captivity, close confinement does not slow the reptile down appreciably and this fact is well illustrated by the following incident which Deraniyagala (1939) says took place at the zoo in Colombo, Sri Lanka:

'A large *C. porosus* 4·3 m [*14 ft 1¼ in*] long was kept for five years in an enclosure with a concrete trough of water about 5 m [*16 ft 6 in*] long, 2 m [*6 ft 6 in*] wide and about 1 m [*3¼ ft*] in depth. In June 1936 a swift-moving adult Langur monkey (*Semnopithecus priam*) escaped from its cage and dashed into this enclosure when the crocodile, which was lying in the trough, hurled itself out of the water with its head and shoulders in the air and seized the monkey as it sprang. A crunch and a gulp, and the crocodile once more lay idly in the water.'

Lengths in excess of 6·09 m *20 ft* have also been reported for the very long-snouted Gharial or Gavial (*Gavialis gangeticus*) of the Ganges, Indus, Brahmaputra and Irrawaddy Rivers of India. The average adult length is 3·65–4·57 m *12–15 ft*. A gharial measuring 6·55 m *21 ft 6 in* was killed in the Gogra River at Fyzabad, United Provinces, in August 1920 (Pitman, 1925), and a 21-footer *6·40 m* was shot in the Cheko River, Buxa Division, Jalpaigur in 1934. Other specimens have been credited with unsubstantiated measurements up to 7·62 m *25 ft*.

The Orinoco crocodile (*Crocodylus intermedius*) has also been credited with great size. Two specimens killed by Alexander von Humboldt and his companion Aime Bonpland in 1800 while they were exploring the course of the Orinoco River reportedly measured 6·80 m *22 ft 4 in* and 5·24 m *17 ft 2¾ in*, but these must have been outsized freaks because the average length is only about 3 m *9 ft 10 in*. According to Neill (1971) the largest *C. intermedius* measured this century was just under 3·96 m *13 ft* in length.

The American crocodile (*Crocodylus acutus*) of the SE USA, Central America, the West Indies and northern South America is somewhat larger than *C. intermedius*, bulls averaging 3·50 m *11 ft. 5½ in*. The largest specimen to be accurately measured was probably a bull shot by Jackson and Hornaday at Arch Creek, Biscayne Bay, Florida, which measured 4·62 m *15 ft 2 in* in length, with an estimated 154 mm *6 in* of tail missing. Another large bull killed by an American alligator (*Alligator mississipiensis*) in 1952 after it had strayed into its pen at the famous Ross Allen Reptile Institute in Silver Springs, Florida, measured exactly 4·26 m *14 ft* in length. There is also an old record of a *C. acutus* killed in Venezuela which measured 7·01 m *23 ft*, but Dr Wilfred T Neill has queried the accuracy of this report.

The American alligator (*Alligator mississipiensis*) of the SE USA is longer and bulkier than *C. acutus*, adult bulls averaging 3·66 m *12 ft*. E H McIlnenny (1934) shot two 18-footers *5·48 m* and another one which taped 5·84 m *19 ft 2 in* in Louisiana, and this latter measurement constitutes a record for this species.

The skin of a 5·94 m 19 ft 6 in *Estuarine crocodile killed in New Guinea (from* To Catch a Crocodile, *published by Angus & Robertson)*

According to Henry Walter Bates (1892) the Black caiman (*Melanosuchus niger*) of the Amazon Basin 'grows to a length of 5·50 or 6·10 metres, and attains an enormous bulk', but Neill (1971) says this species, the largest of all the caimans, does not exceed the American alligator in length. One large individual photographed at Leticia, Colombia, measured just over 3·96 m *13 ft*, and the maximum length reached by this crocodilian is probably about 4·57 m *15 ft*.

The maximum length attained by the Nile crocodile (*Crocodylus niloticus*) of Africa and the Malagasy Republic is a matter of considerable dispute.

In ancient Egypt, where crocodiles – and rulers – were mummified, the word 'great' was applied to any saurian which measured over 7 *Egyptian cubits* 4·01 m *13 ft* $2\frac{1}{3}$ *in* in length. So far, however, no mummified crocodiles exceeding 4·57 m *15 ft* have been found by archaeologists, so it is not surprising that Ditmars (1922), Schmidt (1944) and Pope (1957) all state that this species (average length 3·35–3·65 m *11–12 ft*) never exceeds 4·87 m *16 ft*. Dr Hugh B Cott (1961), on the other hand, begs to differ, and lists several reliable records from Central and Southern Africa of crocodiles measuring between 4·87 m *16 ft* and 5·79 m *19 ft*. They include a 5·05 m *16 ft 7 in* specimen shot by L E Vaughan, a senior game-warden in Northern Rhodesia; one measuring just over 5·31 m *17 ft 5 in* shot by W Hubbard in the Kafue River, also in Northern Rhodesia; a 5·53 m *18 ft 2 in* individual killed by a game-ranger on the Semliki River, Uganda, in 1950; a female (?) measuring 5·59 m *18 ft 4 in* taken on the same river in June 1954; and one measuring 5·74 m *18 ft 10 in* shot by C C Yiannakis near Chipoko, Malawi.

T Murray Smith (1963) obtained a crocodile on Lake Tanganyika which measured 5·50 m *18ft* $0\frac{3}{4}$ *in*, and the naturalist C W Hobley shot one measuring 5·64 m *18 ft 6 in* in the Miriu River.

Another huge saurian shot by a professional hunter named Erich Novotny near Nungwe in the Emin Pasha Gulf of Lake Victoria, Tanganyika (now Tanzania) in 1948 measured exactly 6·40 m *21 ft*, and he claims he saw another one in the same area which was even larger (6·70 m *22 ft*). The Juba River in Somaliland also had a reputation for big crocodiles at one time, and Douglas Jones shot one which was just over 6·40 m *21 ft* long.

According to Foran (1958) the largest of the 1406 crocodiles taken by the commercial hunters Glover and Van Bart on the Kafue River, Northern Rhodesia, in 1950–51 measured 5·38 m *17 ft 8 in* when dry and 5·48 m *18 ft* when landed, but the average length was only 2·89 m *9 ft 6 in*. Other authentic records over 5·18 m *17 ft* quoted by the same writer include a 5·79 m *19 ft* crocodile killed in the Cunene River, South West Africa by Major Trollope in 1952 and one measuring 5·26 m *17 ft 3 in* shot by a policeman at Bentiu, in the Sudan, in September 1947.

Jack Bousfield, who had a hand in killing 45 000 crocodiles in Lake Rukwa, Tanganyika (Tanzania), said the largest example measured 5·31 m *17 ft 4½ in*, and the largest of 500 specimens collected by Graham & Beard (1973) on Lake Rudolf, Kenya, in 1965 was 4·80 m *15 ft 9 in* long and weighed 680 kg *1500 lb*.

In November 1968 a semi-mythical crocodile known as 'Kwena' was brought down by a professional huner, Bobby Wilmot, in the Okavango Swamp, Botswana. This huge bull measured 5·68 m *19 ft 3 in* between pegs and had a maximum girth of 2·13 m *7 ft*. Its weight was estimated at between 793 and 816 kg *1750 and 1800 lb*, and the head alone weighed 165 kg *365 lb*. Inside the stomach were found two goats, half a donkey and the still-clothed trunk of a native woman.

The largest accurately measured Nile crocodile on record was probably one shot by the Duke of Mecklenberg in 1905 near Mwanza, Tanzania, which measured exactly 6·5 m *21 ft 3¾ in* in length (Hubbard, 1927). It was not weighed, but a Nile crocodile this length would be expected to scale somewhere between 1043 and 1088 kg *2300* and *2400 lb* if it was in good condition.

The biggest shot 'officially' was a monster taken on the Semliki River by a member of the Uganda Game and Fisheries Dept in 1953 which measured 5·94 m *19 ft 6 in* between pegs with a belly skin measuring 1·45 m *4 ft 9 in* across.

The smallest known crocodilian is the Congo dwarf crocodile (*Osteolaemus osborni*) which does not exceed 1·14 m *3 ft 9 in* in length. Cuvier's smooth-fronted caiman (*Paleosuchus palpebrosus*) of north-eastern South America has a maximum recorded length of 1·45 m *4 ft 9 in* (Neill, 1971).

At least six of the 23 recognised species in the family Crocodylidae will attack and eat man if given the opportunity, and several of the others are large enough to inflict serious injury or even death.

The most notorious man-eater of them all is undoubtedly the estuarine crocodile. Although no statistics are available, this saurian probably kills well over 2000 people annually.

In December 1975 42 people were reportedly attacked and eaten by estuarine crocodiles in a boating tragedy in Central Celebes, Indonesia. According to a news agency story about 100 people were on a holiday trip when their boat sank in the crocodile-infested Malili River.

The Nile crocodile also has a very bad reputation and probably kills nearly 1000 people (mostly women and children) annually, although at one time the figure may have been as high as 3000. One 4·64 m *15 ft 3 in* bull shot in the Kihange River, Central Africa, by a professional hunter allegedly killed 400 people over the years (Clarke, 1969), but Guggisberg (1972) thinks this figure was highly exaggerated and was only quoted 'to justify the mass slaughter of crocodiles'.

In 1974 the health authorities at Manzini, Swaziland, announced that man-eating crocodiles were being captured and put to work in the local sewage works gobbling up condemned meat and other refuse so that the local natives could cross the near-by river in safety. Whether the project was a success has not yet been ascertained, but the compiler is expecting to hear of staff vacancies at the works any time now!

The strength of the crocodile is quite appalling. Deraniyagala (1939) mentions a crocodile in northern Australia which once seized and dragged into a river a magnificent 1016 kg *1 ton* Suffolk dray horse which had recently been imported from England, despite the fact that this breed can exert a pull of more than 2032 kg *2 ton*, and there is at least one record of a full-grown black rhinoceros losing a tug-of-war with a big crocodile.

Frank Lane (1955) says that tests carried out in France to determine the jaw strength of a 54·43 kg *120 lb* crocodile revealed that this saurian could exert a crushing pressure of 698 kg *1540 lb*. On this basis, a crocodile weighing 1 tonne/*ton* could exert a force of nearly 13 tonne/*ton* (cf. human jaws which can exert a 227 kg *500 lb* crushing pressure).

Very little information has been published on the longevity of crocodilians, but some species probably live longer than 50 years in the wild. On the strength of its comparatively slow growth rate Cott is of the opinion that very large Nile crocodiles i.e. 5·48 m *18 ft* or over, must be at least 100 years old, and that the largest recorded examples may be as much as 200 years old, but further research is needed before any definite statements can be made. According to Neill (1971) toothlessness and other signs of senility are always evident in alligators that have lived for about 50 years; he puts the maximum age as 'not much beyond this', and there is no reason to believe that crocodiles are longer-lived than alligators.

The greatest authentic age recorded for a crocodilian is 56+ years for an American alligator (*Alligator mississipiensis*) which was received at the Dresden Zoo, East Germany, in 1880 and was still alive in September 1936 (Flower, 1937). This ancient saurian may have survived until 13/14th February 1945 when the city was virtually wiped out by RAF saturation bombing.

Another example known as 'Jean-qui-rit' ('Laughing John'), reputedly lived in the Menagerie du Jardin des Plantes, Paris, for 85 years (1852–4 April 1937), but Flower has questioned the validity of this record. The same authority also quotes a reliable record of 50 years for a Chinese alligator (*Alligator sinensis*) which was still living in the Frankfurt Zoo, West Germany, on 8 September 1936 (probably also killed during the Second World War), and Lederer (1941) says another one lived in the Leipzig Zoo, East Germany, for 52 years.

London Zoo's famous American alligator 'George' was reputedly about 60 years old at the time of his death, but only 41 years were spent in captivity. When he was received in 1912 he was described as sexually mature, measuring 1·83 m *6 ft* in length, but in this species sexual maturity is reached at the age of *c.* six years. In other words, 'George' was probably nearer 47 years of age than 60.

The famous Nile crocodile 'Lutembe', who lived for a number of years in a small bay in the Murchison Gulf, Lake Victoria, and used to come ashore to be fed with fish when called, reputedly started her career in the 19th century as royal executioner to the Kings of Uganda, but this report was never substantiated. When she became world-famous in the 1920s, however, she was already a big animal (length *c.* 4·25 m *14 ft*) and must have been at least 25 years old. There is no record of her death, but Guggisberg says she 'disappeared' during the 1940s.

Today the commercial hunting of crocodilians for their skins, and man's irrational hatred of them, has reduced the world's saurian population to such an extent that 16 of the 23 different species have now been brought to the brink of extinction.

The rarest crocodilian in the world is believed to be the Cuban crocodile (*Crocodylus rhombifer*) which, through over-hunting, is now confined to the Cienaga de Zapata, a tiny swamp in Las Villas province, central Cuba. In 1965 there were only an estimated 300 of these reptiles left, and the area has since been declared a sanctuary (Fisher *et al*, 1969).

This heavily muscled crocodile is also the most aggressive member of the Crocodylidae despite its small size (maximum length 2·74 m *9 ft*) and Neill says a 1·83 m *6 ft* specimen will chase an American crocodile half as long again round a pen and badly maul the larger animal if given the opportunity. It will also fight furiously when handled by a herpetologist, and the outwardly tilted teeth can still do a lot of damage even after the jaws have been bound.

The largest reptile (excluding snakes) found in Britain is the Slow-worm (*Anguis fragilis*), which is widely distributed over England, Wales and Scotland. A female example collected in Midhurst, Sussex, measured 460 mm *18·11 in* in length (head and body 215 mm *8·46 in*, tail 245 mm *9·64 in*) and a male from Dorset 427 mm *16·81 in*.

The smallest species of reptile is believed to be *Sphaerodactylus parthenopion*, a tiny gecko found only on the island of Virgin Gorda, one of the British Virgin Islands, in the West Indies. It is known from only 15 specimens, including some gravid females, collected by Mr Richard Thomas of Miami, Florida, USA, and a colleague between 10 and 16 August 1964. The three largest mature females measured 18 mm *0·71 in* from snout to vent, with a tail of approximately the same length (Thomas, 1965).

The smallest reptile found in Britain is the widely distributed Common lizard (*Lacerta vivipara*). Adult males measure 118–170 mm *4·64–6·69 in* in total length and adult females 121–178 mm *4·76–7·0 in* (Smith, 1951).

The highest speed ever recorded for a reptile on land is 29 km/h *18 miles/h* for a Six-lined race-runner (*Cnemidophorus sexlineatus*) pursued by J Southgate Hoyt (1941) in a car near McCormick, South Carolina, USA. This lizard maintained its speed on all four legs for more than a minute before darting off the clay road into the undergrowth.

The largest of all lizards is the Komodo monitor or Ora (*Varanus komodo*), a dragon-like reptile found on the Indonesian islands of Rintja, Padar and Flores. Adult males average 2·59 m *8 ft 6 in* in total length and weigh 79–91 kg *175–200 lb*, and adult females 2·28 m *7 ft 6 in* and 68–73 kg *150–160 lb*.

As with crocodiles, the size attained by this lizard has been much exaggerated in literature.

Major P A Ouwens, the Curator of the Botanical Gardens at Buitenzorg, Java, who first described this giant lizard in 1912, was informed by J K van Steyn van Hensbroek, Governor of Flores, that a 'boeaya-darat' (land crocodile) measuring 4 m *13 ft 1½ in* and weighing over 182 kg *401 lb 4 oz* had been shot by a policeman on near-by Komodo, and the Governor also said that two Dutchmen working for a pearl-fishing company on the same island had told him that they had killed several specimens measuring between 6 and 7 m *19 ft 8 in* and *23 ft*. Another ferocious monster seen by a Swedish zoologist on the shores of Komodo in 1937 was estimated by him to have 'measured 7 m *23 ft*, and the following year an American journalist reported that he had seen one measuring 4·41 m *14 ft 6 in*.

All of these statements, however, were based on estimates (the great girths of older specimens create an impression of enormous size), or referred to estuarine crocodiles which are also found in the area.

According to Ouwens the type specimen, now mounted in the Museum at Buitenzorg, measured 2·9 m *9 ft 6⅛ in* between pegs. Another one collected by De Rooij (1915) on the west coast of Flores in 1915 measured 2·66 m *8 ft 8¾ in*, and the largest of the four monitors collected by the Duke of Mecklenburg in 1923 was just under 3 m *9 ft 10 in*. Of the 54 specimens collected by the Douglas Burden Expedition to Komodo in 1926, the largest measured 2·75 m *9 ft 0½ in*, and the biggest male collected by the Dutch zoologist De Jong (1937) on his second visit to the island measured exactly 2·74 m *9 ft*.

The largest accurately measured Komodo monitor on record was probably a male which was put on exhibition in the St Louis Zoological Park, Missouri, USA, for a short period in *c.* 1937. This specimen measured 3·10 m *10 ft 2 in* in length and tipped the scales at a staggering 165·5 kg *365 lb* (Dr Osmond P Breland, pers. comm. 24 April 1966). Nothing is known of this giant

*The Leatherback turtle, the world's heaviest chelonian (*Underwater Naturalist Magazine*)*

lizard's earlier history, but it seems more than likely that it was the same animal as the one that was presented to an American zoologist by the Sultan of Bima in 1928 which measured 3·05 m *10 ft 0¾ in.*

Two other 'dragons' collected by Dutch and German expeditions to Komodo before 1927 and presented to European zoos were both allegedly over 3·35 m *11 ft* in length (Wendt, 1956), but these measurements have not been substantiated. (A male specimen received at Berlin Zoo on 11 June 1927 measured 2·41 m *7 ft 11 in.*)

Although the Komodo monitor is strictly protected by the Indonesian Government, it is now a threatened species, mainly as a result of local hunting. (The flesh is apparently very palatable.) At the present time the total population is estimated at *c.* 2000 individuals, 900 of which are on Komodo itself and the rest in special reserves on Rintja and Padar.

The heaviest lizard after the Komodo monitor is probably a giant variety of the seaweed-eating Marine iguana (*Amblyrhynchus cristatus*) found in the Galapagos Islands, eastern central Pacific.

Mr Henry C Combe (pers. comm. 27 May 1966) of Langport, Somerset, gave details of an exceptionally large marine iguana he had helped to capture five years previously: 'In September 1961 I was a member of the crew of the brigantine *Yankee* and we were anchored off a small cove on the western shore of Santa Isabella Island. On some very inaccessible rocks below a cliff we came upon about 20 iguanas far bigger than any we had seen during the month we had been in the islands. The Third Mate, Mr Leo Devitt from Bakersfield, California, and myself got on to these rocks and with great difficulty managed to capture one of these monsters. It was not the biggest by any means, but the others all leapt into the sea and escaped. We took it back to the ship, where we "hypnotised" it by turning it on its back and gently stroking its stomach; this puts an iguana to sleep. Then Mr Harvey Segal, of Boston, Massachusetts, a fully qualified zoologist, weighed and measured it. It was 61 in long and weighed almost exactly 100 lb. We then returned it to the water where it dived, and stayed down for over three minutes.'

The longest lizard in the world is the slender Salvator monitor (*Varanus salvator*) of Malaysia, Indonesia and the southern part of New Guinea. Measurements up to 4·56 m *15 ft* have been claimed for this species, but the greatest authenticated length was 3·53 m *11 ft 6½ in* for a male killed in Papua. In February 1961 an unconfirmed measurement of 3·35 m *11 ft* was reported for a Salvator monitor which attacked two RAF policemen on the island of Song Song off the Malayan coast. After a terrific struggle the airmen managed to kill the prehistoric monster with a machete, but the Alsatian dog with them was badly bitten.

The greatest authentic age recorded for a lizard is 'more than 54 years' for a male Slow-worm (*Anguis fragilis*) kept in the Zoological Museum in Copenhagen, Denmark, from 1892 until 1946. At the age of 45+ this specimen mated with a female known to be at least 20 years old (Dr Curry Lindahl, pers. comm. 14 November 1967).

No other species of lizard lives anywhere near as long. The closest approach is probably made by the Komodo monitor which lives 20–30 years in the wild, but the greatest authentic age recorded for a specimen in captivity is 9 years 3 months (Flower, 1937).

The only known venomous lizards are the Gila monster (*Heloderma suspectum*) and the Bearded lizard (*H. horridum*) of the SW USA and parts of Mexico, which both carry an extremely painful neurotoxic venom.

In one series of 34 cases of people bitten by these desert reptiles, which have very powerful jaws and hang on like grim death once they get a grip, there were eight fatalities. The majority of these attacks, however, took place in zoos or laboratories, and most of the dead victims were in bad health – or drunk at the time of the incident (Bogart & Del Campo, 1956). In the wild state these lizards are peaceful animals and rarely intrude into territory occupied by man.

The largest of all chelonians is the Pacific leatherback turtle (*Dermochelys coriacea schlegelii*), also known as the 'Leathery turtle' or 'Luth', which ranges through the Pacific and Indian Oceans from British Columbia to Chile and west to Japan and eastern Africa.

The average adult measures 1·83–2·13 m *6–7 ft* from the tip of the beak to the tip of the tail (length of carapace 1·21–1·52 m *4–5 ft*), about 2·13 m *7 ft* across the front flippers and weighs anything up to 453 kg *1000 lb.*

The greatest authentic weight recorded for a Pacific leatherback turtle is 865 kg *1908 lb* for a male captured *alive* in Monterey Bay, California, USA, on 29 August 1961. The huge chelonian was lassoed cowboy-style by Jerry Lemon, skipper of the fishing-boat *Sea Otter* off Point Pinos. When the noose tightened the marine giant immediately sounded to the length of the 12·19 m *40 ft* long rope, but after a while it ceased to struggle and allowed itself to be towed ashore. The turtle was purchased by Nelson (Bill) Hyler, owner of the Wharf Aquarium, Fisherman's Wharf, Monterey 'for several hundred dollars' and placed in a tank measuring 6·09 × 3·04 × 1·21 m *20 × 10 × 4 ft*, but the unfortunate creature succumbed shortly afterwards. (Captive leatherbacks invariably thresh about and batter themselves to death against the sides of the tank.) This specimen measured 2·54 m *8 ft 4 in* in over-all length (L B Bowhay, pers. comm. 20 April 1963).

Another huge turtle measuring 2·74 m *9 ft* in over-all length and weighing 863 kg *1902 lb 8 oz* was caught in a fisherman's net near San Diego, California, USA, on 20 June 1907.

The Atlantic leatherback turtle (*Dermochelys coriacea coriacea*), which ranges through the Atlantic Ocean, the Gulf of Mexico, and the Caribbean Sea from Newfoundland to the British Isles and south to Argentina and the Cape of Good Hope, is smaller than its Pacific cousin and rarely exceeds 363 kg *800 lb*.

The greatest authentic weight recorded for an Atlantic leatherback turtle is 485 kg *1069 lb 4 oz* for a male found dead at Ameland, one of the West Frisian Islands, in the Netherlands on 4 August 1968. This individual measured 2·44 m *8 ft* in over-all length and had a 1·58 m *5 ft 2¼ in* long carapace. On 5 June 1928 a gravid female weighing 480 kg *1058 lb* was captured off Tazones, NW Spain, and later purchased by the Museu Nacional de Ciencias Naturales, Madrid. It measured 2·18 m *7 ft 2 in* in over-all length. Another exceptionally large example preserved in the Museu Bocage, Lisbon, Portugal, had a total length of 2·46 m *8 ft 0⅞ in* and weighed 421·3 kg *928 lb* in the flesh. (It was caught off Peniche, some 80 km *50 miles* north of Lisbon between 1792 and 1807 and measured 2·64 m *8 ft 8 in* across the front flippers) (Brongersma, 1972).

Some large examples have also been taken in British waters. One of the heaviest actually to be weighed was a male captured alive near Crail Harbour, Scotland, on 27 November 1967 after it had become entangled in the rope of a lobster-creel. It was towed ashore but died soon afterwards. A model of this turtle, in the Royal Scottish Museum, Edinburgh, measures 1·93 m *6 ft 4 in* in over-all length (carapace length over the curve 1·64 m *4 ft 6½ in*) and the animal scaled 350 kg *772 lb* in the flesh. Another one caught off Land's End, Cornwall, on 2 July 1756 reportedly weighed 343 kg *6¾ cwt 756 lb* after it was bled to death, which means it must have scaled just over 363 kg *800 lb* when alive. On 8 May 1958 a turtle measuring more than 2·13 m *7 ft* in over-all length and scaling 452 kg *996 lb* was caught by a French fishing-trawler in the English Channel. On 11 November 1913 a weight of 935 kg *18½ cwt 2062 lb* was reported for a turtle captured alive off Lowestoft, Suffolk, but Brongersma says the actual weight was between 152 and 203 kg *3 cwt* and *4 cwt 336* and *448 lb*.

On 10 July 1962 what may have been the largest turtle ever recorded in British waters was captured by the crew of the ring-netter *Castle Moil* out of Skye in their nets off South Minch, Barra, Outer Hebrides. The crew quickly hitched a nylon rope round the marine giant and then tried to hoist it aboard with a power winch. Unfortunately, however, the rope snapped and the turtle escaped. The men estimated the over-all length of the animal at 3·65 m *12 ft*. This was obviously exaggerated, but any chelonian heavy enough to snap a rope with a 1 tonne/*ton* breaking strain must be something of a superlative!

The largest freshwater turtle in the world is the Alligator snapping turtle (*Macroclemys temminckii*) of the SE USA. The normal upper weight limit for this species is about 91 kg *200 lb*, but one monster caught in the Neosho River in Cherokee County, Kansas, in 1937 reputedly scaled 183 kg *403 lb* (Hall & Smith, 1947).

The smallest marine turtle in the world is the Atlantic ridley (*Lepidochelys kempii*) which measures from 500 to 700 mm *19·68* to *27·55 in* shell length and weighs anything up to 36·28 kg *80 lb*.

The smallest known freshwater turtles are the Mud and Musk turtles (family Kinosternidae) of North and Central America. The Striped mud turtle (*Konosternon baurii baurii*) of the Lower Florida Keys, USA, has a maximum shell length of only 97 mm *3·82 in* (Oliver, 1955).

Several species of turtle are renowned for their aggressive behaviour towards man, but poor eyesight may have something to do with this! Probably the most vicious of them all is the Atlantic loggerhead (*Caretta caretta*), which occasionally enters the Mediterranean.

During the Second World War an RAF plane was shot down off Malta and the crew took to a dinghy. Suddenly a huge loggerhead reared up before them and attempted to demolish the rubber craft with its strong, curved beak, and it was only after a desperate struggle that the airmen managed to drive off the chelonian.

The Pacific loggerhead turtle (*Lepidochelys olivacea*) also has a bad reputation, and in Sri Lanka it is known as the 'Nai amai' or 'dog turtle' because it bites savagely when caught (Deraniyagala, 1939).

In May 1972 a skin-diver was attacked by an amorous short-sighted Atlantic green turtle (*Chelonia mydas mydas*) off the coast of Florida, USA, during the breeding season and was lucky to escape with his life – let alone his honour! – because males of this sub-species go in for 'love bites' which leave deep, bleeding wounds.

The Pacific leatherback turtle (*Dermochelys coriacea coriacea*) is another dangerous chelonian on account of its great size, and there is a record of one biting off the big toe of a fisherman near Sydney, New South Wales, Australia.

The Alligator snapping turtle (*Macroclemys temminckii*) also has an evil temper and is well named. Raymond L Ditmars (1936) complained that he could never take a photograph of this species without being confronted by a pair of enormously wide jaws, and on one occasion an angry specimen bit a leg off his tripod.

This species, incidentally, is basically a scavenger and has an important role to play in nature. One of the most gruesome stories concerning the carrion-feeding propensities of this animal comes from Indiana. Apparently an elderly Indian famous for his ability to locate the bodies of drowned persons when all other methods had failed, used to keep a large *M. temminckii* which he employed like a bloodhound. The chelonian would be taken to the lake being searched and then released from the boat with a long wire attached to it. After a while the 'leash' would be followed up and the turtle found with the body (Schmidt & Inger, 1957).

The greatest authentic age recorded for a turtle is 58 years 9 months 1 day for an Alligator snapping turtle (*Macroclemys temminckii*) which was accidentally killed in Philadelphia Zoological Garden, Pennsylvania, USA, on 7 February 1949. This specimen was one of two alligator snapping turtles received at the zoo on 6 May 1890. The other example died on 10 December 1937 after living in captivity for 47 years 7 months and 4 days (Roger Conant, pers. comm. 10 August 1972).

The Stinkpot (*Sternotherus odoratus*) of the SE USA is another long-lived turtle, and there is a record of one living for more than 53 years 3 months in the Philadelphia Zoological Garden (Conant & Hudson, 1949).

The maximum life potential of marine turtles is not known, but a small group of Atlantic loggerheads (*Caretta caretta*) lived in the Aquario Vasco da Gama, Lisbon, Portugal, from 1898 to 1931. Major Stanley Flower (1937) says they all died suddenly during an exceptional heat-wave.

The leatherback turtles probably live up to 50 years in the natural state, but no large specimens have survived more than a few weeks in captivity.

The rarest marine turtle is the now-protected Atlantic ridley (*Lepidochelys kempii*), which has been practically exterminated by over-fishing, nest-robbing and the slaughter of nesting females. Today nesting ridleys are found only on the coast of the States of Tamaulipas and Vera Cruz, Mexico, and Padre Island, Texas, USA, and the total population has been estimated at 5000–10000 individuals.

The fastest-swimming tetrapods (four-legged animals) are the powerful marine turtles. Speeds up to 35·2 km/h *22 miles/h* have been reported for both the Pacific leatherback turtle (*Dermochelys coriacea schlegelii*) and the Atlantic green turtle (*Chelonia m. mydas*) when frightened, but the normal cruising speed is probably less than 6·4 km/h *4 miles/h*.

Some marine turtles also migrate over great distances. There is a record of a female *C. m. mydas* tagged at Tortuguero, Costa Rica, Central America, being recovered near Campeche, Mexico, having swum 1950 km *1219 miles* in 275 days (Carr & Hirth, 1962), and in September 1972 another female tagged at Surinam, NE South America by the World Wildlife Fund was found on a beach in Ghana, West Africa, having travelled 5920 km *3700 miles*. The Atlantic loggerhead (*Caretta caretta*) is another confirmed wanderer, and one female tagged at Mon Repos, Australia, was recovered 63 days later in the Trobriand Islands near New Guinea, having swum *c.*3200 km *2000 miles* if the coastline was followed (Bustard & Limpus, 1970).

The largest living tortoise is *Geochelone gigantea* of the Indian Ocean Islands of Aldabra, Mauritius and the Seychelles (introduced in 1874).

The normal upper weight limit for males is *c.* 181 kg *400 lb*, but at one time, before heavy exploitation, this giant tortoise sometimes attained weights in excess of 227 kg *500 lb*. A huge male preserved in the Rothschild Museum at Tring, Hertfordshire, has a carapace measuring 1·18 m *46·5 in* in length. Lord Rothschild (1915) says this chelonian weighed 269 kg *593 lb* when alive, and this probably constitutes a weight record for the species.

Some of the giant tortoises (eleven surviving sub-species) of the Galapagos Islands, 960 km *600 miles* west of Ecuador, are also of comparable size, the largest examples weighing between 159 and 181 kg *350* and *400 lb*. In 1708 two of Woodes Rogers's crewmen came across a tortoise which they estimated to weigh over 317 kg *700 lb*, and others were seen over 1·21 m *4 ft* high (Noel-Hume & Noel-Hume, 1958). Another specimen collected in 1830 was so heavy that it required the efforts of six men to lift it into a boat, and Captain Porter examined one (*G. e. porteri*) on Indefatigable Island (= Santa Cruz) which had a carapace measuring (over the curve?) 1·67 m *5 ft 6 in* in length and 1·37 m *4 ft 6 in* in width. There is also a record of a *Geochelone e. vicina* on Albemarle (= Isabela) Island in the same group increasing its weight from 10·3 kg *22·7 lb* to 200·4 kg *442 lb* in only 15 years (Gaymer, 1968).

Although tortoises are the longest-lived of all vertebrates, including man, reports that they sometimes survive for 250 or even 300 years can be discounted. Many exaggerated claims have been based on the mistaken beliefs that (1) tortoises have a very slow rate of growth, and (2) scarred and rubbed shells are reliable evidence of great age; in reality, however, size is not a trustworthy indication of extreme longevity in reptiles, and chelonian shells (in nature) usually become badly worn after 20 years or so.

On 19 May 1966 the death was reported of 'Tu'imalilia' or 'Tui Malela' ('King of the Malila'), the famous but much-battered Madagascar radiated tortoise (*Testudo radiata*) reputedly presented to the King of Tonga by Captain James Cook (1728–79) in 1773, but this record was probably a composite of two (or more) specimens whose periods of residence on the island overlapped.

The giant tortoises (*Geochelone* sp.) have also been credited with extreme life-spans on account of their great size, but since it has been discovered that males can attain a weight of 204 kg *450 lb* in only 15 years – one individual increased its weight from 13 to 159 kg *29* to *350 lb* in seven years! – there is no reason to believe that they live longer than smaller species of tortoise.

In September 1969 'Samir', the oldest giant tortoise at the Giza Zoological Gardens, Cairo, Egypt, allegedly celebrated his 269th birthday (*sic*). According to the story the tortoise, born in France, was presented to the Khedive, Isma'il Pasha, Turkish ruler of Egypt, by the Empress Eugénie, wife of Napoleon III, on the occasion of the inauguration of the Suez Canal on the 16 November 1869. Shortly afterwards the Khedive presented the chelonian, which in Egypt is considered a symbol of good luck, to Giza Zoological Gardens. Thus, up to 1969, the animal had been continuously observed for only 100 years.

In 1975 an unconfirmed age of 180 years was reported for a Giant tortoise (*Geochelone gigantea*) living in the Pamplemousses Royal Botanical Gardens, Mauritius.

The greatest authentic age recorded for a tortoise is 152+ years for a male Marion's tortoise (*Testudo sumeirii*) brought from the Seychelles to Mauritius in 1766 by the Chevalier de Fresne, the French explorer, who presented it to the Port Louis army garrison. When the British captured Mauritius in 1810 the tortoise was officially handed over to the British troops by the surrendering French forces. In 1908 the ancient chelonian went blind, and ten years later it was accidentally killed when it fell through a gun emplacement. The specimen is now preserved in the British Museum (Natural History), London, and as the tortoise was fully mature at the time of its capture Schmidt & Inger (1957) say its actual age 'may be estimated at not less than 180 years'.

Other proven records over 100 years include a Common box tortoise (*T. carolina*) of 138 years (Oliver, 1955) and a European pond-tortoise (*Emys orbicularis*) of 120+ years (Rollinant, 1934).

A Mediterranean spur-thighed tortoise (*T. gracea*) lived in the gardens of Lambeth Palace, London, from 1633 to 1753, which means it was at least 120 years old (Noel-Hume & Noel-Hume, 1958). In 1957 another *T. gracea* named 'Panchard' died in Paignton Zoological & Botanical Gardens, Devon, aged 116+ years. The tortoise was bought at a fair in 1851, at which time it was already fully adult (W E Francis, pers. comm. 1 August 1972).

With the possible exception of chameleons (family Chamaeleonidae) tortoises move slower than any other reptile. Tests carried out on a giant tortoise (*Geochelone gigantea*). in Mauritius revealed that even when hungry and enticed by a cabbage it could not cover more than 4·57 m *5 yd* in a minute (0·27 km/h *0·17 mile/h*). Over longer distances its speed was greatly reduced.

Freshwater turtles usually move faster on land than tortoises, but there is a record of two Common snapping turtles (*Chelydra s. serpentina*) covering a distance of 557 m *1830 ft* in 2 hours 30 minutes, which means they were walking at a rate of 0·216 km/h *0·138 mile/h*.

The rarest tortoise in the world is the protected Short-necked tortoise (*Pseudomydura umbrina*), which is confined to a 209 ha *516 acres* area of swamp known as the Ellen Brook Reserve about 32 km *20 miles* NE of Perth, Western Australia. In 1973 the total population was estimated at 150 animals, including half a dozen specimens being studied at Perth Zoological Gardens. Today each specimen in the reserve carries a homing device so that scientists can keep an eye on their movements.

The longest species of snake in terms of greatest average length is the Reticulated python (*Python reticulatus*) of SE Asia, Indonesia and the Philippines which regularly exceeds 6·09 m *20 ft*. In May 1970 one measuring 7·31 m *24 ft* in length and weighing 83 kg *182 lb* was run over and killed by a taxi near Manila in the Philippines.

One of the largest pythons on record was killed in Penang, Malaya, in October 1859: 'A monster boa-constrictor was killed one morning this week by the overseer of convicts at Bayam Lepas, on the road to Telo' Kumbar. His attention was attracted by the squealing of a pig, and on going to the place he found it in the coils of the snake. A few blows from the changkolks of the convicts served to despatch the reptile, and on uncoiling him, he was found to be 28 ft in length and 32 in in girth' (*The Times*, 1 November 1859).

The Short-necked tortoise, the rarest living chelonian (West Australian Newspapers Ltd)

Another python shot near Taiping, Perak, Malaya, in the late 1800s measured 8·22 m *27 ft* in length and yielded a 10·05 m *33 ft* skin which is now preserved in the British Museum (Natural History), London, and Oliver (1958) says a snake reputedly measuring 9·14 m *30 ft* was killed in Penang in 1844. There is also an unconfirmed record of a 10·05 m *33 ft* python from Java.

The greatest authentic length recorded for a reticulated python is exactly 10 m *32 ft 9½ in* for a monster shot near a mining camp on the north coast of Celebes in the Malay Archipelago in 1912 (Raven, 1946).

Dr F Wayne King (pers. comm. 18 April 1974), Curator of Herpetology at the New York Zoological Park (Bronx Zoo), said some two or three years previously he had received a report from Celebes about the capture of a python measuring 11 m *36 ft* in length – 'but by the time Indonesian officials got to the locality the snake had been killed and burned'.

The semi-aquatic Anaconda (*Eunectes murinus*) of the swamps and slow-moving rivers of tropical South America and Trinidad has also been credited with the title of 'longest snake in the world', but although it is the heaviest of the giant snakes – a 5·18 m *17 ft* anaconda will scale as much as a 7·31 m *24 ft* reticulated python – great bulk can often be misleading when it comes to estimating length.

To prove this point A Hyatt Verrill (1937) once asked the other members of an animal-collecting expedition he was leading in Guayana to estimate the length of an anaconda they had spotted curled up on a rock. The estimates varied from 6·09 m *20 ft* to 18·28 m *60 ft* but, when the snake was shot and straightened out, it proved to be exactly 5·94 m *19 ft 6 in*!

Also complicating the matter is the fact that the anaconda has probably been the subject of more exaggerated claims concerning its size than any other living animal. The early Spanish settlers called it 'matatora' ('bull-killer') and spoke of individuals measuring 18·28–24·38 m *60–80 ft*, but even larger examples have been reported. In 1948, for instance, a 'sucuriju gigante' ('giant boa') measuring 35·05 m *115 ft* in length (*sic*) was allegedly killed by army machine-gun fire at Fort Abunda, in the Guąpore District, SW Brazil (Heuvelmans, 1958), and six years later a measurement of 36·57m *120 ft* was reported for another colossus killed by a Brazilian army patrol at Amapa on the French Guiana border (Gregory, 1962). Needless to say, no fragments of skeleton or skin were preserved to substantiate these claims!

Fortunately some of the early explorers and naturalists who visited South America were more critical in their accounts of this giant snake. Captain J G Stedman (1913), for example, who travelled extensively in the Guianas between 1772 and 1777, says the largest anaconda killed by him measured just over 6·70 m *22 ft*. Robert Schomburgk (1847–8) collected one in Guayana which taped exactly 5·48 m *18 ft*, and Alfred Wallace (1876) who explored large parts of Amazonia, wrote that he never saw an anaconda over 6·09 m *20 ft*, although he was told by his native bearers that they were sometimes 18·28–24·38 m *60–80 ft*.

Nicholas Guppy (1963) shot an anaconda in the Yampari River, Guayana which measured 5·25 m *17 ft 3 in* and had a maximum girth of 711 mm *28 in* (it had just killed an 2·43 m *8 ft* alligator), and on another occasion he shot a much larger specimen on the banks of the Kassikaityu River in the extreme south of that country. Unfortunately this one fell into very deep water and could not be recovered, but Guppy and the rest of his party all agreed that the snake must have been at least 8·22 m *27 ft* long.

Many extreme measurements have been based on the length of skins, but these records are unreliable. According to Oliver (1958) it is impossible to remove the skin of a snake without stretching it, and in the case of the anaconda and other heavy snakes he says the *unintentional* stretching may add as much as 25 per cent to the original length.

The American Consul at Iquitos, Peru, told Leonard Clark, the American explorer, that many of the anaconda skins brought to the city by traders each year measured 12·19 m *40 ft*, and Thomas Barbour, the American naturalist, claims he saw a 13·71 m *45 ft* skin (Perry, 1970).

In the 1920s Raymond L Ditmars, Curator of Reptiles at New York Zoological Park (Bronx Zoo) personally offered $1000 to anyone who could supply him with an anaconda skin measuring over 12·19 m *40 ft* in length, but the money was never claimed. (The largest skin submitted to him measured 6·50 m *21 ft 4 in* and came from a snake killed in the Province of Minas Geraes, Brazil.)

One of the longest anacondas ever taken in the field was a tremendously bulky specimen shot by W L Schurz (1962) in Brazil which measured 8·46 m *27 ft 9 in* in length and had a maximum bodily girth of 1·11 m *44 in*. Its weight was not recorded, but a snake of these dimensions would probably scale about 181 kg *400 lb*. Col Leonard Clark killed another one near Iquitos, NE Peru, which taped 8·16 m *26 ft 9 in*, and Ditmars (1931) mentions another Brazilian example which was 'a few inches in excess of 25 ft'. (The skin of the latter reptile may be the one in the Bhutantan Institute.)

According to some herpetologists the largest accurately measured anaconda on record was a 37½-footer *11·43 m* shot by Roberto Lamon, a petroleum geologist working for the Richmond Oil Co, and other members of his team while they were exploring for oil in the steppes of the Upper Orinoco River near the Colombia-Venezuela boundary. It appears the men had sat down for lunch on the bank of a slow-moving river one day when suddenly one of them noticed the huge body of a snake in the water. The party immediately opened fire with their ·45 automatics and, when all writhings had ceased, they dragged the serpentine colossus up on to the bank where they measured it with a steel surveyor's tape; later, however, when they went back to skin their prize they found it had gone! Apparently the bullets had only stunned it (Dunn, 1944).

The colossal snake shot by Col P H Fawcett (1953) on the Rio Abunda not far from the

The skin of a 6·70 m 22 ft *long Anaconda (*World Wide Magazine*)*

confluence of the Rio Negro, western Brazil in 1907 was stated by him to have measured 18·89 m *62 ft* (13·71 m *45 ft* out of the water and 5·18 m *17 ft* in it), but as this anaconda had a maximum diameter of only 304 mm *12 in* = circumference of 957 mm *37·7 in*, this length must be considered more than excessive. (A 24-footer *7·31 m* shot by Paul Fountain in Brazil had a maximum girth of 1·06 m *42 in*.)

Dr Afranio do Amaral, Director of the Bhutantan Institute in São Paulo, and a distinguished herpetologist of international reputation, who reviewed many of the reports of giant anacondas in his paper 'Serpentes Gigantes' (1948) accepted a Brazilian record of a 11·28 m *37 ft* specimen, and he said another snake killed in southern Brazil in 1913 by a small group of Indians measured just over 11·58 m *38 ft*. After a thorough study of the evidence (written) he came to the conclusion that the maximum length attained by the anaconda was somewhere between 12 and 14 m *39 ft 4½ in* and *45 ft 11 in*.

Clifford Pope (1961) has suggested that unusually large anacondas might belong to a sub-species with a limited range. He could be right but, until an actual body is produced and measured by a reputable scientist, the upper size limit for the anaconda must remain in the region of 9·14 m *30 ft* rather than 13·71 m *45 ft*.

The African rock python (*Python sebae*) of tropical Africa is another large constrictor whose size has been grossly exaggerated in literature. According to the historians of ancient Rome the army of Attilus Regulus, while laying siege to Carthage (near modern Tunis) was attacked by an enormous serpent which was destroyed only after a tremendous battle. The skin of this monster, measuring 36·57 m *120 ft* in length, was sent to Rome where it was preserved in one of the temples.

Suetonius says another giant python measuring 50 cubits 22·86 m *75 ft* in length was exhibited in front of the Comitium in Rome, and one captured alive in Ethiopia for Ptolemy II and put on view at the royal palace in Alexandria was credited with a length of 13·71 m *45 ft*.

As recently as 1932 a python allegedly measuring 39·62 m *130 ft* in length was killed in the Semliki Valley, Central Africa, but the snake was conveniently converted into stew by the local Bwambwa tribe before any reliable measurements could be taken! In reality the adult African rock python averages 3·96 m *13 ft* to 4·87 m *16 ft*, but individuals measuring over 6·09 m *20 ft* have been reliably reported.

Arthur Loveridge (1929) measured the freshly removed skin of a python speared by natives on the banks of the Nigeri River near Morogoro, Tanzania and found that it was exactly 9·14 m *30 ft* in length. He estimated that the snake must have taped *c*. 7·62 m *25 ft* in the flesh and said he was quite prepared to believe that this species sometimes reached 9·14 m *30 ft*. Three years later he was proved right when a python measuring an incredible 9·81 m *32 ft 2¼ in* was shot in the grounds of a school in Bingerville, Ivory Coast, West Africa, after being found in a hedge of bougainvillia (Pope, 1961). This must have been an outsized freak, however, because no examples approaching this size have been collected since.

On 14 May 1963 what was described as the 'grand-daddy' of all African rock pythons attacked and killed a three-month-old native baby near Lobatsi, Botswana, while its mother was collecting firewood. The snake, which had lived in the area for years and was something of a legend, was estimated by Africans who had seen it to measure nearly 10·66 m *35 ft* in length which, allowing for some distortion, means that the reptile must have been at least 7·62 m *25 ft* long.

In the late 1940s the New York Zoological Society offered a reward of $5000 for a *living* example of any snake 9·14 m *30 ft* or more in length, but the money has never been claimed. 'As more and more time goes by', writes Dr F Wayne King, 'I doubt that our reward for the giant snake will ever be collected. Too many constrictors are killed as vermin every time they are found by local people. This argues rather persuasively that the really large specimens are becoming fewer and fewer in numbers. This is also supported by the size of skins that are reaching the leather industry from all parts of the world.'

Another problem with captive giant snakes is that there is no really practical method of obtaining a 100 per cent accurate measurement, although Oliver (1958) thinks the best method is one which utilises a lot of manpower which can 'straighten the snake out and hold it in that position until it relaxes sufficiently to stretch to its approximate maximum length'.

The longest snake ever held in a zoo was probably a male reticulated python named 'Agamemnon', received at London Zoo on 20 December 1935. This specimen was 6·40 m *21 ft* long on arrival and grew to just over 8·53 m *28 ft* by the time of his death on 22 November 1942. Another male reticulated python named

The 8·38 m 27·5 ft *long Reticulated python 'Cassius' being moved to new quarters at Knaresborough Zoo (Camera Press)*

'Praggers' presented to London Zoo by the Prince of Wales (the late Duke of Windsor) on 22 May 1922 was credited with a measurement of 7·31–7·62 m *24–25 ft* in 1935 and a length of *c.* 9·14 m *30 ft* at the time of his demise on 31 August 1942, but the latter figure was an estimate. Both pythons were collected in Malaysia.

The male reticulated python 'Rex' shown at the First International Snake Exposition in New York in 1936 measured 8·30 m *27 ft 3 in* and scaled 87 kg *191 lb 1 oz* (Pope, 1962), and the same authority says the largest specimen exhibited in the National Zoological Park, Washington, DC, USA, was 7·62 m *25 ft* long and weighed 138 kg *305 lb*. In 1957 a male reticulated python measuring 7·62 *25 ft* in length and weighing 100 kg *220 lb* died in Bangkok Reptile Grove, Thailand, from forced feeding.

Another reticulated python caught by Charles Mayer, the French animal-collector, in the State of Negri Sembilan, Malaya, and brought back by him to Europe was stated to have measured 9·75 m *32 ft*. This snake later came into the hands of Henry Trefflich, the American animal-dealer, who confirmed this length after running a string along the curve of the body. (It took eight men to hold and straighten out this snake on the floor.) Later, however, he told Oliver that he sold it as a 34-footer *10·36 m* to compensate for contraction of the body. This measurement may well be reliable, but unfortunately it will never be proved 100 per cent because nobody seems to know what happened to this giant after it was sold.

The heaviest snake ever held in a zoo was a female reticulated python named 'Colossus', who died in Highland Park Zoological Gardens, Pennsylvania, USA, on 15 April 1963. She arrived at the zoo on 10 August 1949 after being shipped to the USA from Singapore and was probably of Malayan origin. Her length was 6·70 m *22 ft*. In June 1951 she measured 7·08 m *23 ft 3 in*, and by February 1954 her length had increased to 8·28 m *27 ft* (weight 134 kg *295 lb*). On 15 November 1956 she measured 8·68 m *28 ft 6 in*, when she was said to be still growing at the

The huge bulk of the Reticulated python is shown in this picture. The snake measures 6·70 m 22 ft *in length (Camera Press)*

rate of just over 152 mm *6 in* a year. Her maximum bodily girth before a feed was measured at 914 mm *36 in* on 2 March 1955 and she scaled 145 kg *320 lb* on 12 June 1957 (Barton & Allen, 1961). Shortly before her death on 15 April 1963, the length of this snake was estimated to be in excess of 9·14 m *30 ft*, but this measurement was not borne out by posthumous examination.

William B Allen, Jr, (pers. comm. 23 April 1966), Curator of Reptiles at Highland Park Zoo, said: 'The snake was measured after its death but it was fairly hard to get a good measurement. It was stiffened up and vertebrae had pulled together shrinking the snake. We had a measurement of over 24 ft, but this being put on the same ratio as a smaller snake dying and shrinking, we could add the difference that would have given us our 28 ft alive. . . . It was not weighed because of its stiffened condition, but it scaled over 200 lb, as it took several men all they could do to move it, by dragging and pulling.'

An autopsy revealed that several segments of the vertebrae were eaten almost completely through, along with several rib sections, by reptilian tuberculosis, and it was this factor, along with a lung infection and possibly old age, that contributed to this snake's death.

Because anacondas are more difficult to keep in captivity than other constrictors, very few large specimens have been exhibited in zoos. The largest specimen ever held in captivity was probably one in Highland Park Zoological Gardens, Pennsylvania, USA, for which a length of 6·28 m *20 ft 7 in* was reported on 10 July 1960. Sadly this snake died shortly afterwards without increasing its growth. Another one in the Rio de Janeiro Zoo, Brazil, was credited with a length of 8 m *26 ft 3 in* in 1971, but Mr P J Davey (pers. comm. 14 May 1975), British Vice-Consul in Rio de Janeiro, said the snake in question actually measured 5 m *16 ft 5 in*.

The longest snake held captive anywhere in the world today is a female reticulated python named 'Cassius' owned by Mr Adrian Nyoka, owner of Knaresborough Zoo, Yorkshire, which measures *c.* 8·38 m *27 ft 6 in* and tips the scales at 109 kg *240 lb*. It was collected in Malaysia in 1972 and was purchased as a 26-footer *7·92 m*. Another exceptionally large python in the Kebin Binatang Zoo, Djakarta, Indonesia, is reported to be very close to 9 m *29 ft 5¼ in* in length, but Dr F Wayne King, who has seen this snake thinks 7 m *23 ft* is a more realistic figure.

The longest snake found in Britain (three species) is the Grass snake (*Natrix natrix*), also known as the 'Ringed snake', which is widely distributed throughout England and Wales. Adult males average 660 mm *26 in* in length and adult females 760 mm *29·92 in* (Appleby, 1971).

The longest accurately measured specimen on record was probably a female killed in South Wales in 1887 which measured 1775 mm *5 ft 10 in*, and another female also collected in South Wales measured 1750 mm *5 ft 9 in* (Leighton, 1901).

The longest venomous snake in the world is the King cobra (*Ophiophagus hannah*), also called the 'Hamadryad', of eastern India, China, the Malay Archipelago and the Philippines. Adult

The King cobra, the longest venomous snake in the world (Fox Photos)

individuals average 3·65–3·92 m *12–13 ft* and weigh about 6·8 kg *15 lb*. In April 1937 an enormous example measuring 5·53 m *18 ft 2 in* in length was captured alive near Port Dickson, in the State of Negri Sembilan, Malaya, and kept in captivity for a time by a Mr J Leonard of Ruthkin Estate (Gibson-Hill, 1948). Later it was sent to London Zoo where a measurement of 5·56 m *18 ft 4 in* was recorded for it shortly after arrival. At the time of its death a few days before the outbreak of the Second World War – all the venomous snakes at the Zoo were ordered by the Government to be killed as a safety precaution – it measured a magnificent 5·70 m *18 ft 9 in* (David J Ball, pers. comm. 16 March 1971).

The longest known sea snake (50 species) is *Hydrophis spiralis*, which ranges from the Persian Gulf to the Malay Peninsula and Archipelago. The average adult length is *c.* 1·83 *6 ft*, but individuals measuring 2·74 m *9 ft* and even 3·04 m *10 ft* have been reliably reported.

The heaviest living snake is the anaconda, which has already been discussed. On the basis of the weight/length relationship of captive specimens measuring less than 6·09 m *20 ft* in length Dowling (1961) has calculated that a 30-footer *9·14 m* in good condition would probably weigh over 272 kg *600 lb*, which means it would be nearly twice as heavy as a reticulated python of the same length.

The heaviest venomous snake is the Eastern diamondback rattlesnake (*Crotalus adamenteus*) of the SE USA. Adult examples average 1·52–1·83 m *5–6 ft* in length and scale 5·5–7 kg *12–15 lb*. One huge individual killed by Rutledge (1946) measured 2·36 m *7 ft 9 in* in length and tipped the scales at 15·5 kg *34 lb*. Another one killed by two boys in the Big Santee Swamp, South Carolina, allegedly measured 2·43 m *8 ft* in length and weighed 18 kg *40 lb* (Wallace, 1950), but Klauber (1956) has queried the accuracy of this claim.

In 1919 a length of 3·45 m *11 ft 4 in* was reported for an eastern diamondback killed at West Palm Beach, Florida, but this measurement is much too extreme to be acceptable. John L Behler (1975), Associate Curator of Herpetology at the New York Zoological Society, who investigated this claim, managed to come up with a photograph but nothing in the way of physical evidence. The slightly out of focus picture shows about 95 per cent of the not

The Eastern diamondback rattlesnake – the heaviest venomous snake

unduly bulky snake being supported by five men standing shoulder to shoulder with their arms held away from their bodies. This pose obviously exaggerates the true size of the reptile inasmuch as it brings the carcass some 460 mm *18 in* nearer to the camera, and if we take this into consideration then the 'monster' as shown probably did not tape much more than 2·43 m *8 ft* (2·45 m *8 ft 1 in* perhaps?).

The Gaboon viper (*Bitis gabonica*) of the tropical rain forests of Africa south of the Sahara is even bulkier than *Crotalus*, but this species does not exceed 1·83 m *6 ft*. A 1·74 m *5 ft 8½ in* female killed in the Mabira Forest, Uganda, scaled 8 kg *18 lb* with an empty stomach (Pitman, 1938), and a weight of 9 kg *20 lb* has been reported for another individual which measured exactly 1·83 m *6 ft*.

Mention should also be made of the King cobra (*Ophiophagus hannah*), although this is a relatively slender species. In 1951 a weight of 12 kg *26 lb 8 oz* was recorded for a 4·74 m *15 ft 7 in* hamadryad captured alive on the golf-course at the Royal Island Club on Singapore Island and taken to the Raffles Museum. It is interesting to note that the two men who caught this snake did so by *hand* because they had mistaken it for a python due to its great size; fortunately when they seized it by the tail the cobra offered no effective resistance. (It had probably just eaten a heavy meal and was drowsy.) On 26 February 1973 the death was reported of 'Junior', a 4·39 m *14 ft 5 in* king cobra at New York Zoological Park (Bronx Zoo) after establishing a longevity record for the species of 15 years 7 months. The posthumous weight of this snake was given as 12·75 kg *28 lb* despite the fact that it had suffered from a prolonged illness (Dresner, 1973).

The shortest snake in the world is the very rare thread snake *Leptotyphlops bilineata*, which is known only from the islands of Martinique, Barbados and St Lucia in the West Indies. Of eight specimens examined by Richard Thomas (1965) the two longest (from Martinique) both measured 108 mm *4¼ in*.

Some of the blind snakes (*Typhlops*) of Africa are almost as diminutive, the East African blind snake (*T. fornasinii*), *T. caecatus* of West Africa, the Angola blind snake (*T. anchietae*) and Hallowell's blind snake (*T. hallowelli*) of West Africa all measuring 101–152 mm *4–6 in* (Isemonger, 1968).

The shortest venomous snake in the world is the Striped dwarf garter snake (*Elaps dorsalis*) of South Africa, adults averaging 152 mm *6 in* in length (maximum 304 mm *12 in*). Peringueyi's desert adder (*Bitis peringueyi*) of South West Africa is also of comparable size, adults averaging 203 mm *8 in* (maximum *304* mm *12 in*).

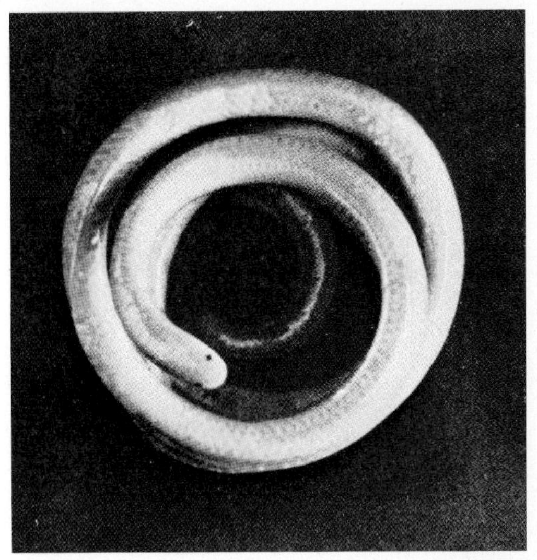

The diminutive thread snake (G Scortecci)

When food is plentiful snakes often eat prodigious meals. They are able to do this because their jaws are highly elastic and can be stretched to a remarkable degree. James Oliver (1958) says he once owned a 355 mm *14 in* long Cottonmouth moccasin (*Ancistrodon piscivorous*) which swallowed a very slender 736 mm *29 in* Ribbon snake (*Thamnophis sauritus*) sharing the same case, and he quotes another case of a 965 mm *38 in* King snake (*Lampropeltis getulus*) disposing of an 203 mm *8 in* DeKay's snake (*Storeria dekayi*), a 381 mm *15 in* grass lizard and a 1016 mm *40 in* Corn snake (*Elephe guttata*) in the space of one day!

When it comes to eating *heavy* meals, however, the giant constrictors are the undisputed champions, although the actual size of the prey they swallow is sometimes exaggerated.

The largest animal on record to be swallowed by a snake was a 59 kg *130 lb* Impala (*Aepyceros melampus*) which was removed from a 4·87 m *16 ft* African rock python (Rose, 1955). The weight of the snake was not recorded, but it was probably not much heavier than its victim! Gustav Lederer (1944) once induced a 7·31 m *24 ft* reticulated python to swallow a 54·5 kg *120 lb* pig, and a Collared peccary (*Tayassu angulatus*) weighing an estimated 45 kg *100 lb* was removed from an anaconda measuring 7·81 m *25 ft 8 in*. On

2 February 1973 Adrian Nyoka's huge female reticulated python Cassius swallowed a 38·5 kg *85 lb* pig with comparative ease. It took her less than an hour to perform this feat. Dr Walter Rose believes the maximum capacity of a really big constrictor is about 68 kg *150 lb*, but a very large African rock python (i.e. 7·62–9·14 m *25–30 ft*) could probably surpass this poundage.

Although many stories have been published of giant snakes attacking and swallowing human beings, only a very few have been authenticated. In most of the cases the victims were children and the constrictors reticulated pythons.

Felix Kopstein (1927), a student of East Indian reptile fauna, cites the case of a 14-year-old boy who was attacked and swallowed by a 5·18 m *17 ft* python on Selebaboe Island in the Talaud group south of the Philippines, and he also says there is an authentic record of an adult woman being devoured by a python reportedly over 9·14 m *30 ft* long.

In May 1972 an eight-year-old boy was swallowed by a 6·09 m *20 ft* python in the jungles of Lower Burma. The snake was eaten by the local villagers in revenge.

It is extremely doubtful whether even the largest python could swallow an average-sized man (70 kg *154 lb*) because the shoulders would probably get stuck in the gullet, but Arthur Loveridge (cited by Oliver, 1958), says there is one reliable record of a python swallowing a Burmese of small stature in 1927.

The man, Maung Chit Chine, who worked for a firm of European jewellers, had apparently set out with four friends on a hunting expedition in the Thaton district when they were suddenly overtaken by a thunderstorm. Chine sought refuge under a large tree, and when he failed to rejoin his colleagues they started searching for him. Eventually they found his hat and his shoes near a gorged python measuring between 6·09 and 9·14 m *20* and *30 ft* in length.

'They killed the snake', writes Loveridge, 'and on opening it found the body of their late companion inside; he had been swallowed feet first!'

There are no authentic records of the anaconda, the Indian python or the amethystine python swallowing human beings, although in March 1953 a three-month-old baby narrowly missed ending up in the stomach of a huge amethystine python in Western Australia. The child was seized by the cheek while her parents were sleeping, and if her mother hadn't suddenly awoken at the crucial moment and raised the alarm she would surely have been eaten.

At the other end of the scale the giant snakes are also great fasters and there are a number of records of individuals going twelve months or more without food. One female reticulated python at the Frankfurt Zoo, West Germany, fasted 570 days, took food for a time and then fasted another 415 days before eating, and a much larger example at the same zoo reportedly went 679 days without food although it drank regularly (Lederer, 1944).

The most venomous snake (300 species) in the world is the sea snake *Hydrophis belcheri* (*H. melanocephalus*?), which abounds around Ashmore Reef in the Timor Sea, 320 km *200 miles* NW of Darwin, Northern Territory, Australia. This species was not discovered until 1946 and is named after Sir Edward Belcher, who headed a survey of the area that year.

In 1973 an eleven-man international team of specialists interested in sea snakes took part in a six-week cruise of the Timor Sea aboard the American research vessel *Alpha Helix* and collected 100 examples of *H. belcheri*. Later one of the scientists, Dr Noburo Tamiya (1974), Head of Department of Biochemistry at Tohoku University, Japan, carried out a number of MLD (minimum lethal dose) tests on live mice under laboratory conditions and revealed that this species (maximum length 1·52 m *5 ft*) has a venom 100 times more toxic than any other known sea or land snake.

Prof William A Dunson (pers, comm. 20 November 1974), leader of the expedition, said later: 'Only small amounts of the venom were available . . . and the extreme potency hardly makes this a highly dangerous species since the head is small and the snake is not aggressive. It probably does not inject much of the venom when it bites, or as most sea snakes, does not often inject any venom.'

Apparently this sea snake has developed a small head in order to enter the burrows of small eels on which it feeds, and this reduction in size probably accounts for the very high toxicity of the venom.

The most venomous land snake is the Tiger snake (*Notechis scutatus*) of southern Australia and Tasmania, which grows to a length of 1·21–1·52 m *4–5 ft*. F G Morgan (1956) of the Commonwealth Serum Laboratories in Melbourne found that the *dried* neurotoxic venom of twelve black specimens (possibly a sub-species) collected on Reevesby Island in the Spencer Gulf had a lethal dose of 0·00087 mg per 100 g *3·54 oz* of body-weight when injected subcutaneously in guinea-pigs, which means that the fatal dose for an average 70 kg *154 lb* man would be 0·609 mg *0·000021 oz*. This extremely high potency is exceptional, however, and is more than twice as toxic as the venom found in mainland tiger snakes (i.e. 0·0002 mg per 100 g of body-weight). The bite of the tiger snake causes vomiting and excessive sweating, followed by respiratory paralysis, which usually results in death in two to three hours. In one series of 45 victims 18 died, giving a fatality rate of 40 per cent (Fairley, 1929), but since the development of an effective antidote the death toll has dropped considerably. The venom is four times as toxic as that of the Death adder (*Acanthopis antarticus*), also found in Australia, 20 times that of the Asiatic cobra (*Naja naja*), and 100 times that of Russell's viper (*Vipera russelli*) of SE Asia.

The Kraits (genus *Bungarus*) of SE Asia and the Malay Archipelago are also highly venomous. The Indian krait (*B. caeruleus*) is particularly nasty (lethal dose for man 2–3 mg *0·000070–0·000105 oz*) and about 50 per cent of bites are fatal, even with anti-venom treatment (Minton & Minton, 1969). The venom of the Javan krait (*B. javanicus*) is reported to be even more toxic – Kopstein says a father and son bitten in quick succession by this snake in SW Yunnan Province both died soon afterwards – but this species is only known from the type specimen and more information is needed.

The most dangerous snake in the world is the Saw-scaled viper (*Echis carinatus*), which ranges across Africa north of the Equator, through the Middle East to the Indian subcontinent and north Sri Lanka. It is an extremely prolific species and its venom is unusually toxic for man. Minton & Minton (1969) report that the lethal dose may be as small as 3–5 mg *0·000105–0·000175 oz* and this, coupled with the fact that the snake becomes extremely aggressive when frightened or disturbed, makes it a truly formidable adversary and one greatly feared by rural populations in Africa and India. These fears were substantiated in 1973 when Dr H Alistair Reid, of the Liverpool School of Tropical Medicine, made a short research visit to rural parts of northern Nigeria.

'In three weeks I saw 60 victims of *E. carinatus* bite. Fortunately we had an effective anti-venom and therefore there were no deaths, but without anti-venom the mortality rate in medical literature is quoted as 10–20 per cent' (pers. comm. 1 April 1975).

The King cobra (*Ophiophagus hannah*), the Bushmaster (*Lachesis muta*) of tropical Central and South America, and the Taipan (*Oxyuranus scutellatus*) of northern Australia and SE New Guinea have also been credited with the title of 'most dangerous snake' by various authors, but these species have a much more limited range than *E. carinatus* and are nothing like as abundant; in fact, the taipan is comparatively rare. In addition these snakes are not naturally aggressive, despite many claims to the contrary, and even when frightened they would probably try to escape.

Although sea snakes are extremely venomous they are not generally aggressive and there are relatively few records of these creatures attacking human beings deliberately in the water. Dr H A Reid (1956) gives details of two bathing fatalities at Penang, Malaysia, both off the same beach on the same day, but these cases were exceptional. One of the victims, an eight-year-old boy, died 13 hours after being bitten, and the other, a 26-year-old man, lived for 72 hours.

Sea snakes present a much bigger hazard to the native fishermen of SE Asia who often catch them in their nets for use as food and general medicine, and there have been a number of fatal accidents. The trouble with the sea snake is that the bite is painless and the venom very slow acting, so by the time symptoms develop like general debility and intense thirst, first-aid measures are too late. According to Reid (1959) the mortality rate due to sea-snake bites is about 5 per cent of those bitten, and three-quarters of these die within 24 hours. The reason why the figure is so low is because the amount of venom injected is usually very small.

The largest yield of venom ever recorded for a snake is 1530 mg *0·05 oz* for a Jararacucu (*Bothops jararacussu*) of Brazil (Schottler, 1951), but this was exceptional (average 150–200 mg *0·005–0·006 oz* dry weight).

The Gaboon viper (*Bitis gabonica*) probably has the highest *average* yield. Three specimens 'milked' at the serpentarium in Brazzaville, Congo, gave a combined weight of 2970 mg *0·10 oz* (Grasset, 1946). The Eastern diamondback rattlesnake (*Crotalus adamanteus*) is another big producer, with a normal range of 370–720 mg *0·01–0·02 oz*. The yield of the Western diamondback rattlesnake (*C. atrox*) is usually half that amount (i.e. 175–325 mg *0·006–0·01 oz*), but Dr Laurence Klauber (1956), the leading authority on rattlesnakes, 'milked' one example which yielded 1145 mg *0·04 oz* or sufficient to kill 45 people! There is also a record of a Cottonmouth moccasin (*Agkistrodon piscivorous*) from the SE

USA producing 1094 mg *0·038 oz* (Wolff & Githens, 1939), compared to 300–500 mg *0·010–0·019 oz* for the Bushmaster (*Lachesis mutus*) and 350–450 mg *0·012–0·015 oz* for the King cobra (*Ophiophagus hannah*). The highly venomous Eastern coral snake (*Micrurus fulvius*), on the other hand, has a yield of only 2–6 mg *0·000070–0·00210 oz* and the even more toxic Banded small-headed snake (*Hydrophis fasciatus*) less than 1 mg *0·000035 oz*.

If we exclude China, the USSR and Central Europe, it is estimated that between 30000 and 40000 people die from snakebite each year – 10000 to 15000 of them in India, where the rate is 5·4 deaths per 100000 population. In 1955 alone, 1300 people reportedly died of snakebite in the former Bombay State (Deoras, 1959).

The country with the highest mortality rate from snakebite is Burma, the average rate being 15·4 deaths per 100000 population, but Swaroop & Grab (1956) claim the rate in some districts is as high as 30. (In Sagaing district the rate was 36·8!) During the period 1935–40 there were 12733 reported deaths in Burma, but Caras (1964) says he finds this figure difficult to accept for a country with a population of only 14500000, although he does admit that 'Burma is one of those unfortunate areas where there are large numbers and many species of truly dangerous snakes.'

The only venomous snake in Britain is the Adder (*Vipera berus*), which is widely distributed throughout England, Wales and Scotland. Adult males average 500 mm *19·68 in* in length and adult females 550 mm *21·65 in* (Appleby, 1971). The largest accurately measured specimen on record was probably a female killed on Walberswick Common, east Suffolk, on 8 July 1971 which measured 965 mm *38 in*. Another adder killed in a garden at Kingstone, Hertfordshire, on 18 June 1968 reportedly measured 927 mm *36½ in*, but this snake was not preserved for scientific examination.

Since 1890 ten people have died from snakebite in Britain, including six children, but there have only been two fatalities since 1950. One of them was a 14-year-old boy who was bitten on the right hand at Carey Camp, Dorset, on 13 May 1957. He was rushed to Poole Hospital, but died three hours later from cardio-respiratory failure due to anaphylactic shock. On 1 July 1975 a five-year-old boy died in a Glasgow hospital 44 hours after being bitten on the ankle by an adder during a family outing to the hills near Callander, Perthshire. It was the first death from an adder-bite in Scotland since records began 100 years ago.

The longest-fanged snake in the world is the highly venomous Gaboon viper (*Bitis gabonica*) of tropical Africa. In a 1·30 m *4 ft 3 in* long example they measured 29 mm *1·14 in*, and in a 6-footer *1·83 m* the length was 50 mm *1·96 in* (Minton & Minton, 1971). The only other venomous snake with fangs of comparable size is the much longer Bushmaster (*Lachesis mutus*). Those of an 3·44 m *11 ft 4 in* individual measured 35 mm *1·37 in* (Cochran, 1943).

Although venomous snakes are said to be able to strike 'like lightning', the speed of the movement is actually quite slow; in fact, a human hand can snatch an object quite a bit faster than most snakes can strike.

In 1974 Mr N Schaefer, a herpetologist at the Port Elizabeth Museum, South Africa, photographed with a cinécamera, different species of snakes lunging at moving mice, and was able to calculate their striking speeds. He discovered that the African egg-eater (*Dasypeltis scabra scabra*) had a striking speed of 123 cm/s *45·5 in/s*, the Horned adder (*Bitis caudalis*) 205 cm/s *80·7 in/s*, the Puff adder (*Bitis arientans*) 236 cm/s *92·5 in/s*, the Gaboon viper (*Bitis gabonica*) 247 cm/s *97·2 in/s* and the Shield-nose snake (*Aspidelaps scutatus*) 271 cm/s *106·6 in/s*.

The huge fangs of the Gaboon viper (Philadelphia Zoological Garden)

The Boa constrictor 'Popeye' – the oldest snake on record (Philadelphia Zoological Garden)

The human hand by comparison recorded a speed of 328 cm/s *129 in/s* in a quick snatch. Mr Schaefer concluded that the unexpectedness of a snake's striking movement over a relatively short distance was the reason why observers usually over-estimated its speed.

Like growth, the longevity of snakes has been greatly exaggerated and this largely stems from the ancient belief that periodic sloughing of the skin was a form of rejuvenation. This is why, in mythology, snakes are often 1000 years old.

The greatest irrefutable age recorded for a snake is 39+ years for a male Boa constrictor (*Boa constrictor*) named 'Popeye' who was still living in Philadelphia Zoological Garden, Pennsylvania, USA, in 1975 after spending approximately 38 years 6 months in captivity. The snake was purchased by Mrs Eugenia S Shorrock of Massachusetts from a London dealer in December 1936 and later presented to the zoo (J Kevin Bowler, pers. comm. 4 April 1975).

An Indian python (*Python molurus*) purchased by Mrs Shorrock from the same dealer on 9 December 1936 died in Philadelphia Zoological Garden on 20 February 1971 after being in captivity for 34 years 2 months and 11 days (Roger Conant, pers. comm. 10 August 1972).

There is also an authentic record of a female Anaconda (*Eunectes murinus*) living for more than 31 years. This specimen was received at Basle Zoo, Switzerland, on 18 August 1930 and died on 8 May 1962 (Wackernagel, pers. comm. 25 May 1966).

The oldest venomous snake on record was probably a Northern copperhead (*Agkistrodon contortrix mokeson*) collected in the Blue Hills near Boston, Massachusetts, in May 1941. It died in Philadelphia Zoological Garden on 7 April 1971 after having been in captivity for approximately 29 years 11 months. An African black-lipped cobra (*Naja melanoleuca*) hatched in San Diego Zoological Garden, California, USA, on 1 October 1928 died there on 12 November 1957 aged 29 years 42 days (Charles E Shaw, pers. comm. 13 January 1971).

Although some of the slender species of snakes give the impression of fast movement as they glide smoothly over the ground, this is in fact an illusion combined with ignorance of their metabolism. Except for short bursts the vast majority of snakes cannot keep pace with a man walking at a normal brisk pace i.e. 6·4 km/h *4 miles/h*, and their type of heart and circulation are such that they tire very rapidly.

In the 1930s Dr Walter Mosauer (1935) carried out some speed tests on several species of desert snakes in California. The results were extremely disappointing. The maximum velocity recorded was 5·7 km/h *3·6 miles/h* for a Colorado desert whip snake (*Masticophis flagellus piceus*), while

The slender Black mamba, the fastest moving land snake (Artco)

The deadly Black and yellow sea snake, one of the most venomous members of the Hydrophiidae (Artco)

the allegedly swift Sidewinder rattlesnake (*Crotalus cerastes*) could only notch up 3·26 km/h *2·04 miles/h*. The slowest of the six snakes tested was the Rosy boa (*Lichanura roseofusca*) with a miserable 0·358 km/h *0·224 mile/h*.

On another occasion Dr James Oliver and his wife released a Black racer (*Coluber constrictor*) on rough ground in Florida and invited a group of people to guess its speed. The estimates ranged from 16 to 24 km/h *10* to *15 miles/h*, but the best stopwatch reading was 5·9 km/h *3·7 miles/h*.

In another series of speed trials carried out in Australia with racers (family Elapidae), the maximum velocity recorded was 5·6 km/h *3·5 miles/h* (Kinghorn, 1956).

Even if we make due allowance for the fact that these reptiles rarely travel in a straight line between two points – a snake's meandering course can add 25–35 per cent to the measured distance it has to travel – it is still doubtful whether the faster serpents like the whip snakes and the racers ever exceed 8–9·6 km/h *5–6 miles/h* – even in an emergency!

The fastest-moving land snake in the world is probably the slender, agile Black mamba (*Dendroaspis polylepis*), which is found on the African continent south of the Sahara Desert and is principally a tree snake.

It has been claimed that this snake can keep up with a galloping horse (*c.* 40–48 km/h *25–30 miles/h*), but this belief is based solely on the evidence of a few cases of men who have been bitten as high up as the thigh when mounted on horseback – and trotting.

On 23 April 1906 Col Richard Meinertzhagen (1955) timed an angry black mamba over a distance of 43 m *47 yd* near Mbuyuni on the Serengeti Plains, Kenya, after he and his men had baited it with clods of earth, and recorded a speed of 11·2 km/h *7 miles/h*. This specimen – shot when the man it was pursuing tripped and fell over – was only 1·70 m *5 ft 7 in* long, however, and as this snake is said to be fastest about the 2·43–2·74 m *8–9 ft* mark (maximum length 4·26 m *14 ft*), it is more than likely that an angry or frightened mamba this size could reach 16 km/h *10 miles/h* on level ground for a short distance (24 km/h *15 miles/h* when rushing downhill to escape from a bush fire).

The British grass snake (*Natrix natrix*) has been timed at 6·7 km/h *4·2 miles/h*, which classifies it as one of the 'sprinters' of the serpentine world.

The fastest-swimming snakes are the sea snakes (Hydrophidae) of the Indo-Pacific region. Their tails have been modified into paddles and they can propel themselves through the water at considerable speed.

The swiftest member of this family is the widely distributed Black and yellow sea snake (*Pelamis platurus*) of the Pacific which has an extremely streamlined (flattened) body. No speed records have been published for this 914 mm *3 ft* long serpent, but the velocity may be as high as 16 km/h *10 miles/h* for short bursts when pursuing prey.

Section IV

AMPHIBIANS

(class Amphibia)

An amphibian is a cold-blooded, air-breathing vertebrate which lives both in the water and on land. It normally has four legs and is distinguished from reptiles by its naked and moist skin which is used in respiration. Throughout the class the brain is of a very low type. The majority of young are hatched from eggs deposited in water and breathe by means of external gills during the larval stage. At the completion of metamorphosis the gills close up and breathing is transferred to the skin and lungs so that the animal can live on land. In cold weather it burrows into the earth or mud at the bottom of a lake or pool where it passes into a state of hibernation.

The earliest known amphibian and the first quadruped was *Ishthyostega* which lived about 350000000 years ago. Its remains have been discovered in Greenland.

There are about 2000 living species of amphibian and the class is divided into three orders. These are the *Salientia* (frogs and toads); the *Urodela* (newts and salamanders); and the *Gymnophiona* (the wormlike caecilians). The largest order is *Salientia*, which contains about 1800 species or 90 per cent of the total number.

The largest amphibian in the world is the Chinese giant salamander (*Megalobatrachus davidianus*), which is found in the cold mountain streams and marshy areas of north-eastern, central and southern China. Adult specimens of both sexes average 1 m *39·37 in* in total length and scale 11–13 kg *24·25–28·66 lb*, the weight depending on the amount of water contained in the body.

The largest accurately measured giant salamander on record was a huge individual collected in Kweichow (Guizhou) Province, southern China in *c.* 1923 and described by Sowerby (1925), who says it measured 1·52 m *5 ft* snout to tail between pegs, and 1·75 m *5 ft 9 in* along the curve of the body. Unfortunately this animal was not weighed, but it must have scaled nearly 45·35 kg 100 lb.

The Japanese giant salamander (*Megalobatrachus japonicus*), found in the cold mountain streams of southern Japan proper, has also been credited with the title of 'largest living amphibian', but in reality this species 'bulks out' slightly smaller than its Chinese counterpart because its tail is proportionately longer. Adult specimens of both sexes average 1 m *39·37 in* in total length (Che'eng-Chao Liu, 1950).

According to Flower (1936) a very large Japanese giant salamander which died in the Leipzig Zoo, East Germany, on 31 May 1930 measured 1·44 m *4 ft 8¾ in* in a straight line and 1·64 m *5 ft 4½ in* along the curve of the body. This individual weighed 40 kg *88·18 lb* when alive and 45 kg *99·20 lb* after death, the body having absorbed water from the aquarium.

The only other salamander which approaches *Megalobatrachus* in terms of length is the eel-like Three-toed amphiuma (*Amphiuma means tri-*

A goliath frog alongside a falcon for size comparison.

dactylum) of the USA which has been measured up to 1·01 m *39·76 in* (Bishop, 1943). The Hellbender (*Cryptobranchus alleganiensis*), also of the USA, is not so long (maximum length 685 mm *27 in*), but is a bulkier creature than *A. m. tridactylum* (Oliver, 1965), and probably runs second to *Megalobatrachus* for weight.

The largest frog in the world is the rare Goliath frog (*Rana goliath* = *Conraua goliath*) of Cameroun and Spanish Guinea, West Africa. On 23 August 1960 a female weighing 3·305 kg *7 lb 4½ oz* was caught in the cataracts of the River Mbia, Spanish Guinea. According to Dr Jorge Sabater Pi, (pers. comm. 15 August 1967), Curator of the Centro de Ikunde in Bata, Rio Muni, this monster had an over-all length of 815 mm *32·08 in* and measured 340 mm *13·38 in* snout-vent. An even larger female with a snout-vent length of 356 mm *14·01 in* was collected by Dr Zahl (1967) in Spanish Guinea in December 1966 (now preserved in the National Geographic Society, Washington, DC), but this individual weighed 3·1 kg *6 lb 13¾ oz*.

Weights up to 5·89 kg *13 lb* and snout-vent lengths up to 609 mm *2 ft* have been claimed for this species, but Dr Sabater Pi considers these figures to be exaggerated.

The largest species of tree frog is *Hyla vasta*, found only on the island of Hispaniola, West Indies. The average snout-vent length is about 90 mm *3·54 in*, but a female collected from the San Juan River, Dominican Republic, by G S Miller, Jr for the US National Museum, Washington, DC, in March 1928 measured 143 mm *5·63 in* (Cochran, pers. comm. 1 June 1967).

The largest frog found in Britain is the Marsh frog (*Rana ridibunda*), which was introduced into this country in February 1935. (Twelve specimens from Debreczen, Hungary, were released on the edge of the Romney Marshes, Kent). Adult males have been measured up to 96 mm *3·77 in* and adult females up to 126 mm *4·96 in* snout-vent.

Mr D Stevenson of Eastbourne, Sussex, (pers. comm. 22 November 1972) said he and two friends caught a gigantic female on Romney Marsh in June/July 1966 which had a snout-vent length of approximately 133 mm *5·25 in*.

The largest native British frog is the Common frog (*Rana temporaria*) which has been measured up to 95 mm *3·74 in* snout-vent in Scotland (Boulenger, 1893). The average snout-vent length is 40–45 mm *1·57–1·77 in*.

The Edible frog (*Rana esculenta*) has been credited with snout-vent measurements up to 106 mm *4 in*, but this species is probably not indigenous.

The largest toad in the world is the Marine toad (*Bufo marinus*) which – thanks to man – is probably the most widely distributed amphibian living today.

An enormous female collected at Miraflores, Colombia, on 24 November 1965 and later exhibited in the reptile house at New York Zoological Park (Bronx Zoo) had a snout-vent length of 238 mm *9·37 in* and weighed 1302 g *2 lb 14 oz* shortly before its death. (Dr Hutchinson, pers. comm. 5 October 1967.) This specimen

A Marine toad with a Common toad on its back

The smallest toad in the world – Bufo taitanus beiranus

is now preserved in the American Museum of Natural History in New York.

This size is closely matched by the Blomberg toad (*Bufo blombergi*) of Colombia. The type specimen was collected by Rolf Blomberg in the vicinity of Nachao, SW Colombia, on 11 September 1950. It measured 207 mm *8·14 in* snout-vent and weighed 1000 g *2·20 lb*. Another one collected by Blomberg in the same area in May 1951 was slightly larger, measuring 215 mm *8·46 in* snout-vent (Myers & Funkhouser, 1951), but its weight was not recorded. Both these dreadnoughts were presented to the New York Zoological Society (Bronx Zoo), and Prof Myers (pers. comm. 13 September 1967), of Stanford University, California, thinks they 'grew a few millimetres larger before they died'.

The largest toad found in Britain is the Common toad (*Bufo vulgaris*). Adult males have an average snout-vent length of 60–65 mm *2·36–2·55 in* and females 75–80 mm *2·96–3·14 in*, but measurements up to 70 mm *2·75 in* and 102 mm *4·01 in* respectively have been recorded (Smith, 1951).

The smallest known amphibian is the Arrow-poison frog *Sminthillus limbatus* of Cuba. Eleven specimens (sex unknown) with snout-vent lengths ranging from 8·5 to 12·4 mm *0·33* to *0·48 in* were described by Schwartz & Ogren (1956).

A Hoffman (1944), describing a *new* species of frog (*Phrynobatrachus chitialaensis*) he had collected in Malawi, says the snout-vent length of the type specimen was only 13 mm *0·51 in*, but this amphibian was not mature. When Loveridge (1953) examined the material later, he realised that *P. chitialaensis* was in fact the subspecies *P. ugkingensis mababiensis* (maximum snout-vent length 22·2 mm *0·87 in*), and Prof Stewart (pers. comm. 9 August 1967) of the State University of New York, Albany, NY, has since obtained the following measurements: juveniles 9·8–16 mm *0·38–0·62 in*; adult males 14–18 mm *0·55–0·70 in*; adult females 16–22 mm *0·62–0·87 in*.

The smallest frogs in Africa are actually *Microbatrechella capensis*, which is restricted to the Cape Flats near Cape Town, South Africa, and *Cacosternum nanum parvum* of the Drakensberg mountains of Cape Province and Natal, both having a maximum snout-vent length of 16 mm *0·62 in* (Poynton, 1964).

The smallest tree frog in the world is the Least tree frog (*Hyla ocularis*) of the SE USA, which also has a maximum snout-vent length of 16 mm *0·62 in* (Oliver, 1955).

The smallest toad in the world is *Bufo taitanus beiranus*, which was originally discovered in the Beira area of Mozambique, Africa (Loveridge, 1932). The largest of the 13 specimens examined by Dr Poynton (1964) of the University of Natal had a snout-vent length of only 24 mm *0·94 in*. Four other examples have since been collected at Nova Freixo, N Mozambique, by Dr Broadley (pers. comm. 1 December 1967), Curator of the Umtali Museum in Rhodesia. The largest specimen had a snout-vent length of 21·5 mm *0·84 in*.

The smallest amphibian found in Britain is the Palmate newt (*Triturus helveticus*). Mature examples measure 75–88 mm *2·95–3·45 in* in total length and the weight ranges from 1·50 g *0·052 oz* to 2·39 g *0·082 oz* (Boulenger, 1894; Evans, 1894).

The greatest authentic age recorded for an amphibian is 51 years 7 months 3 weeks 2 days for a male Japanese giant salamander (*Megalobatrachus japonicus*) which was born in the Aquarium at Amsterdam Zoo, Netherlands, on 10 November 1903 and died there on 6 July 1955. Another male which died in the same aquarium on 3 June 1881 was allegedly 55 years old; Flower (1936) says it was brought to Europe in 1829 and lived in the Royal Museum for Natural History in Leiden up to about 1840, when it was transferred to the zoo. Mr Graaf (pers. comm. 28 September 1972), Curator of the Aquarium, said, however, that he doubted the authenticity of this longevity claim and that he preferred the more recent record. According to Schneider (1932) a female Japanese giant salamander which died in the Leipzig Zoo, East Germany, on 31 May 1930 had spent 65 years in captivity having been received at Berlin Zoo in 1865, but Flower refused to accept this record. He claims this specimen was one of two received at the zoo in 1884, which means it was about 46 years old at the time of its death.

Toads, too, are long-lived creatures and probably run second to *Megalobatrachus*. T Pennant (1776) says one *adult* female Common toad (*Bufo vulgaris*) was kept under observation by an English family for 36 years and became so tame that it allowed itself to be put on a table and fed by hand. Even then, it didn't die of old age – the family's tame raven suddenly took a dislike to the ancient toad – and Malcolm Smith (1951) estimates that it must have been at least 40 years old at the time of its death. (The females of this species take four years to reach adulthood.)

The popular belief that toads can survive for decades – or even for a century or more – imprisoned in cavities in wood or solid rock without food, water or air is a myth.

This was neatly proved by Dr William Buckland, the famous English naturalist who, on 26 November 1825, enclosed two dozen common toads of different sizes in separate airtight compartments made of compact sandstone or porous limestone which he buried in his garden. When he examined them again on 10 December 1826 he found that all the toads in the sandstone compartments were dead and very much decayed, while most of the others in the larger cells of porous limestone were still alive although emaciated. The survivors were buried again for a further year but, at the end of this period, they were all found to be dead.

The toads in the sandstone compartments probably died within a few hours of being buried because they cannot live in dry air. The others in the porous limestone compartments only survived because they were getting a constant source of moisture and air, but in the end they died of starvation.

On another occasion four toads of the same species were placed in holes cut into the trunk of an apple tree which were then sealed. A year later, however, when the seals were removed the toads were all found to be dead and very much emaciated.

In May 1919 a miner working in Netherseal Colliery was holing in the dirt beneath the coal when his pick broke into a 'pocket' and a live toad came rolling out.

'The creature', said a newspaper of the day, 'is about three inches in length and has a skin of dirty brown colour. Its eyes were open, but it was obvious that it could not see at first. Two days afterwards it gave indications of returning to normal. The toad has no mouth, but there are evidences that it once possessed this useful member. On the same day as the sight began to return the toad also started to leap about in a clumsy manner. The webbed feet differ from present well-known varieties. The spot where it was found is only 200 yd below the surface. The toad has been sent to Birmingham University.'

In July 1933 a live toad with eyes but no mouth was found in the solid chalk face of a greystone lime pit at Dorking, Kent. It was exposed after blasting operations 15·24 m *50 ft* from the surface and 1·22 m *4 ft* into the face of the cliff. The amphibian was about 31 mm *1¼ in* long. When first captured it was pink in colour, but gradually got darker. It was later taken to the British Museum (Natural History), London, for examination.

According to Dr Cochran, Curator of Reptiles & Amphibians at the US National Museum, Washington, DC, who has made a special study of this subject, toad eggs often filter down through the soil and rock with surface water and get lodged in limestone, sandstone and coal cavities where the temperature is sufficient for them to hatch. The young animals feed on the insects which accumulate in these hollows until eventually they become too big to get out through the narrow apertures they had entered. Similarly, toads hatched above ground regularly hide in cool, moist holes and crevices during the summer in an effort to escape the heat and dryness, and the entrance to a mine or the bottom of a deep excavation must seem a very attractive proposition.

Other amphibians reliably credited with a maximum life-span in excess of 20 years include the Hellbender (*Cryptobranchus alleganiensis*) 29 years; the Three-toed amphiuma (*Amphiuma means tridactylum*) 27 years; the Great siren (*Siren lacertina*), the Spotted salamander (*Ambystoma maculatum*) and *Amphiuma punctatum* 25 years; and the European or Fire salamander (*Salamandra salamandra*) 24 years (Walterstorff, 1928; Koch, 1952; Oliver, 1955).

Dr Oliver (1955) gives the maximum life-span of the American toad (*Bufo terrestris americanus*) as 31 years, but this statement was obviously made tongue-in-cheek. The animal he was referring to – 'Old Rip' – was removed alive from the cornerstone of the old courthouse in Eastland, Texas, in 1928 after a former county clerk alleged he had imprisoned a toad there in 1897!

Frogs are generally not as long-lived as toads. The American bullfrog (*Rana catesbeiana*) has been known to live up to 16 years in captivity (Durham & Bennett, 1963), and the Edible frog (*Rana esculenta*) for 16+ years (Sebesta, 1935), but these ages were exceptional. The rare Goliath frog (*Rana goliath*) may live much longer, but more information is needed on the subject. (The monstrous Marine toad has a maximum life-span of *c.* 15 years.)

In a letter published in the magazine *African Wild Life* (Vol. 10, no. 4, December 1956) D Cairncross described how he kept a live specimen of the very large Bullfrog *Pyxicephalus adspersus* for nearly 17 years, having acquired it in Pretoria West in November 1939 when it was already fully grown.

'In January 1940 a tadpole of this species metamorphosed in my aquarium and remained in my care for seven years when it unfortunately escaped. If this may be taken as a guide, it would appear that the animal requires about 28 years to reach its full size; and that the large specimen obtained in November 1939 hatched from an egg laid not later than November 1911. This would make it 45 years of age.'

The finest jumpers among amphibians are frogs, the abilities of which are largely dependent on body-weight, hind-leg length and the surface from which they take off.

The jumper *par excellence*, not only for distance covered, but also in terms of snout-vent length, is the 51 mm *2 in* long South African sharp-nosed frog (*Rana oxyrhyncha*). In 1950 Dr W Rose of Cape Town collected a small number of these frogs in Zululand and decided to test out the jumping ability of a 55 mm *2·16 in* long female. Her best leap was 2·98 m *9 ft 10 in*, and in three consecutive jumps she covered a distance of 7·74 m *25 ft 5 in*. (In 1950 the American world record stood at 4·47 m *14 ft 8 in*.) This incredible performance, however, was over very uneven ground and the jumps were not quite in a straight line. Some four years later, on 18 January 1954, a Frog Olympics was held on Green Point Racing Track, Cape Town, in the presence of some 5000 spectators. Most of the contestants were examples of *Rana fuscigula*, which is about half as large again as *R. oxyrhyncha*, and only two of them managed a three-jump total in excess of 2·74 m *9 ft*. It was then the turn of a specimen of *R. oxyrhyncha* named 'Leaping Lena' (later discovered to be a male) which had been hurriedly flown in from Durban to do his stuff.

'Seeing the tiny chap squatting there, the crowd began to laugh', says Dr Rose, 'then, as he stayed blinking in the unaccustomed light, they commenced to jeer. Our hearts alternated between our mouth and our boots, for we were in effect his sponsor. Then, to our joy, the little fellow braced himself together, inflated his little chest, and leapt; once, twice, three times in a dead straight line. At the first leap the crowd gasped; at the second it cried out in amazement and at the third tumult broke out.'

The actual distance covered by this frog in three consecutive leaps was 7·40 m *24 ft 3½ in*, and shortly afterwards Dr Rose was a witness when 'Leaping Lena' achieved a three consecutive leap of 9·82 m *32 ft 3 in* under the same carefully observed conditions. After the competition this 'perfect little jumping machine' was released, and the record still stands today.

At the annual Calaveras County Jumping Frog Jubilee at Angels Camp, California, USA, in May 1955 another male of the same species made an unofficial *single* leap of over 4·57 m *15 ft* when being retrieved for placement in its container.

The greatest number of consecutive jumps attributed to a frog is 120 for a freshly caught adult Spring peeper (*Hyla crucifer crucifer*) of the USA which was placed on a grassy lawn. Not unnaturally the distance between each hop gradually decreased (Rand, 1952).

The leaping ability of toads is not nearly so impressive, although they probably have more stamina than frogs. Rand tested a small number of Fowler's toads (*Bufo woodhousi fowleri*) on the same lawn and found that they jumped from 304 mm *12 in* to 571 mm *22½ in* per hop, or up to 7·8 times snout-vent length; on sand, however, the distance covered was only 152–368 mm *6–14½ in*.

One of the strongest contenders for the title of 'poorest jumper' among anurans is the diminutive Greenhouse frog (*Eleutherodactylus ricordi planirostris*) of the USA. One specimen tested by Dr Oliver (1955) with a snout-vent length of 31 mm *1¼ in* could only manage a personal best of 120 mm *4¾ in* in five trials, or just under four times its own snout-vent length.

The greatest altitude at which an amphibian has been found is 8000 m *26 246 ft* for a Common toad (*Bufo bufo*) collected in the Himalayas. This species has also been found at a depth of 340 m *1115 ft* in a coal-mine.

Although many amphibians carry at least a trace of poison in their body or secrete this substance from glands in the skin, Caras (1964) says most of the members of this class 'are more colourful and musical than dangerous'. There are exceptions, however, and some of the Arrow-poison frogs (family Dendrobatidae) of Central and South America produce the most potent biotoxins known to man.

The most active poison is the batrachotoxin derived from the skin secretions of the Kokoi arrow-poison frog *Phyllobates bicolor*, which is found in the dense forests near the headwaters of the Rio San Juan and its tributaries, NW Colombia, South America. Only about 0·0001 g *0·0000004 oz* is sufficient to kill an average-sized man, which means that *28·3 g* 1 oz *of this poison would be enough to wipe out 2 500 000 people!*

According to Marki & Witkop (1963) the Choco Indians of the region capture these frogs by the following ingenious method: 'They imitate the frog's peeping, which sounds like fiu-fiu-fiu, with great skill, by whistling and at the same time beating their cheeks with their fingers. Their imitation is so perfect that a frog present not too far away usually answers the call and thus can be located. Trying to find these small frogs (snout-vent length 20–30 mm *0·78–1·18 in* weight *c.* 1 g *0·035 oz*), which live hidden among the plants near the ground, by any other means would seem hopeless.'

When the amphibians have been collected and taken back to the village, they are pierced through the mouth and body with a specially cut stick called siuru kida ('bamboo tooth') and then held over an open fire. The heat and pain contract the skin and force out a milky secretion, especially on the back of the creature. The tips of the arrows are dipped in the secretion and then left to dry in the shade. One tiny frog produces enough poison to tip up to 50 arrows.

Today the Choco Indians use the arrows for hunting game such as the jaguar, deer, monkeys and birds but, in earlier times, Wassen (1957) says the blowguns were used with deadly effect against neighbouring hostile tribes.

'An animal struck by a poisoned arrow', write Marki & Witkop, 'becomes paralyzed almost immediately and dies within a short time. The Indians then cut out the arrow from the flesh together with a small piece of meat immediately surrounding it. This is only done as a precaution, since the kokoi venom – like curare – is usually completely harmless when taken orally. A small scratch in the mouth, however, or an ulcer in the digestive tract of a person eating such meat may, quite obviously, cause a dangerous situation.'

The poison of the kokoi is ten times more powerful than the deadly tetrodotoxin produced in the body of the Japanese puffer fish (*Tetraodon hispidus*) (see page 157) which causes rapid respiratory paralysis in humans and has no known antidote (Halstead, 1956).

The tetrodotoxin produced by the abundant and well-developed glands of the California newt (*Taricha torosus*) is also particularly virulent and experiments have shown that 9 mg *0·0003 oz* is sufficient to kill 7000 mice.

The longest gestation period of any terrestrial animal is that of the Alpine salamander (*Salamandra atra*). According to Hafeli (1971) females living above an altitude of 1400 m *4593 ft* in the Swiss Alps produce two young in the fourth summer after fertilisation, which means the gestation period must be an incredible 3 years 1–2 months (cf. a maximum 2 years 1 month for the Asiatic elephant). Lower down, however (this species is not found under 600 m *1968 ft*), the young are usually born in the third summer, i.e. 2 years 1–2 months (Goodhart, pers. comm. 19 November 1971).

Section V

FISHES

(classes Agnatha, Chondrichthyes, Osteichthyes)

A fish is a cold-blooded vertebrate which lives in water and takes in oxygen by means of gills. Usually it has a muscular, streamlined body covered with scales and limbs modified into paired fins for swimming, but in some species the skin is unprotected or concealed by bony plates, and others have no recognisable fins at all. A few primitive fish also move on land and breathe with lungs. The brain is basically a primitive structure, but bony fish are capable of association and show a capacity for learning. The majority of young are hatched from eggs deposited in the water.

The earliest known fish, and the first vertebrate, was *Agnathans*, a jawless fish, which lived about 480 000 000 years ago. Its remains have been found near Leningrad, USSR.

There are about 30 000 living species of fish, 2300 of them freshwater, and they are divided into three very distinct classes. These are the *Agnatha* (lampreys and hagfishes); the *Chondrichthyes* (sharks, rays and chimaerids) and the *Osteichthyes* (higher bony fishes). The largest class is *Osteichthyes* which contains over 90 per cent of living species.

The largest fish in the world is the rare plankton-feeding Whale shark (*Rhineodon typus*), which is found in the warmer areas of the Atlantic, Pacific and Indian Oceans. The 5·03 m *16 ft 6 in* type specimen, one of the smallest on record, was harpooned in April 1828 by fishermen in Table Bay, Cape of Good Hope, South Africa, after they had noticed its unusual coloration (greenish grey with white spots). The shark was examined by Dr Andrew Smith, a military surgeon attached to the British troops stationed at Cape Town, who published a brief description the following year, and a more detailed one in 1849. The dried skin, which Dr Smith purchased for Sterling £6, is now preserved in the Museum d'Histoire Naturelle, Paris.

The enormous Whale shark, the largest fish in the world (Associated Press)

More than 40 years elapsed before *Rhineodon* was heard of again. In 1868 a young Irish naturalist named E Perceval Wright spent six months in the Seychelles, a group of islands in the western Indian Ocean. During his stay there he heard of a monstrous fish called the 'Chagrin' and offered a reward of $12 for the first example to be harpooned and delivered to him onshore. Eventually two individuals measuring 5·48 and 6·09 m *18* and *20 ft* respectively were secured, both of which he photographed and dissected. Two years later, in Dublin, he wrote: 'I have seen specimens that I believe to have exceeded 50 ft in length, and many trustworthy men, accustomed to calculate the length of the Sperm whale [one of the most important stations for this cetacean is off Ile Denis, one of the Seychelles group], have told me of specimens measuring upwards of 70 ft in length.'

He also said that Mr Swinburne Ward, Civil Administrator of the Islands, had informed him that he had measured one whale shark personally at slightly over 13·71 m *45 ft* in length.

Since then more than 100 whale sharks of varying sizes have been stranded or rammed by ships, but only a few of them have ever been scientifically examined.

The largest whale shark on record about which there is reliable information was a monstrous specimen which got wedged in the entrance of a bamboo stake-trap set in 15·24 m *50 ft* of water off Koh Chik (=Chick Island), on the east side of the Gulf of Siam, in the early part of 1919.

'The fish', writes Dr Hugh M Smith (1925), Fisheries Adviser to the Siamese Government, 'remained stuck for 7 days, during which time all fishing had to be suspended. It was finally killed with rifle bullets and hauled out of the trap, but the combined efforts of the local fishermen were insufficient to drag it ashore. The fishermen are quite familiar with sharks, which are caught almost daily in the bamboo traps set in the offshore waters of this section, but none of them had ever seen or heard of a shark of this size or kind. From the descriptions of its shape, colour, mouth and teeth given to me by eye-witnesses, I have no doubt the fish was a *Rhineodon*.'

The stuffed body of the 11·58 m 38 ft *Whale shark killed in Southern Florida (W. Gudger)*

The shark was not measured, but Dr Smith said its actual length was known by its position alongside the 'leader' as it lay jammed in the narrow entrance of the trap.

'From several independent sources I have learnt that the length of the monster was determined by the fishermen to be over 10 wa. The wa is the Siamese fathom and originally represented the full stretch of a man's outspread arms; in recent years it has been stabilised and adopted by the Royal Survey Dept as the equivalent to 2 metres. Therefore, whether we regard the wa as being the somewhat elastic measure of the Siamese fishermen, with say 1·7 to 1·8 m as an approximate average, or as being a full 2 metres, it would seem that in the fish in question we have rather more than the maximum length that has heretofore been ascribed to the whale shark.'

On the evidence of Dr Smith's statement this immense fish must have measured between 17 m *55 ft 9 in* and 20 m *65 ft 7 in* in length and scaled at least 40 tonne/*ton*.

In September 1934 the liner *Maurganui* collided with a whale shark 96 km *60 miles* NNE of Tikehau Atoll in the South Pacific and 'cut so deeply into the body of the fish that it was literally impaled on the bow'. Gudger (1940) says about 4·57 m *15 ft* of the shark hung on one side of the bow, and another 12·19 m *40 ft* on the other, making a total length of 16·76 m *55 ft*.

During the summer of 1926 Mack Sennett, one of the pioneers of screen comedy, organised a fishing expedition in the Gulf of California and took along with him a newly invented underwater camera to film the marine life in the area. While the party was in Los Frales Bay, about 64 km *40 miles* inside the Gulf, an enormous whale shark suddenly came into camera range, and swam round the fishing vessel for several minutes. Sennett, who later showed the film he had shot to Gudger (1927) in New York, said the shark had measured an estimated 19·81 m *65 ft* in length and 3·04 m *10 ft* across.

The most publicised whale shark on record was a 38-footer *11·58 m* killed by Captain Charles Thompson and some local fishermen just below Knight's Key, southern Florida,

A large Basking shark caught in a herring weir near Weymouth, Nova Scotia (Bob Brooks)

USA, in May 1912. They took nine hours to beach this leviathan, and it only succumbed after a piece had been cut out of its head and the small brain pierced by a knife attached to a long pole. The huge carcass was then towed to Miami, where it was hauled out of the water and placed on a railway flat-car – which promptly collapsed under the weight! The fish was later purchased by an enterprising promoter who, after having it skinned and stuffed, (a job taking several months) took it on a very successful tour of the eastern USA.

Gudger worked out the weight of this shark as 11·8 tonne/*ton* 26 594 lb, basing his calculations on a length of 11·58 m *38 ft* and a circumference of 5·48 m *18 ft* ('a 20 ft line put around his body for the purpose of anchoring him to the sand-bar, lapped over 2 ft so that we judged he was about 18 ft in circumference'), but this measurement was taken over the first gill-slit instead of directly behind the pectoral fins. Another whale shark killed near Marathon in the Florida Keys on 9 June 1923, after a fight lasting 54 hours, measured 9·60 m *31 ft 6 in* in length and had a circumference of 7·01 m *23 ft* just behind the pectorals, and 5·33 m *17 ft 6 in* over the first gill-slit. Another 11·58 m *38 ft* whale shark captured by fishermen off the coast of Karachi, Pakistan, also had a circumference of 7·01 m *23 ft* just behind the pectorals and must have weighed – using Gudger's formula – about 19·3 tonne/*ton*.

The largest fish ever held in captivity was a whale shark weighing several thousand pounds which was exhibited at the Mito Aquarium, Japan, for several months many years ago. The fish was kept in a small bay separated from the open sea by a net septum (Hiyama, cited by Gilbert, 1963).

The only other fish which compares in size with the whale shark is the Basking shark (*Cetorhinus maximus*), another plankton-feeder, which is found in all temperate waters of the world, although it is most common in the North Atlantic. Adult specimens average 7·92–8·83 m *26–29 ft* in length and weigh 5·25–6·5 tonne/*ton*, but much larger individuals have been reported.

In 1806 a basking shark measuring 11·12 m *36 ft* in length and weighing an estimated 8 tonne/*ton* was washed ashore at Brighton, Sussex. Another example measuring 10 m *32 ft 10 in* was captured near Brown's Point, Raritan Bay, New Jersey, USA, in 1821 (Lesueur, 1882), and one 12·27 m *40 ft 3 in* long was trapped in a herring gill-net in Musquash Harbour, New Brunswick, Canada, in 1851 (Perley, 1852). In 1865 a 40-footer *12·19 m* was captured at Povoa de Varzim, Portugal, and three years later a specimen measuring 10·66 m *35 ft* was stranded at Eastport, Maine, USA. The ten largest basking sharks taken on the Norwegian coast during the period 1884–1905 measured 13·71 m *45 ft* (the largest basking shark on record), 12·19 m *40 ft* (three), 10·97 m *36 ft*, 9·80 m *32 ft 2 in*, 9·75 m

The decomposed remains of a Basking shark, the source of many 'monster on the beach' stories

32 ft, 9·44 m *31 ft*, 9·29 m *30 ft 6 in*, and 9·22 m *30 ft 3 in* (Collett, 1905).

Other reliable records over 9·14 m *30 ft* include: a 9·29 m *30 ft 6 in* shark stranded at Portland, Victoria, Australia, in November 1883 (McCoy, 1885); a 34-footer *10·36 m* washed up dead at Auckland, New Zealand, in 1885; another the same length stranded near San Francisco, California, USA, in *c.* 1896 (this shark was embalmed and put on exhibition); a 36-footer *10·97 m* taken in Monterey Bay, California, USA, in *c.* 1915 and purchased by the owner of a travelling museum (Gudger examined this specimen personally); one measuring 11·58 m *38 ft* captured near Concarneau, NW France in 1917; a 9·75 m *32 ft* specimen caught in a cod-trap at Petty Harbour, near St John's, Newfoundland, Canada, in 1934; a 12·19 m *40 ft* individual taken off Portland, Maine, USA, the same year; a 31-footer *9·44 m* taken at Long Point, near Provincetown, Massachusetts, USA; one measuring 10·66 m *35 ft* (head missing) washed up on Chesil Beach, Portland, Dorset, on 1 October 1935; a 40-footer *12·19 m* trapped in a herring weir on Grand Manan Island, New Brunswick, Canada, in July 1958; one measuring 10·66–12·19 m *35–40 ft* caught in a salmon gill-net in Portugal Cove, Conception Bay, Newfoundland, on 15 July 1961; a 10·05 m *33 ft* individual killed in Monterey Bay, California, USA, in 1948; a 32-footer *9·75 m* entangled in a net at Grand Bruit, Newfoundland, in July 1962; and one measuring 12·19 m *40 ft* caught at Burgeo, Newfoundland, the same year.

The largest basking shark seen by Gavin Maxwell (1952) at his shark fishery on Soay in the Inner Hebrides, Scotland, between 1945 and 1949 measured 9·57 m *31 ft 5 in*, but he says he narrowly missed capturing another one between Uisenish and Lochboisdale, South Uist, Outer Hebrides, which was 'a full 40 ft'.

The rotting carcass of a basking shark provides the basis for more than 90 per cent of the 'monster on the beach' stories that appear in newspapers. This is not altogether surprising, because when the body of one of these great selachians is found in an advanced state of decomposition, it does *look* more like a sea serpent than a basking shark. This is because the gill apparatus and the massive lower jaw have long dropped off and been washed away by the sea, leaving behind a tiny box-like cranium and a fleshy backbone, which could pass as a small head on a long slender neck (shades of Nessie!). The often reported 'white mane' is also easily explained: it is nothing more than thousands of threads from the disintegrating muscle fibres. (See photograph page 141.)

As recently as November 1970 a 9·14 m *30 ft* long 'unknown marine animal' described by local police as a sea serpent was washed up on a beach near Scituate, Massachusetts, USA. This bizarre creature reportedly weighed between 15 and 20 tonne/*ton* (*sic*) and looked like a giant camel without legs, but later it was positively identified as – yes, a basking shark!

The largest carnivorous fish (excluding plankton-eaters) is the comparatively rare Great white shark (*Carcharodon carcharias*), also called the 'Man-eater', which ranges from tropical waters to cool, temperate zones. Adult specimens (females are slightly larger than males) average 4·26–4·57 m *14–15 ft* in length and generally scale between 521 kg *1150 lb* and 771 kg *1700 lb*, but much heavier weights have been reported for carcharodons in this measurement range. For instance, one tremendously bulky 15-footer *4·57 m* captured at San Miguel Island off the coast of California, USA, weighed more than 1360 kg *3000 lb* (Kenyon, 1959).

In 1964 professional shark fisherman Captain Frank Mundus – the man on whom Peter Benchley based the character of 'Quint' in his best-selling novel *Jaws* – harpooned a huge carcharodon off Montauk Point, at the tip of Long Island, New York, NY, USA. The shark, measuring 5·33 m *17 ft 6 in* in length, 3·96 m *13 ft* in maximum girth and weighing 2063 kg *4550 lb*, was boated after a five-hour battle. On another occasion Captain Mundus killed a 5·03 m *16 ft 6 in* long great white shark which reportedly weighed 1360 kg *3000 lb*.

A 'blue pointer', as it is called in South Africa, harpooned from a whale catcher about 160 km *100 miles* off the coast of Durban, Natal, in July 1952 measured 5 m *16 ft 5 in* in length and weighed 1500 kg *3306 lb*. (The jaws of this shark are preserved in Durban Aquarium.)

In July 1965 a 5·63 m *18 ft 6 in* long female carcharodon with a maximum girth of 3·27 m *10 ft 9 in* and weighing an estimated 1587 kg *3500 lb* was killed by rifle fire at Fremantle, Western Australia, after it had eaten two 1·52 m *5 ft* sharks hooked on lines and was in the process of swallowing a third (Andrews, pers. comm. 17 July 1965). In August 1975 sailors aboard an oil company boat harpooned a great white shark measuring 4·56 m *15 ft* and weighing 1090 kg *2400 lb* in the waters off California.

These marked variations in poundage reflect the physical condition of the shark.

The rare Great white shark, the largest of the true man-eaters (Government of Western Australia)

Norman & Fraser (1948) cite another record of a gravid carcharodon caught at Agamy near Alexandria, Egypt, shortly before the Second World War which measured 4·26 m *14 ft* in length and weighed 2540 kg *5600 lb* (*sic*), but this avoirdupois is impossible for a shark – or even a whale – of this length. (The nine fully developed young in this female were all credited with a length of 609 mm *2 ft* and a weight of 49 kg *108 lb*, but the latter figure must obviously mean the *combined* poundage.)

Although the general literature on sharks abounds with stories of enormous carcharodons, there are very few authentic records of specimens reaching or exceeding 6·09 m *20 ft* in length. In 1758 a great white shark measuring 6·09 m *20 ft* in length, 2·74 m *9 ft* across the pectoral fins and weighing 1778 kg *3924 lb* was harpooned from a French frigate in the Mediterranean after swallowing a sailor who had fallen overboard.

In October 1906 a 20 ft 'white pointer' was harpooned by whaleman Archer Davidson from a 3·04 m *10 ft* dinghy at Eden, New South Wales, and killed with a boat spade (Caldwell, 1937).

Another man-eater caught by fishermen near Aix, southern France, in 1829 measured 6·70 m *22 ft* and weighed over 1814 kg *4000 lb* (Smith, 1833). This length was exceeded by a carcharodon trapped in a herring weir at Harbour de Loutre, Campobello Island, New Brunswick, Canada, in November 1932 which measured 7·92 m *26 ft* (Piers, 1934).

William Travis (1963) says a great white shark caught by him off the Seychelles and raised to within 6·09 m *20 ft* of the surface was only 914 mm *3 ft* shorter than his 9·75 m *32 ft* shark cutter, but the line holding this giant eventually snapped under the tremendous strain and the fish escaped.

Jordan & Evermann (1896) mention a 9·14 m *30 ft* man-eater taken off Soquel, California, in 1880 which had a 45 kg *100 lb* sea-lion in its stomach, but Dr Stewart Springer of the US Fish and Wildlife Service (pers. comm. 7 July 1964) does not consider this record to be reliable: 'I can find no confirmation of the measurement of this shark and rather doubt that Jordan saw it. Perhaps he accepted the report on hearsay, feeling that if it could swallow a hundred pound sea-lion entire, it was really very large. But of course a 14–16 ft White shark can swallow a hundred-pound sea-lion, or two for that matter.'

Another man-eater caught in Santa Monica Bay, California, after 1950 was credited with a length of 9·75 m *32 ft*, but this measurement was put out by local practical jokers. John E Fitch, Research Director of the Dept of Fish and Game at Long Beach, (pers. comm. 29 April 1974) said that the shark in question measured 4·26 m *14 ft*.

Roedel & Ripley (1950) report that a great white shark measuring 7·01 m *23 ft* in length was caught in Bodega Bay, California, in 1943, but this unconfirmed claim was based upon a newspaper account and a picture which gave no idea as to size.

The largest authenticated carcharodon ever taken in Californian waters was one measuring 5·48 m *18 ft* in length and weighing 1882 kg *4150 lb* harpooned by Larry Mansur, a commerical fisherman, near Los Angeles in July 1976. Another outsized specimen – 4·87 m *16 ft* and 1406 kg *3100 lb* – caught by him in the same area two days previously had two *whole* sea-lions in its stomach which weighed 79 kg *175 lb* and 57 kg *125 lb* respectively!

The popularly held view is that the largest great white shark on record of which there is evidence was a 11·12 m *36 ft 6 in* monster taken off Port Fairy, Victoria, Australia, in 1842, the jaws of which are preserved in the British Museum (Natural History), London (Gunther, 1870). This length, however, has also been queried by Dr Springer, who had an opportunity to examine the jaws several years ago. He said he was 'greatly surprised' to find that the shark's teeth were about the same size as those of a 16-footer *4·87 m* he had dissected. (The largest tooth in the upper jaw has an enamel length of 57 mm *2·24 in*.) 'I think it likely that the specimen really was somewhere around 16 ft long, perhaps 18 ft or even 20 ft, but certainly not 36½ ft', he concluded.

This opinion is also shared by John E Randall (1973), an ichthyologist at the Bernice B Bishop Museum, Honolulu, Hawaii, who after carrying out his own examination of the jaws, calculated that the Port Fairy shark must have measured about 5·4 m *17 ft 6½ in* in life. (An Australian ichthyologist has suggested that the 36 ft 6 in was a printer's error and that the figure should have read 16 ft 6 in.)

The size of bite inflicted by the great white shark is another good guide to approximation of length, a 4·87 m *16 ft* individual producing a wound area measuring 279 × 330 mm *11 × 13 in*. C Ostle of the Dept of Fisheries and Fauna at Albany, Western Australia, told Randall that a whale killed in South Australian waters on 26 May 1972 had five enormous bites on its body measuring 483 × 610 mm *19 × 24 in* when it was found the next day, from which it was deduced that the carcharodon must have measured about 7·62–7·92 m *25–26 ft* in length. (Man-eaters this size can exert a crushing pressure of at least 20 tonne/*ton* per square inch.) Even larger bites were seen on a whale in 1968, but Ostle said they were not measured.

According to Vladykov & McKenzie (1935) the largest great white shark ever recorded in Canadian waters was a 11·27 m *37 ft* monster found trapped in a herring weir at White Head Island, New Brunswick, in June 1930.

Both these men are still alive, and in December 1972 the compiler made contact with Mr R A McKenzie, long since retired and living in St Andrews, New Brunswick, who said that this particular record was supplied by Dr Vadim Vladykov. He also revealed that his co-author did not arrive on the Atlantic coast of Canada until the summer of 1931, which meant that he did not examine the carcharodon personally, but obtained his information second-hand. The important question is: would Dr Vladykov, a man of high reputation, have included hearsay evidence unless he was positive it was accurate? In other words, did he have proof in the way of teeth, the jaws or even a photograph to substantiate this extreme measurement?

Unfortunately we may never know the answer for, despite strenuous efforts, the compiler has been unable to contact Dr Vladykov.

The huge fish may, of course, have been a wrongly identified basking shark, which is the only really large selachian found in the area, but it is difficult to believe that a fisherman – presumably the source of Dr Vladykov's information – could have made such a mistake, because even if the carcass was in a bad state of decomposition the huge, serrated teeth would have been an immediate means of identification.

In another letter (8 January 1973) Mr McKenzie said: 'There have been many more White Sharks taken on the Canadian Atlantic coast since our account of the Nova Scotian fishes was prepared. The most recent was not too far from here and was around 16–17 ft long. However, a few years ago a fantastic story re a White Shark was relayed to us by a fisheries officer, and if the fish was only half as long as indicated, it was still a big shark. Since none of our men saw it or measured it, however, nothing has been published about it.'

Captain J S Elkington of Queensland told David Stead (1963) that one day in 1894 an enormous carcharodon drew alongside his launch just outside Townsville breakwater and lay there virtually motionless for half an hour. He claimed the shark was at least 1·21 m *4 ft* longer than his 10·66 m *35 ft* launch.

Lawrence Green (1958), the popular South African writer, goes even further by stating that a 13·10 m *43 ft* great white shark ran aground in False Bay, Cape Province, many years ago after it had followed a ship with plague on board. Later (pers. comm. 6 July 1966) he said: I gathered a great deal of information on whaling and shark fishing in those waters from a Tristan islander named George Cotton who settled at Simonstown towards the end of the last century. He died only a few years ago, almost a centenarian.

'Cotton was very vivid and accurate in his descriptions, but vague about dates. Thus it would be almost impossible to give the date of the 43 footer. In any case the newspapers of those days gave little or no space to such events.'

The largest carcharodon actually to be weighed was a 6·40 m *21 ft* female caught by fishermen near Havana, Cuba, in May 1948 which tipped the scales at a massive 3312 kg *7302 lb*. The shark was mounted and put on display but was later discarded when it started to fall to pieces (L Howell Rivero, pers. comm. 9 September 1964).

The largest fish ever caught on a rod was a great white shark weighing 1208 kg *2664 lb* and measuring 5·13 m *16 ft 10 in* taken on a 59 kg *130 lb* test line by Alf Dean at Denial Bay, near Ceduna, South Australia, on 21 April 1959. Mr Dean has also caught five other carcharodons weighing over 1 tonne/*ton*. They scaled 1150 kg *2536 lb* (5·11 m *16 ft 9 in*), 1076 kg *2372 lb* (4·85 m *15 ft 11 in*), 1063 kg *2344 lb* (4·57 m *15 ft*), 1057 kg *2333 lb* (4·95 m *16 ft 3 in*) and 1048 kg *2312 lb* (4·90 m *16 ft 1 in*) respectively. In 1952 he nearly caught another one nicknamed 'Barnacle Lil' which he estimated measured 6·09 m *20 ft* and weighed 1360 kg *3000 lb* (probably much heavier), and in 1954 he played a carcharodon weighing an estimated 1814 kg *4000 lb* for 5 hr 30 min before losing it.

Apart from its very impressive size, the great white shark is also the man-eater *par excellence*.

According to Rondelet (1554) whole men in armour have been found in large individuals caught off Nice and Marseilles in the Mediterranean, and he says 'two tunny and a fully-clothed sailor' were removed from the stomach of another man-eater.

Thomas Pennant (1776), in his account of this species, writes: 'This grows to a very great bulk, Gillius says to the weight of four thousand pounds: and that in the belly of one was found a human corpse entire, which is far from incredible considering their vast greediness after human flesh.'

On 2 October 1954 the body of a 13-year-old boy was found intact in the stomach of a 907 kg *2000 lb* great white shark caught near Nagasaki, Japan (Baldridge, 1974).

In 1931 a carcharodon was reponsible for a temporary breakdown in the quaint mail service of Niuafoo ('Tin Can Island') in the extreme north of the Tonga Archipelago, central Pacific Ocean.

This island has very steep shores which make landing very difficult, and for years a native had swum out to meet the monthly mail ship. The man would carry with him outgoing letters sealed in biscuit tins and bring the incoming mail back to shore. On this occasion, however, the aquatic postman ran into a hungry man-eater on the return trip and, ever since then, the journey has been made by canoe.

Although Britain's coasts should fall comfortably within the range of the great white shark this species, until recently, had only been recorded three times – in 1769, 1813 and 1863. Since the 1960s, however, carcharodons have been *positively* identified off the Irish coast, and there have been a small number of unconfirmed sightings off SW Cornwall.

Fortunately there is not one single authentic case of a bather being mauled or killed by a shark in British waters, but fishermen have suffered injuries after carelessly handling landed sharks. (In June 1960 a trawlerman was treated at Wick, Caithness, Scotland, for an arm injury after being bitten by a small blue shark accidentally brought up in the catch.)

In June 1971 Jimmy Johnson, 32, a member of the Dolphins Sub-Aqua Club, Aldershot, Hampshire, was reportedly attacked by a large shark some 46 m *50 yd* off the beach at Bessand, near Dartmouth, S Devon.

'The shark was about 10 or 12 ft long. At first I thought it was basking, but after looking closer I decided it wasn't. It circled me once and then suddenly came straight at me. I had my lobster hook with me and held that out towards it, hitting it with a defensive action on the forehead. It passed a couple of feet beneath me.'

Exactly a year later the leader of a two-man team of naval clearance divers was allegedly attacked and mauled by a shark while surfacing from a 55 m *180 ft* dive in Falmouth Bay, Cornwall. The man, Lt John Rayner, from Portsmouth, said he was waiting at the 6 m *19 ft 6 in* mark to recompress when the shark – estimated length 2·43 m *8 ft* – suddenly zoomed in and sunk its teeth into his arm just above the wrist. Fortunately his companion, Petty Officer Barry Brand, was near and drove off the fish with his knife, but the wound required 19 stitches.

The story was later shot down in flames by a naval spokesman at Portsmouth who said the whole thing was a hoax. 'The man was not attacked by a shark. He just fell overboard and cut his arm. Also, he's a rating, not a lieutenant.'

After the official denial the two sailors, both of whom had been using fictitious names, were taken back to Portsmouth to be dealt with by their commanding officer.

The only other species of carnivorous sharks which have been reliably measured over 6·09 m *20 ft* are the Tiger shark (*Galeocerdo cuvieri*) of tropical and sub-tropical waters, and the Greenland shark (*Somniosus microcephalus*) of the arctic and northern seas.

Measurements up to 9·14 m *30 ft* have been claimed for the tiger shark, but the longest accurately measured specimen on record was one caught off the island of Taboga in the Gulf of Panama in 1922 which was 6·32 m *20 ft 9 in* long and weighed a very low 798 kg *1760 lb* (Hedges, 1923). An 18-footer *5·48 m* caught in a shark net at Newcastle, New South Wales, Australia, in early 1954 was much bulkier, scaling 1524 kg *3360 lb* (Grant, 1972; P R Wilson, pers. comm. 4 April 1975). Another tiger shark (originally thought to be a carcharodon) which attacked and completely devoured a skin-diver hunting for abalone off La Jolla Cove, California, on 14 June 1959 was estimated by his diving companion to have measured at least 6·09 m *20 ft* in length (Limbaugh, 1959; Baldridge, 1974).

The Greenland or Sleeper shark rarely exceeds 4·26 m *14 ft*, but one huge individual caught off May Island in the Firth of Forth, E Scotland, in 1895 was 6·40 m *21 ft* long and scaled 1020 kg *2250 lb*. Another one found entangled in cod-nets in the Moray Firth, NE Scotland, in the spring of 1929 was 4·57 m *15 ft* and weighed 1066 kg *2352 lb*.

In 1966 Dr C Lafond of the US Navy's Electronics Laboratory, and deep-diving pilot Joseph Thompson, were probing and photographing the 1219 m *4000 ft* deep bottom of the San Diego Trough in the 5·48 m *18 ft* research submarine 'Deepstar 4000' when suddenly they found an enormous shark looking into the lens of their camera. 'The eyes were as big as dinner plates', said Thompson afterwards. 'Then came the huge pectoral fins and finally the tail.'

By comparing its size with the known distance between scientific instruments placed on the sea-bed the two men calculated that the fish was about 9·14 m *30 ft* long.

The shark was later identified as a Greenland shark by Dr Carl Hubbs, Professor of Marine Biology at the Scripps Institution of Oceanography, after studying photographic evidence.

Excepting the Selachii (sharks, rays and chimaeras), the only other fish which has been reliably measured over 6·09 m *20 ft* is the Oar or Ribbon fish (*Regalecus glesne*), which has a worldwide distribution. In *c.* 1885 a specimen measuring 7·62 m *25 ft* in length was caught by fishermen off Pemaquid, Maine, USA.

On 18 July 1963 a ribbon fish measuring an estimated 15·24 m *50 ft* was seen swimming off Asbury Park, New Jersey, by scientists aboard the 25·9 m *85 ft* long Sandy Hook Marine Laboratory research vessel *Challenger*, but the huge creature was not captured.

In 1866 a ribbon fish measuring 4·57 m *15 ft* in length was caught near Hartlepool, Durham, and is now preserved in the Dorman Museum, Middlesborough (C Thornton, pers. comm. 17 April 1975).

Although no reliable statistics are available, it is estimated that *at least* 1000 people are killed annually by sharks, 70–80 per cent of them off the coasts of Africa, South America and Asia, where fatalities often go unrecorded. Of the 40–50 shark attacks actually reported worldwide each year, mostly in warmer waters, about 60 per cent survive the ordeal, although possibly with fewer appendages than they had originally.

The waters around Australia are particularly dangerous, and since 1898 at least 420 people have died after being attacked.

If we look at it another way, though, Man is a far greater threat to sharks than vice versa inasmuch as he kills hundreds of thousands of them annually for use as food, fertilisers, etcetera.

Probably the greatest mass attack on man by any large animal on record occured on 28 November 1942 when a German U-boat fired a salvo of torpedos into the hull of the Liverpool steamer *Nova Scotia* (6796 tonne/*ton*) some 48 m *30 miles* off the coast of Zululand, Natal, South Africa. On board were 900 men, including 765 Italian prisoners of war bound for colonial work camps.

The ship went down in seven minutes, and as the hundreds of men in the water thrashed round or clung desperately to pieces of wreckage (only one lifeboat was launched) packs of hungry sharks moved in for the kill.

When a Portuguese rescue ship eventually arrived on the scene, there were only 192 survivors, the rest having been eaten by sharks in a bloody orgy of horror (Davis, 1964).

In March 1975 at least 50 people were drowned or killed by sharks when a ferry with 190 passengers on board capsized near Sandwip Island in the Ganges-Brahmaputra Delta, Bangladesh.

If we exclude the Greenland shark, the largest carnivorous shark ever recorded in British waters was probably a 4·14 m *13 ft 7 in* Common hammerhead (*Sphyrna zygaena*) found trapped among the rocks at Ilfracombe, Devon, on 31 July 1865 (Day, 1880). The fish was not weighed, but a 4·29 m *14 ft 2½ in* specimen caught at Acapulco, Mexico, in January 1940 weighed 474 kg *1047 lb*.

The largest fish ever killed underwater by a spear fisherman was a 4·26 m *14 ft* 725 kg *1600 lb* great white shark taken in Australian waters in the 1960s by a skin-diver using an explosive head fitted to a hand spear (Baldridge, 1974). In 1955 a 4·19 m *13 ft 9 in* long 635 kg *1400 lb* basking shark was killed by Robert Lorenz off Santa Monica, California, USA.

The smallest known shark is *Squaliolus laticaudus* of the Gulf of Mexico, which does not exceed 152 mm *6 in*.

The heaviest bony or 'true' fish in the world is the Ocean sunfish (*Mola mola*), which is found in all tropical, sub-tropical and temperate waters. Adult specimens average 1·83 m *6 ft* from tip of snout to tip of tail fin, 2·43 m *8 ft* between the dorsal and anal fins (vertical length) and weigh up to 1 tonne/*ton*.

The largest Ocean sunfish ever recorded was one accidentally struck by the S.S. *Fiona* shortly after 1 pm on 18 September 1908 off Bird Island, some 64 km *40 miles* from Sydney, New South Wales, Australia. A full account of this incident appeared in the 'Wide World' magazine for 10 December 1910:

'. . . all hands were alarmed by a sudden shock, as though the steamer had struck a solid substance or wreckage. The result was strange and remarkable, for the port engine was brought up all standing. The starboard engine was quickly stopped and a boat lowered and sent to investigate. On getting under the steamer's counter the boat's crew were astonished to find that a huge sun-fish had become securely fixed in the bracket of the port propeller. One blade was completely embedded in the creature's flesh, jamming the monster firmly against the stern-

A 1587 kg 3500 lb *Ocean sun fish caught off the coast of California (American Museum of Natural History)*

post of the vessel. It was impossible to extricate the fish at sea, so the boat was hoisted on board again and the steamer proceeded on her passage to Sydney with the starboard engine only working.

'On reaching Port Jackson, the *Fiona* was anchored in Mosman Bay, where all hands were set to work to remove the fish. After much difficulty and with the aid of the steamer's winch, the sun-fish was hoisted clear and swung on board.'

The fish was later taken to a nearby wharf and put on a weighbridge where it registered 2235 kg *4928 lb*. It measured 3·04 m *10 ft* in length and 4·26 m *14 ft* between the dorsal and anal fins.

The largest ocean sunfish ever recorded in British waters was one weighing 363 kg *800 lb* which ran aground at a salmon fishing station at Montrose, Scotland, on 14 December 1960. It was sent to the Marine Research Institute in Aberdeen. Another one measuring 1·98 m *6 ft 6 in* between the anal and dorsal fins and weighing 305 kg *672 lb* was washed ashore at Kessingland near Lowestoft, Suffolk, on 19 December 1948 after a storm.

The largest bony fish in the world is the Russian sturgeon (*Acipenser huso*), also called the 'Beluga', which is found in the temperate areas of the Adriatic, Black and Caspian Seas but enters large rivers like the Volga and Danube for spawning.

The average adult female measures about 2·13 m *7 ft* in length and weighs 140–163 kg *310–360 lb.* (the average weight of 1977 adult males caught in the Don River in 1934 was 75–90 kg *165–198 lb*), but exceptional fish may measure up to 4·26 m *14 ft* in length and weigh 907 kg *2000 lb*.

In former times when it was more abundant this species attained a much greater size. Seeley (1886) says sturgeons measuring 6·31 m *24 ft* and upwards in length were quite common (*sic*) in the Danube and fishermen sometimes caught fish 'so large that they were unable to drag them from the river'. One female caught at Saratov on the west bank of the Volga in 1869 scaled 1251 kg *2760 lb*.

According to Dr Leo S Berg (1948), the Russian ichthyologist, the largest beluga on record was a gravid female taken in the estuary of the Volga in 1827 which measured 7·31 m *24 ft* in length and weighed 1474·2 kg *3249 lb*. Another gravid female caught in the Caspian Sea in 1836 scaled 1460 kg *3218 lb*, and one weighing 1451 kg *3200 lb* was taken at Sarepta (now called Krasnoarmeisk) on the Volga in 1813.

On 11 May 1922 a female weighing 1200 kg *2645 lb* was caught in the estuary of the Volga. The head weighed 288 kg *635 lb*, the body 667 kg *1470 lb* and the eggs or caviare 146·5 kg *323 lb*. Another female weighing 1228 kg *2706 lb* caught in the Tikhaya Sosna River, in the Biruch'ya Kosa region, in 1924 had 7700000 eggs weighing 245 kg *541 lb* in her ovaries (Babushkin, 1926).

The Kaluga or Daurian sturgeon (*Huso dauricus*) of the Amur River and adjacent lakes of eastern Siberia also reaches a great size. The largest one listed by Soldatov (1915) measured 4·18 m *13 ft 8½ in* and weighed 541 kg *1193 lb*, but Berg (1932) quotes weights of 820 kg *1807 lb* and 1140 kg *2513 lb* for two others.

A White sturgeon (*Acipenser transmontanus*) caught in the Columbia River at Astoria, Oregon, USA, in 1892 and exhibited at the World's Fair in Chicago the following year, was stated to have weighed more than 907 kg *2000 lb*, but Gudger (1934) says he was unable to confirm this weight. There are also two claims for a 680 kg *1500 lb* white sturgeon on record: one reportedly taken from the Weiser River, Washington, in 1898, after it had blundered into a salmon gill-net, and the other from Snake River, Oregon, in 1911 (Brown, 1962).

The official record is held by a 3·81 m *12 ft 6 in* fish taken in the Columbia River near Vancouver, Washington, in May/June 1912 which weighed 583 kg *1285 lb* and was described as the largest 'stickleback' in the world.

The largest sturgeon found in British waters is the slender Common sturgeon (*A. sturio*). The heaviest recorded specimen was a female measuring 3·18 m *10 ft 5 in* and weighing 317 kg *700 lb* netted by the trawler *Ben Urie* off the

Orkneys and landed at Aberdeen on 18 October 1956. Another female measuring 3·65 m *12 ft* and weighing 305 kg *672 lb* was caught by the trawler *King Athelstan* in the North Sea and landed at Lowestoft, Suffolk, on 4 April 1937. The largest river specimen was a 2·74 m *9 ft* male weighing 230 kg *507 lb 8 oz* accidentally netted in the River Severn at Lydney, Glos, on 1 June 1937. Its captor, James Legg, a Blakeney lave net salmon fisherman, said the fish had a severe wound on its back which was thought to have been inflicted by the propeller of a boat (S E V Jones, pers. comm. 18 October 1974).

The largest fish which spends its whole life in fresh or brackish water is the rare Pa beuk or Pla buk (*Pangasianodon gigas*), a giant catfish found in the deep waters of the Mekong River of Laos and Thailand. According to Seidenfaden (1923) this fish attains a length of up to 3 m *9 ft 10 in* and a weight of 240 kg *529 lb*, and he says he saw one personally which measured 2·5 m *8 ft 2¼ in* in length, 1·70 m *5 ft 7 in* in maximum bodily girth and weighed 180 kg *397 lb*.

In June 1974 Mr A E Davidson, the British Ambassador in Vientiane, Laos, and a keen amateur ichthyologist, served up fillet of Pa beuk ('Pa beuk de ban Sai') at a banquet given in honour of the Laotian Prime Minister, Prince Souvanna Phouma, but there were few takers. The 49 kg *108 lb* head of this 170 kg *375 lb* fish was later presented to the British Museum (Natural History).

Mr Davidson (pers. comm. 18 July 1974) told the compiler that the season for the Pa beuk fishery at Ban Houei Sai was usually in April and May, and that between 20 and 30 fish were caught in a normal year. In 1974, however, only 14 examples were caught in a very short season lasting just four weeks. Of this total only two were females, and they weighed 165 kg *364 lb* and 160 kg *353 lb* respectively. The twelve males ranged from 135 kg *297 lb* to 200 kg *441 lb*. In this series of fish the lengths were not recorded, but the measurements probably ranged between 2·28 m *7 ft 6 in* and 2·59 m *8 ft 6 in*.

The Common sturgeon – the largest fish found in British rivers (A L Allen and Edwin Lewis)

In earlier times, before it was over-fished, the European catfish or Wels (*Silurus glanis*) was considered the largest freshwater fish in the world, but today this species is no longer in contention.

According to Kessler (1856) the catfish found in the Dnieper River reached a length of 2·43–4·26 m *8–14 ft* and weighed up to 270 kg *600 lb*, and he says the largest one on record measured 4·57 m *15 ft* and tipped the scales at 326 kg *720 lb*.

Seeley (1886) writes: 'It is often from six to nine or ten ft long, and occasionally reaches a length of thirteen ft. In the Danube it often attains a weight of 400 or 500 lb, and in South Russia may exceed 600 lb. With age it increases chiefly in circumference, and sometimes is as much as two men can span [*sic*].'

The largest accurately measured wels of which there is reliable evidence was a 3 m *9 ft 10 in* female caught in the Danube in Rumania which is now preserved in the Museum d'Histoire Naturelle, Paris (Pellegrin, 1931). Another exceptionally large example from Rumania measured 2·85 m *9 ft 4¼ in* in length and weighed 170 kg *375 lb*.

The heaviest Russian example listed by Berg (1949) was taken in the Desna River 9·6 km *6 miles* from Chernigou in the Ukraine in September 1918. It weighed 256·7 kg *565 lb*. Another one caught in the Dnieper River near Kremenchug allegedly weighed 300 kg *661 lb*, but further details are lacking. Berg says catfish weighing 200 kg *441 lb* are not a rarity in the Syr-Darya and Chu Rivers, but this statement would not stand up today.

This species, incidentally, has been dubbed a man-eater by some of the earlier writers. Grossinger (1793), for instance, has been quoted as saying that two young girls in Hungary who had gone to fetch water from a river were devoured by great catfish, and the same naturalist also mentions another wels caught by a fisherman near the Turkish frontier with Europe which contained the body of a young woman. Gudger (1945), on the other hand, although admitting that a large wels has a mouth and gullet big enough to swallow a small child, believes that most human parts recovered from giant catfish have been taken from rotting corpses, particularly of children 'which in some way – by drowning or some other accident – have gotten into rivers'.

There are also several huge catfishes in South America. The longest – although not the heaviest – is the Lau-lau (*Brachyplatystoma filamentosum*), found in the rivers of Guayana as well as the Amazon, which has been credited with a maximum length of 3·65 m *12 ft* (Hillhouse, 1825), but 2·43 m *8 ft* is probably a more realistic figure. Dr William Beebe and John Tee-Van told Gudger (1943) that the two largest specimens taken by them at Kartabo at the junction of the Cuyuni and Mazaruni Rivers in Guayana in 1927 measured 1·88 m *6 ft 2 in* and 1·92 m *6 ft 3½ in* in length. (These measurements did not include the tail fin which would account for about another 304 mm *12 in*.) Eleven years earlier Beebe had measured another lau-lau caught at Kartabo which taped 2·10 m *6 ft 11 in* (about 2·43 m *8 ft* with the tail fin).

The Pirahyba (*Piratinga piraiba*) of the Amazon River, which is closely related to the lau-lau, has been called the goliath of catfishes, but very little information has been published on its size and weight. Ex-President Theodore Roosevelt (1914) was told by the doctor attached to his Brazilian expedition that while he was working at Itacoatira, a small town at the mouth of the Madeira River, he had seen a pirahyba measuring 3 m *9 ft 10 in*. It had been killed by two men with machetes after it had attacked their canoe.

Another specimen collected by Dr J D Haseman (cited by Gudger, 1943) in the Guapore tributary of the Mamore River in 1909–10 for the Carnegie Museum, Pennsylvania, USA, was so heavy that it required the efforts of four sturdy men to pull it up on to a sandbar. This fish measured 1·85 m *6 ft 1 in* and weighed an

North America's huge Alligator gar (Dwight Franklin)

estimated 159 kg *350 lb*. Dr Haseman put the maximum length and weight of this species at 2·13 m *7 ft* and 181 kg *400 lb*.

Africa's only giant freshwater fish is the Nile perch (*Lates niloticus*). Large specimens normally measure 1·21–1·37 m *4–4½ ft*, but Boulenger (1907) says one measuring 1·85 m *6 ft 1 in* in length, 1·39 m *4 ft 7 in* in maximum girth and weighing 121 kg *266¾ lb* was taken a few miles up the Sobat River, a southern tributary of the Nile. Another huge example caught in Lake No at the junction of the Bahr al Ghazal with the Bahr el Jebel or Nile proper weighed 115 kg *255 lb*, and Gudger (1944) says another one from the same lake reportedly weighed 127 kg *280 lb*. There is also a record of a 246-pounder *111 kg* being taken in Lake Rudolf, northern Kenya, and one said to weigh 163 kg *360 lb* was caught in a seine net on Lake Albert (Gordon, 1969).

Some writers claim that the largest freshwater fish in the world is the pike-like Arapaima (*Arapaima gigas*), also called the 'Pirarucu', which is found chiefly in the Amazon drainage of Brazil and Peru, but the size attained by this fish has been much exaggerated in the literature. When Schomburgk (1841) visited Brazil in 1836 the natives of the Rio Negro told him that they had caught pirarucu measuring 4·57 m *15 ft* in length and weighing 181 kg *400 lb*, but the naturalist-explorer did not see any fish remotely approaching that size. The two longest specimens collected by him measured 2·46 m *8 ft 1½ in* and 2·13 m *7 ft* respectively.

Paul Fountain (1914) claims he killed an arapaima on the Rio Negro which scaled 285 kg *628 lb* piecemeal, and weights up to 453 kg *1000 lb* have been reported elsewhere (Norwood, 1964), but these poundages are much too extreme to be acceptable.

Dr Haseman told Gudger (1943) that he measured an 2·43 m *8 ft* arapaima at Meura on the Rio Negro which scaled 120 kg *264 lb*, and heard from the local fishermen that they had caught a 10-footer *3·04 m* weighing 200 kg *441 lb* in the same spot a few days earlier.

'Haseman did not see the fish', says Gudger, 'but he saw the head. This was so much larger than the head of the one he measured that he thought the report of length and weight "reasonably accurate".'

Three individuals caught by Edward McTurk, a rancher at Karanambu, Guayana, on 13 May 1947 were 2·12 m *6 ft 11¾ in* and 92 kg *203 lb*, 1·93 m *6 ft 8 in* and 67 kg *148 lb* and 1·81 m *5 ft 11 in* and 50 kg *110 lb* respectively. McTurk, who has probably seen more arapaima than any other white man, told McCormick that he once killed a fish measuring 2·74 m *9 ft*, but it was not weighed.

The largest freshwater fish found in North America is the Alligator gar (*Lepisosteus spatula*) of the Mississippi River and its tributaries. The largest specimen listed by Gudger (1942) was a female taken in Belle Island Lake, Vermilion Parish, Louisiana, in *c.* 1925 which measured 2·95 m *9 ft 8½ in* in length and weighed 137 kg *302 lb*. Another one caught by Dr Henry Thibault in the Arkansas River near Little Rock measured 2·79 m *9 ft 2 in* and scaled 105 kg *232 lb*, and there is also a record of an alligator gar taken in Moon Lake, Mississippi, which measured 3·05 m *10 ft* and scaled 104 kg *230 lb*. The rod and reel record is 126 kg *279 lb* for an exceptionally bulky individual 2·35 m *7 ft 9 in* long caught by Bill Valverde in the Rio Grande River, Texas, on 2 December 1951 (Field & Stream, February 1975).

The largest freshwater fish found in Britain is the Pike (*Esox lucius*), and there are several records of fish exceeding 13·5 kg *30 lb*. The British rod and reel record is 19·5 kg *43 lb* for a monster caught by Roy R Whitehall in Lockwood Reservoir, Walthamstow, Essex, in 1975 and there is also an authentic record of a 42-pounder *10 kg* caught in the River Barrow, County Offaly, Ireland, on 22 March 1964. In 1945 a pike weighing 19·5 kg *43 lb 11 oz* was caught in Loch Lomond, Scotland, and a 53-pounder *24 kg* was landed on a spoon from Lough Conn, Co Mayo, Eire, in 1923. Another pike allegedly weighing 23·5 kg *52 lb* was recovered when Whittlesea Mere, Cambridge, was drained in 1851, and one reputedly weighing 32·5 kg *72 lb* was landed from Loch Ken, Dumfries and Galloway, Scotland, in 1777. There is also an unconfirmed record of a pike weighing in excess of 41·75 kg *92 lb* taken from the River Shannon at Portumna, Co Galway, in *c.* 1796.

Although the Salmon (*Salmo salar*) is classified as a freshwater fish by the National Anglers' Council this is not strictly true, because it feeds in the sea. It has been claimed that this species was originally a freshwater fish that later became semi-marine, but most ichthyologists believe that it started off life as a marine fish.

The shortest known fish, and the shortest of all vertebrates, is the Dwarf pygmy goby (*Pandaka pygmaea*), a colourless and nearly transparent creature found in the streams and lakes of Luzon in the Philippines. Adult males measure 7·5–9·9 mm *0·29–0·38 in* in length and weigh 4–5 mg *0·0014–0·0017 oz*, and adult females 9–11 mm *0·35–0·43 in* and 5–6 mg *0·0017–0·0021 oz*. This fish is so tiny that it has to be studied with a microscope rather than a strong magnifying glass.

The Sinarapan (*Mistichthys luzonensis*), another goby found in Lake Buhi, in southern Luzon, is almost as diminutive. Adult males measure 10–13 mm *0·39–0·51 in* in length and adult females 12–14 mm *0·47–0·55 in* (Smith, 1902).

Surprising as it may seem, the latter fish is much in demand as *food* and has considerable commercial importance. The natives of the region catch them with large close-web nets, pack them tightly into woven baskets until the water drains out leaving a compact mass, and then sell them in dried cake form. (A 0·45 kg *1 lb* cake contains about 70000 fish!)

The shortest recorded marine fishes are the Marshall Islands goby (*Eviota zonura*) measuring 12–16 mm *0·47–0·62 in* and *Schindleria praematurus* from Samoa, measuring 12–19mm *0·47–0·74 in*. Mature specimens of the latter fish, which was not described until 1940, have been known to weigh only 2 mg *0·0007 oz*, equivalent to 14,175 to the ounce – the lightest of all vertebrates and the smallest catch possible for any fisherman.

The smallest fish found in British waters is the goby *Lebutus orca* which grows to a length of 39 mm *1·53 in*. It has been recorded from the waters off SW Cornwall, the west coast of Ireland and the Irish Sea (Wheeler, 1969).

The longest-lived species of fish is difficult to determine because (1) aquaria are of too recent origin and (2) interpretations of age are not uniform. Early indications are that it is the Lake sturgeon (*Acipenser fulvescens*) of North America. Between 1951 and 1954 the ages were assessed of 966 specimens caught in the Lake Winnebago region, Wisconsin, USA, by examination of the growth rings (annuli) in the marginal ray of the pectoral fin. The oldest sturgeon was found to be a male measuring 2·01 m *6 ft 7 in* which gave a reading of 82 years, and was still actively growing. The next oldest was a 6-footer *1·83 m* aged 49 (Probst & Cooper, 1954). Another lake sturgeon measuring 2·05 m *6 ft 9 in* and weighing 97 kg *215 lb* caught in the Lake of the Woods, Ontario, Canada, on 15 July 1953 was believed to be 152 years old based on a growth ring count (Anonymous, 1954), but in the light of published information this figure must be considered excessive. According to Magnan (1966) females live longer than males, and he gives 55 years as the usual maximum age for males and 80 years for females in Quebec, the maximum age being reached in the more northern, slower-growing populations.

A 2·30 m *7 ft 7 in* White sturgeon (*A. transmontanus*) caught in the lower part of the Fraser River, British Columbia, Canada, in the summer of 1962 was believed to be 71 years old based on an annual count (Semakula & Larkin, 1968), and another specimen 3·50 m *11½ ft* long taken from the same river was estimated to be 82 years old.

According to Petrov (1927) a 4·2 m *13 ft 11 in* Russian sturgeon caught on 3 May 1926 in the estuary of the Ural River and weighing over 1000 kg *2200 lb* was about 75 years old, and Solatov (1935) quotes a figure of 55 years for a 656 kg *1446 lb* kaluga.

A Sterlet (*A. ruthenus*) lived in the Royal Zoological Society Aquarium in Amsterdam for 69 years 8 months (Nigrelli, 1959). It was received in 1883 when 380 mm *15 in* long.

Dr Alex Comfort (1964) puts the upper age limit for the largest specimens of sturgeons at about 120 years.

Stories of Methuselah fish are legendary. The most famous of them all is probably the one about an enormous pike which was caught in the Kaiserwag Lake in Wurttemburg, Germany, in 1497. This fish reportedly had a copper ring round its 'neck' with an inscription saying it had been put in the lake in 1230, which made it 267 years old. Its length was given as 5·18 m *17 ft* and its weight as 159 kg 350 lb (*sic*). The *alleged* skeleton of this monstrous fish was later preserved in Mannheim Cathedral. In the 19th century a celebrated German anatomist examined the bones and discovered that they came from several different fish!

Exaggerated claims apart, though, the pike is still one of the longer-lived fishes. Francis Bacon (1645) says it sometimes lives to the 40th year, and this view is also shared by Dr Tate Regan (1911) who remarks that 'it is probable that fish of 60 or 70 lb are at least as many years old'.

Comfort (1964), on the other hand, is much more cautious and refers to a record of a 15 kg *34 lb* pike caught in 1961 which had a scale reading of only 13–14 years, whilst Frost & Kipling (1959) give the maximum scale reading for 5000 pike taken in gill-nets from Lake Windermere from 1944 to 1957 as 'slightly in excess of 15

The Lake sturgeon – the longest lived species of fish (Associated Press)

years'. Although this is a big reduction it is still a very respectable age for a fish.

Recently some incredible ages have also been claimed for the Koi fish of Japan, a form of fancy carp. In 1974, for example, a 7·3 kg *16 lb* specimen named 'Hanako' living in a pool in Higashi Shirakawa Village, Gifu Prefecture, allegedly celebrated his 223rd birthday (confirmed by a scale reading) after having been cared for by seven generations of one family. His owner is Dr Kimiski Koshihara, President of Nagoya Women's College, whose ancestors served as the village headmen for three centuries from the beginning of the Tokygawa Shogunate until the feudal clan system was abolished by the Meiji Restoration. There is also another record of a koi fish living for 250 years. Extreme ages like these, however, are so much in excess of maximums quoted for other species of fishes that the compiler suspects that two or even three contour lines are laid down annually in this fish, as opposed to a single growth ring in the sturgeons.

According to Regan (1911) the carp lives up to 50 years under artificial conditions, but only 12–15 years in the wild state.

'Clarissa', London Zoo's famous 20 kg *44 lb* Common carp (*Cyprinus carpio*), who died in May 1971 after spending 19 years in the aquarium there, was believed to have been about 30 years old when she was received in 1952, which means she must have been *c.* 49 years at the time of her death.

After the sturgeons the European catfish (*Silurus glanis*), the European freshwater eel (*Anquilla anguilla*) and the Halibut (*Hippoglossus hippoglossus*) are probably the longest-lived fishes.

Flower says two European catfish placed in a pond at Woburn, Bedfordshire, in 1874 were still alive on 16 January 1935 aged 60+ years, and he also cites records of 40, 42 and 55 years for the European freshwater eel. In 1957 a female halibut measuring 3·05 m *10 ft* in length and weighing 228 kg *504 lb* was caught in the North Sea and landed at Grimsby. Its age was given as 60+ years on a scale reading. Another one caught off Iceland in October 1963 weighed 267 kg *588 lb*.

The life-span of most sharks is less than 25 years. One exception is the Australian school shark (*Galeorhinus australis*), females of which take 12 years to mature and probably live 30 years. The virtually indestructible Whale shark (*Rhineodon typus*) may live 70 years or more.

A 267 kg 588 lb *Halibut caught off Iceland in 1963 which was believed to be 34 years old*

The highly streamlined Sailfish, the fastest fish in the sea

The exhibition life of a Goldish (*Carassius aurktus)* is normally about 17 years (Comfort, 1964), but much greater ages have been reliably reported. Bateman (1890) cites a record of 29 years 10 months 21 days for an example kept in an aquarium at Woolwich, London, from 20 May 1853 until 11 April 1883, and Mennel (1926) gives details of several others which lived 25 years. In May 1971 a 28-year-old goldfish was reported from Coventry. There is also an unconfirmed claim of a goldfish living in a water-butt for 40 years (Comfort, 1956). In China common goldfish have been known to live for over 40 years.

The shortest-lived fishes are probably certain killifish of the sub-order Cyprinodontei which live about eight months in the wild state. These include *Nothobranchius guentheri*, *N. rachovii* and *N. melanospiius* of Africa, and *Cynolebias bellottii*, *Poterolebias longipinnis* and *Austrofunduls dolichopterus* of South America.

They are found in temporary ponds, drainage ditches and even in the water-filled footprints of large animals. The eggs are laid in the mud at the bottom of the water. When this dries up, the fish die and the eggs, protected from drying out completely by morning dew and the moisture retaining properties of the sub-stratum, aestivate (fall into a state of torpor) until the next wet season, when they hatch, grow at great speed and spawn, dying in the next drought.

Because of the practical difficulties of measurement very little accurate datum has been published on the speeds attained by fishes. A good insight into rate of performance is provided by the shape of the tail and body. Species with deeply forked or crescent-shaped tails and cigar-shaped bodies thickest in the middle are capable of high speeds; slow swimmers, on the other hand, usually have square or round tails and short, laterally compressed bodies.

Most ichthyologists share the view that the fastest fish in the world over a short distance is the highly streamlined Sailfish (*Istiophorus platypterus*), which is found in all tropical waters. The maximum velocity reached by this species is not known for certain, but Hamilton M Wright (cited by Hunt, 1935) says that in a series of speed trials carried out with a stopwatch at the Long Key Fishing Camp, Florida, USA, between 1910 and 1925 one sailfish took out 91 m *100yd* of line in three seconds, which is equivalent to a speed of 109 km/h *68·18 miles/h*. (cf. 96 km/h *60 miles/h* for the cheetah).

'The speed of the sailfish', writes Wright, 'is sometimes such that I have known a man on his first fishing trip to think that there were two fish when only one was on the line, because the fish reappeared on the surface so quickly in another quarter.'

When it is travelling at high speed the long dorsal fin of the sailfish folds back into a slot in the back, and the pectoral and ventral fins are pressed flush against the body to cut down drag to a bare minimum. The same streamlining technique is also used by other fish of the family Istiophoridae.

The Swordfish (*Xiphias gladius*) has also been credited with amazing bursts of speed, but the evidence is based mainly on bills that have been found deeply embedded in ship's timbers. A speed of 50 knots (92 km/h *57·6 miles/h* has been calculated from a penetration of 558 mm *22 in* by a bill into a piece of timber, but this figure has been questioned by some authorities. According to Sir James Gray (1968) a 272 kg *600 lb* swordfish travelling at 16 km/h *10 miles/h* would hit a wooden vessel travelling at the same speed in the opposite direction with a force of about half a tonne/*ton*, all of it concentrated into the 25 mm *1 in* tip of the bill. This, he says, would be equivalent to a 0·45 kg *1 lb* projectile hitting the vessel at a speed of 48 km/h *30 miles/h*.

Most experts put the maximum speed of the swordfish at between 30 and 35 knots (56–64 km/h *35–40 miles/h*). Anything higher, they say, would be impossible because the drag from this fish's non-depressable dorsal fin and long bill would be too great.

Some American fishermen believe that the Bluefin tuna (*Tunnus thynnus*) is the fastest fish in the sea, and speeds up to 104 km/h *65 miles/h* have been claimed. Certainly it is beautifully streamlined for speed, and when it is travelling flat out its first dorsal, pectoral and pelvic fins are withdrawn into grooves in its body to facilitate swimming. (The second dorsal and anal fins remain fixed because they have to act as stabilisers.)

The question of how fast a bluefin tuna can swim was answered by H Earl Thompson off Liverpool, Nova Scotia, Canada, in 1938. Using a device he and a friend had invented called a 'Fish-O-Meter', which consisted of a motorcycle speedometer, a flexible cable and V-pulley mounted on a rod and reel, Thompson hooked a young bluefin weighing 27 kg *59·5 lb* which registered a speed of 69·6 km/h *43·4 miles/h* in a 20 second dash. He said that specimens weighing 227–272 kg *500–600 lb* were the fastest swimmers (Patterson, 1939).

*The Bluefin tuna of the North Atlantic, one of the marine world's top sprinters (*Sports Illustrated*)*

Some sharks are also very swift over short distances. but the statement by Perry (1972) that they can overtake warships travelling at 40 knots (73·6 km/h *46 miles/h*) must be considered excessive. Pelagic sharks normally cruise at 3–5 knots but this speed can be increased four-fold in an emergency. One of the fastest-swimming species is the Mako shark (*Isurus oxyrinchus*). One 12-footer *3·65 m* chased by Thomas Helm (1961) and a colleague in the Florida Keys kept ahead of their speed boat travelling at 27 knots (49·6 km/h *31 miles/h*) for 805 m *880 yd*.

The Great blue shark (*Prionace glauca*) can also reach a very high velocity. In one experiment carried out by Magnan and Sainte-Legue (1928) in which the fish was tethered by an extremely fine thread to a tachometer, a great blue shark measuring 1·98 m *6 ft 6 in* in length and weighing 32 kg *70 lb 8 oz* registered a speed of 21·3 knots (39·2 km/h *24·5 miles/h*). In another experiment using water current a very young great blue shark measuring 609 mm *2 ft* in length and weighing only 0·58 kg *1·3 lb*, held its own against a current of 26 ft/s (39·2 km/h *24·5 miles/h*), and in short bursts of speed reached an astonishing 68·8 km/h *43 miles/h* (Budker, 1971).

The great white shark is also extremely fast. One day Captain Jacques Cousteau and a member of his team were diving off the Cape Verde Islands when a 12·19 m *25 ft* carcharodon appeared out of the gloom some 12·19 m *40 ft* in front of them. Suddenly it saw the two divers. 'His reaction', writes Cousteau (1953), 'was the least conceivable one. In pure fright, the monster voided a cloud of excrement and departed at an incredible speed.'

The Wahoo (*Acanthocybium solandri*) and the Bonefish (*Albula vulpes*) of the tropical Atlantic are also magnificent sprinters. Walters & Fiersteine (1964) timed a wahoo running out a light line in open water at 77·6 km/h *48·5 miles/h*, and Lane (1955) says Zane Grey once hooked a bonefish which dashed off at a calculated speed of 64 km/h *40 miles/h*.

Flying-fish (family Exocoetidae) of tropical waters break the surface of the water at speeds of between 24 and 32 km/h *15* and *20 miles/h* when fleeing from underwater pursuers like the Bonito (*Katsuwonus pelamis*) and the Dolphin-fish or Dorado (*Coryphaena hippurus*). They then 'taxi' along the surface briefly with the lower part of their tail beating in the water up to 50 times a second, and then accelerate to something like 64 km/h *40 miles/h* before rising into the air.

Leonard P Schultz (1948) timed hundreds of these aerial ventures with a stopwatch and found that the majority of them lasted from two to ten seconds. He clocked one wind-assisted flight of 10 seconds and said the longest on record was one of 42 seconds, but this latter time has since been beaten.

On 29 November 1972, while serving aboard the USS *Davis* Lt Stephen J Kuppe (pers. comm. 14 December 1972) watched a flying fish about 254 mm *10 in* long with a 355 mm *14 in* wing-span remain airborne 'for no less than 90 seconds while maintaining the same relative position alongside my ship at a distance of about 30 ft'.

At that particular time the ship was travelling through the Mozambique Channel, East Africa at a speed of 24 knots (44·22 km/h *27·64 miles/h*), which means the flying-fish must have travelled a total airborne distance of 1109 m *1214 yd*.

The swiftest domestic fish is probably the Pike (*Essox lucius*) – but only for very short distances. According to Hertel (1966) this fish can cover 7·5 to 12·5 times its own length in one second when lunging at prey, and Bainbridge (1958) has calculated that fish measuring up to 1 m *39·37 in* should be able to swim ten times their own body length per second. On this basis, a 914 mm *3 ft* pike could cover 6·85–11·43 m *22·5–37·5 ft*, which is equivalent to 24·5–40·8 km/h *15·3–25·5 miles/h*.

The fastest fishes in the world over a sustained distance are the Marlins (genus *Tetrapturus*). Norman & Fraser (1948) state that they can swim at a very rapid rate for several hours, and Bandini (1933) writes: 'One minute there is nothing, the next they are all around you, and a minute afterwards the sea is empty once more.'

The swimming endurance record is held by an 3·50 m *11 ft 6 in* long female Grey nurse shark (*Odontaspis arenarius*) named 'Skipper II' which, during the four-year period 1932–6 clocked up an estimated 168 000 km *105 000 miles* of non-stop swimming in the aquarium at Taronga Park Zoo, Sydney, New South Wales, Australia. She was eventually attacked and blinded by a 150 mm *6 in* Porcupine fish (family Diodontidae) living in the same tank and died three weeks later.

The slowest-moving marine fishes are the Sea horses (family Syngnathidae). Some of the very small species measuring less than 25 mm *1 in* probably never get above 0·016 km/h *0·01 mile/h* – even in an emergency!

The most restricted fish in the world is the diminutive Devil's Hole pupfish (*Cyparinodon diabolis*) which is confined to a small area of water directly above a rock shelf in a spring-fed pool in Ash Meadows, Nevada, USA. Because the pool is located 15·24 m *50 ft* below ground, the population of this species is completely dependent on the amount of sunshine received.

'During the summer', writes Leontine Nappe (1974), 'when the sun shines on the ledge for several hours a day, the population rises to 700 or more. But during the winter when no sunlight can enter the spring the algae, and the small invertebrates upon which the pupfish depend, dwindle; the pupfish numbers then drop to about 200.'

The greatest depth from which a fish has been recovered is 8299 m *27 230 ft* for a 165 mm *6½ in* brotulid, *Bassogigas profundissimus* (only the fifth ever recorded) sledge-trawled by the American research vessel *John Elliott Pillsbury* in the Puerto Rico Trench in April 1970 during a National Geographic Society/University of Miama Deep-Sea Programme. The previous record had been held by another specimen of *B. profundissimus* sledge-trawled by the Royal Danish research vessel *Galathea* in the Sunda Trench, south of Java, at a depth of 7120 m *23 392 ft* in *c.* September 1951 (Brunn, 1956). In 1910 the Norwegian Michael Stars Expedition trawled several Rat-tailed fish (family Macrouridae) from a depth of 4700 m *15 420 ft* in the North Atlantic (Marshall, 1954). There is also a record of a Deep-sea eel (family Synaphobranchus) in a state of metamorphosis being taken at a depth of 4040 m *13 254 ft* in the Indian Ocean.

On 24 January 1960 Dr Jacques Piccard and Lt Don Walsh, US Navy, sighted a sole-like fish about 304 mm *12 in* long (tentatively identified as *Chascanopsetta lugubris*) from the bathyscaphe *Trieste* at a record depth of 10911 m *35 802 ft* in the Challenger Deep (Mariana Trench) in the western Pacific. 'Slowly, very slowly,' relates Piccard (1960), 'this fish . . . moved away from us, swimming half in the bottom ooze, and disappeared into the black night, the eternal night which is its domain.'

At this depth the fish would have been subjected to a pressure of 7257 kg *16 000 lb* to the square inch.

This sighting has been questioned by some ichthyologists, who still regard the brotulids of the genus *Bassogigas* as the deepest-living vertebrates. The two men had earlier sighted and photographed a *Bathypterois*, which is related to lantern-fishes, and a *Haloporphyrus* at 7010 m *23000 ft*.

Some fishes lay an enormous number of eggs, fecundity increasing with weight. A Carp (*Cyprinus* sp.) had 2000000 in its ovaries; a Halibut (*Hippoglossus hippoglossus*) some 2750000; a Cod (*Gadus morhua*) 6652000; a Turbot (*Scophthalmus maximus*) 9000000; a Conger eel (*Conger conger*) 15000000; and a Common ling (*Molva molva*) 28361000. The most fertile fish of them all, however, is the Ocean sunfish (*Mola mola*). According to Johannes Schmidt, the famous Danish marine zoologist, the ovaries of one female contained 300000000 eggs, each of them measuring about 1·27 mm *0·05 in* in diameter. It has also been established that there is a greater size difference between a newly born ocean sunfish (length 2·54 mm *0·1 in*) and an adult than between any other living animal. Gudger (1936) says 'the larval sunfish is to its mother as a 150 lb rowboat is to 60 Queen Marys.'

The bony fish which produces the least number of eggs is probably the Tooth carp (*Jordanella floridae*) of Florida, USA. Mature females deposit about 20 eggs over a period of several days (Innes, 1945). The average yield of the Guppy (*Poecilia reticulatus*) is 40–50, but one female measuring 30 mm *1·2 in* in length had only four in her ovaries, while another measuring 51 mm *2 in* had 100.

The largest egg produced by any living animal is that of the Whale shark (*Rhineodon typus*). On 29 June 1953 Captain Odell Freeze of the shrimp trawler *Doris* was fishing 208 km *130 miles* south of Port Isabel, Texas, USA, at a depth of 31 fathoms (56 m *186 ft*) when he noticed a very large egg case in one haul of the net. 'I saw this thing in the net and, on picking it up, felt something kicking around in it. I opened it with a knife, out flopped this little shark, very much alive!'

The egg case measured 304 × 139 × 88 mm *12 × 5·5 × 3·5 in*, and the embryo was 350 mm *13·78 in* long (Baughman, 1955).

The most venomous fish in the world are the hideous-looking Stonefishes (family Synanceidae) of the tropical waters of the Indo-Pacific region. These demons (up to 381 mm *15 in* and 1·13 kg *2 lb 8 oz*) administer their neurotoxic poison, which disrupts both the circulatory and the nervous systems, through the 13 spines of their fins, and direct contact causes excrutiating pain, followed by delirium and often death.

In January 1950 a stonefish victim at Bundaberg, Queensland, Australia, was saved by inhaling trilene (trichlorethylene) which served to numb the body and counteract the shock.

The most venomous fish found in British waters is the widely distributed Lesser weever (*Trachinus vipera*), which is generally found in sandy, shallow bays. This fish half-buries itself in the sand with its poisonous dorsal fin projecting waiting to snap at passing shrimps and is easily trodden on by unwary bathers. The resultant wound is so agonisingly painful that fishermen at sea have been known to try to throw themselves overboard, and there is at least one case of a man cutting off his own finger to obtain relief! People with heart conditions have been known to die within minutes of being stung.

This venom potency may be matched by that of the Sting ray (*Dasyatis pastinaca*), which is commonly found throughout British waters. In addition, the sting or dagger of this fish can cause severe lacerations which may result in permanent damage to the affected limb. Fatalities, however, are extremely rare.

The most poisonous fish in the world is the Japanese puffer fish (*Arothron tetraodon*), known also as the 'Deadly death puffer'. The internal organs, skin, muscles, bones and even blood of this species contain a nerve biotoxin which is 200000 times more potent than curare, the deadly plant poison used by the natives of South America for their arrow tips, and human victims usually die within two hours.

Extraordinary as it may seem, the flesh of this fish and other puffers of the same family is considered a great delicacy in Japan – connoisseurs claim it produces a tremendously exhilarating feeling – and in the specially licensed restaurants where this 'gourmet's Russian roulette' is served, highly qualified 'fugu' chefs with a three-year apprenticeship behind them (good eyesight is a must) remove the poisonous parts without contaminating the rest of the fish.

Despite these elaborate precautions, however, fugu still accounts for a small number of restaurant victims annually. In 1963 Japanese statistics revealed that of 168 persons taken ill after eating this fish, 82 died, most of them uneducated villagers who served up the fish for dinner and wiped out half the family, and Caras (1964) says there is a proverb in Japan: 'Great is the temptation to eat fugu, but greater is the dread of losing life.'

Dr Bruce Halstead (1956) says there is no known antidote (induced vomiting at a very early stage might help), but the Japanese claim that burying a victim up to their neck in earth acts as a cure.

In September 1975 a Japanese restaurateur in Bayswater, London, announced that he was planning to serve fugu at £6 a portion in his eating house, but shortly afterwards Health Officials moved in and seized 25 kg *55 lb* of the deadly delicacy worth £3000 on the grounds that it was unfit for human consumption.

The most powerful electric fish in the world is the Electric eel (*Electrophorus electricus*) which is found in the rivers of Brazil, Colombia, Venezeula and Peru. An average-sized specimen can discharge 400 V at 1 A, but a measurement of 650 V has been registered for a 90-pounder *41 kg* in the New York Aquarium (Bronx Zoo), (Coates, 1937). This latter discharge would be sufficient to kill a man on contact or stun a horse at a distance of 6·09 m *20 ft*.

In 1941 two Indians in the state of Amazonas, Brazil, died when they accidentally fell into a pool containing a number of electric eels. They were killed instantly (Caras, 1964).

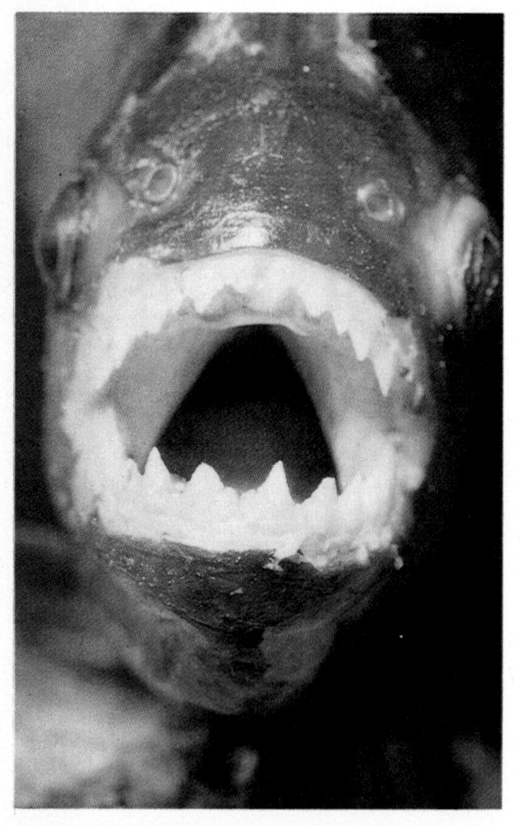

The terrifying jaws of the Piranha, the world's most ferocious freshwater fish (Camera Press).

The Electric catfish (*Malapterurus electricus*) of the rivers and lakes of tropical Africa can also produce a powerful discharge, and measurements up to 350 V at 1 A have been recorded.

The most powerful electric marine fishes are the Torpedo rays (genus *Torpedo*) which are found in all warm and temperate waters. The Black torpedo ray (*T. nobiliana*) of the Mediterranean and adjacent parts of the eastern Atlantic, including the English Channel, normally produces 50–60 V at 1 A, but discharges up to 220 V have been measured. Marine electric fishes do not need very high voltages to stun or kill their prey because salt water is a better conductor of electricity than fresh water.

Some years ago an enterprising longshore fisherman at Brighton, Sussex, made a tidy sum of money out of a large black torpedo ray he found caught in his net. He exhibited the fish on the seaside promenade as 'the heaviest fish on earth' and invited the public to guess its weight for a small fee. Needless to say no one held the fish long enough to gain any idea of its avoirdupois – or win the prize offered in the competition to anyone who could lift the ray above his head. An excellent day's business was done until the fish, showing signs of battery exhaustion, was wisely withdrawn from public exhibition for a period of rest and feeding.

The most ferocious marine fish is the Bluefish (*Pomatomus saltatrix*), which is found in most of the warmer parts of the world. This fast-moving species, which travels in large schools, has been described by Prof Spencer F Baird (1871) of the US Commission for Fish and Fisheries as 'an animated chopping machine whose sole business in life is to cut to pieces and destroy as many other fish as possible in a given space of time'.

It will attack mackerel, weakfish and herring with unbelievable ferocity, leaving behind a trail of oil, blood and pieces of its victims, and often destroys ten times as many fish as it can eat. In fact, its gluttony is so great that when its stomach is full to bursting point (up to 40 fish have been removed from captured specimens) it often disgorges the contents and starts the slaughter all over again just for the sheer joy of it. It has been calculated that an average-sized bluefish weighing 2·26 kg *5 lb* will eat nearly a tonne/*ton* of fish a year.

This raises a very interesting question: would a school of hungry bluefish skeletonise a human swimmer unfortunate enough to fall overboard in their midst? The answer is probably yes if the person was not taken out of the water very quickly.

On 9 April 1976 a large school of bluefish pursuing mullet went on a feeding frenzy along a section of the Florida Gold Coast, USA, injuring at least a dozen swimmers and forcing the closure of 24 km *15 miles* of beach.

The most ferocious freshwater fish in the world are the ever-hungry Piranhas of the genera *Serrasalmus*, *Pygocentrus* and *Pygopristis*, which live in the *sluggish* waters of the large rivers of South America. These utterly fearless cannibals have razor-sharp teeth, and their jaw muscles are so powerful that they can bite off a man's finger or toe like a carrot. They usually move about in large shoals numbering anything up to 1000 strong and will attack with lightning speed any creature, regardless of size, if it is injured or making a commotion in the water. There is a record of a 45 kg *100 lb* capybara being reduced to a skeleton in less than a minute, and a wounded alligator has been stripped of all flesh in under five minutes.

Fortunately only four out of the 16 species of piranhas are considered dangerous to man (*Serrasalmus nattereri* has the most fearsome reputation), and although many fatalities have been attributed to these 'river sharks', most of the experiences have been largely based on hearsay.

When Nicholas Guppy (1963) visited Guayana he did not hear of a single death, but when he went to Apoera on the Courantye River he found most of the adult population had lost fingers, toes or penny-sized chunks out of their arms or legs after bathing or washing clothes. One boy at nearby Orealla had most of his foot bitten off and spent months in hospital.

Harold Schultz (1963) of the Museu Paulista, São Paulo, Brazil, who spent 20 years travelling in the interior, said of all the many thousands of people he met during that time only *seven* had been injured by piranhas, and these were only slight bites. He, himself, however, admitted that he once nearly lost a toe to one of these freshwater devils.

The Amazonian dolphin or Boutu (*Inia geoffrensis*) and the Giant Brazilian otter (*Pteronura brasiliensis*) both feed on this fish with impunity.

Section VI

ECHINODERMS

(phylum Echinodermata)

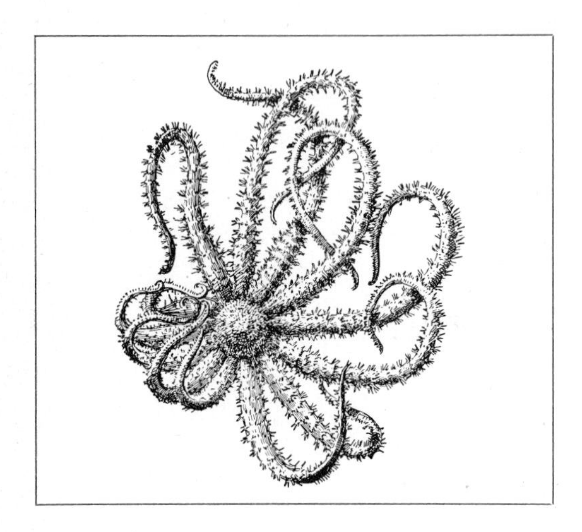

An echinoderm is a spiny-skinned marine invertebrate animal with a hard calcareous internal skeleton sometimes reduced. It occurs in all depths of the sea, and the shape and size of the body is exceedingly varied. The main characteristic is radial symmetry, which means the different parts of the body are arranged round a central axis or disc. Another feature is the water-vascular system, which is connected both with the external sea-water by a siege-plate and with the hollow tube feet and serves as a means of locomotion and respiration. Most species lay eggs in the sea and these hatch into free-swimming planktonic larvae.

Echinoderms are a very ancient order of animal, and their fossil remains have been found in rocks dating back 600000000 years.

There are about 6200 living species and the phylum is divided into five classes. These are: the *Asteroidea* (starfishes), *Ophiuroidea* (brittlestars and basketstars), *Echinoidea* (sea-urchins), *Holothuroidea* (sea-cucumbers) and the *Crinoidea* (sea-lilies). The largest class is *Ophiuroidea*, which contains about 2100 species or 33 per cent of the total number.

The largest of the 1600 known species of starfish in terms of span is the very fragile brisingid *Midgardia xandaros*. A female of this new genus collected in a dredge by the Texas A & M University research vessel *Alaminos* in the southern part of the Gulf of Mexico at a depth of 457 m *1500 ft* on 18 August 1969 measured 1380 mm *54·33 in* from arm tip to arm tip (R = major radius 680 mm *26·77 in*), but the diameter of the disc was only 26 mm *1·02 in*. Shortly after captured weighed 4·53 kg *10 lb* although it only twelve arms and further damage occurred during subsequent handling. The dry weight of the fragmented parts was given as 70 g *2·46 oz* (Downey, 1972).

Midgardia's nearest challenger is the five-armed *Evasterias echinosoma* of the North Pacific. In June 1970 a Russian expedition from the Institute of Marine Biology in Vladivostock collected a huge example in the flooded crater of a volcano in Broughton Bay, Semushir, one of the Kurile Islands. It measured 960 mm *37·79 in* from arm tip to arm tip (R = 505 mm *19·88 in*) – more than twice the width of an average dustbin lid – and weighed just over 5 kg *11 lb* when alive (Lukin, pers. comm. 25 November 1970).

The bulkiest starfish is the Sunflower or Twenty-rayed star (*Pycnopodia helianthoides*), also of the North Pacific. This echinoderm reaches its greatest size in Puget Sound, Washington, USA, and spans up to 1219 mm *4 ft* have been claimed (Williams, 1952). Fisher (1928), however, says the largest recorded *Pycnopodia* had a major radius of 400 mm *15·75 in*, which is equivalent to an arm tip to arm tip measurement of *c.* 780 mm *30·7 in*.

Luidia savignyi also reaches an impressive size. One specimen collected from Mauritius had a

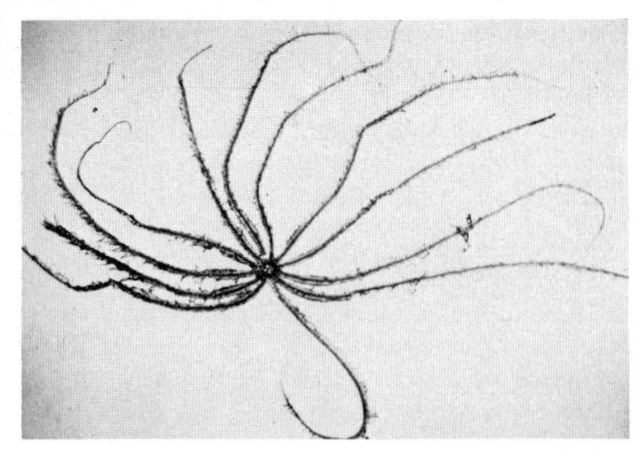

The very fragile brisingid Midgardia xandaros, *which has been measured up to 138 cm* 54·3 in *across the arms (Texas A and M University)*

major radius of 370 mm *14·56 in* (span *c.* 720 mm *28·34 in*) (Bell, 1889), and Downey (1972) mentions a *Luidia magnifica* from Hawaii preserved in the National Museum of Natural History, Smithsonian Institution, which has a major radius of 400 mm *15·74 in* (span *c.* 770 mm *30·31 in*).

The Five-armed starfish (*Pisaster brevispinus*) is another very bulky asteroid. In the 1968 'World Championship Starfish Grapple' in Hood Canal, Puget Sound, the largest specimen captured weighed 4·53 kg *10 lb* although it only measured 457 mm *18 in* from arm tip to arm tip (Furlong & Pill, 1970). The five-armed *Orester reticulatus* (span up to 508 mm *20 in*) of the West Indies is also massively built and is practically all disc.

The largest starfish found in British waters is the Spiny starfish (*Marthasterias glacialis*), which has been measured up to 350 mm *13·78 in* in major radius or 665 mm *26·18 in* arm tip to arm tip (Mortensen, 1927).

The fragile seven- or eight-armed *Luidia ciliaris* is also of comparable size, having been measured up to 300 mm *11·81 in* in major radius or *c.* 570 mm *22·43 in* in diameter (Mortensen, 1927). On 20 April 1973 Mr Steven Cox of Porthcawl, South Wales, caught a specimen off Fowey, southern Cornwall, on a hand-line which had a span of 596 mm *23½ in*. This starfish was presented to Fowey Aquarium but died by fragmentation after a fortnight. According to P Cox (pers. comm. 23 November 1973) the curator told him that *Luidia ciliaris* measuring over 609 mm *24 in* in diameter were not uncommon in that area.

The heaviest starfish found in British waters is probably the Common sunstar (*Crossaster papposus*) which has been measured up to 350 mm *13·78 in* in diameter (Mortensen, 1927). This twelve-armed species has a very large disc and *may* reach a weight of *c.* 2·27 kg *5 lb*. The Purple sunstar (*Solaster endeca*) is even larger in terms of diameter, having been measured up to 400 mm *15·74 in*, but it is not such a bulky animal.

Among the smallest known starfishes is the Mediterranean deep-sea species *Marginaster capreensis* which is not known to exceed *c.* 20 mm *0·78 in* in diameter.

The smallest starfish commonly found in British waters is the Cushion starfish (*Asterina gibbosa*) which measures up to *c.* 60 mm *2·36 in* in diameter but is usually *c.* 25 mm *1 in* across.

Although very little information has been published on the longevity of starfishes (each species has its own average life-span), most probably live less than four years. Notable exceptions include *Marthasterias glacialis* and *Asterias rubens* which do not reach sexual maturity in captivity until they are seven years and five to six years respectively (Comfort, 1964; Bull, 1934), and there is also a record of an *Astropecten irregularis* living for 6 years 6 months (Feder & Christensen, 1966). MacGinitie & MacGinitie (1949) believe that the North Pacific species *Pisaster ochraceus* may live for 20 years in areas where it develops slowly, but this needs confirming. Estimates up to 100 years have been quoted for the Antarctic starfish (*Odontaster validus*) based on extrapolation for very slow larval development (Pearse, 1969), but this duration of life has not yet been proved and is probably much too extreme for any member of this class of echinoderms.

Marthasterias glacialis – *the largest starfish found in British waters*

The West Indian starfish Oreaster reticulatus, *one of the bulkiest members of the Asteroidea*

The greatest depth from which a starfish has been recovered is 7584 m *24872 ft* for a specimen of *Porcellanaster ivanovi* collected by the Russian research vessel *Vityaz* in the Mariana Trench, W Pacific (Belyaev, 1969). Earlier a depth of 7630 m *25032 ft* had been reported for a specimen of *Eremicaster tenebrarius* collected by the Danish research vessel *Galathea* in the Kermadec Trench, central Pacific, in 1951 (Brunn, 1956), but no details of this capture are given in the official report of the Galathea Deep Sea Expedition (Madsen, 1956).

The most destructive starfish in the world is probably the Crown of Thorns (*Acanthaster planci*) of the Indo-Pacific region and the Red Sea. This giant asteroid (304–609 mm *12–24 in* in diameter) feeds on the polyps which make up coral reefs, and a single individual can destroy up to 0·91 m^2 *1* yd^2 of coral a day. Once the polyps are destroyed the empty coral cups become overgrown with algae. The edible fish in the area then move to a new habitat, and with them goes the shallow-water fisheries on which many small islands in the South Pacific depend for a living. In the 1960s some massive populations of this species were reported on parts of the Australian Great Barrier Reef which were thought might endanger the ecology of the largest deposit of coral in the world, but this claim was based on the assumption that such infestations were not localised. According to the findings of Dr Peter Vine (1970), however, who headed a British diving expedition to the Great Barrier Reef between December 1968 and April 1970, large populations of this starfish are often concentrated in small areas of the reef. It now seems likely that large local increases in the numbers do not represent a general plague, but are merely a result of natural fluctuations.

All starfishes can renew arms, but according to MacGinitie & MacGinitie (1949) only one family (Ophidiasteridae) can grow an entirely new starfish from a piece of one arm without retention of the disc or any part of it. In one of their experiments they discovered that a new *Linckia columbiae* (diameter 101 mm *4 in*), a species found in the eastern Pacific, could grow from a section of arm measuring only 10 mm *0·39 in* in length, although they admitted that this regeneration often went haywire (the starfish sometimes ended up with as many as eight arms instead of five).

Most of the brittlestars (Ophiurae) are small creatures averaging less than 12 mm *0·5 in* across the disc and having a span of 101–127 mm *4–5 in*. There are some exceptions, however, and the largest brittlestar is probably the tropical Indo-Pacific variety *Ophiarachna increassata* which has a disc diameter of over 51 mm *2 in* (Clark, 1962), and a span of *c.* 508–609 mm *20–24 in*.

The basketstars (Euryalae) are much larger (up to 101 mm *4 in* across the disc) and most species with arms *fully extended* probably reach 609 mm *24 in* in diameter, but the long branching arms are normally coiled in such a tangle that it is difficult to obtain accurate measurements.

The largest known ophiuroid is the Antarctic basketstar *Astrotoma agassizii*, which has unbranched, elongated arms. One giant specimen with its arms broken off measured 57 mm *2·24 in* across the disc and Fell (1966) says it must have measured about 1000 mm *39·37 in* in diameter when complete. Other individuals collected by Mortensen (1936) at South Georgia had discs measuring up to 60 mm *2·36 in* across, but these also had their arms broken off and the over-all size was not recorded.

The largest ophiuroid found in British waters is the Gorgon's head basketstar (*Gorgonocephalus caput-medusae*). The disc of this species has been measured up to 90 mm *3·54 in* across and the span may reach 700 mm *27·65 in*. If, however, we accept Mortensen's definition of British waters to include the Faroe Channel and the continental slope to the west of Ireland, this brings in the Arctic species *Gorgonocephalus eucnemis*, *G. lamarcki* – and possibly *G. arcticus* –

all of which match or just exceed *G. caput-medusae* in size.

The smallest known ophiuroids are the brittlestars of the genus *Ophiomisidium* which measure only 3–5 mm *0·11–0·19 in* in diameter.

Very little information has been published on the life-spans of ophiuroids. Very small species like the ones just mentioned may live only two to three years, but Fell (1966) believes 'most ophiuroids continue to grow for a period of 8 years at least, and probably for 10 or 15 years'. Buchanan (1964) found that it takes *c.* 15 years for the brittlestar *Amphiura chiajei* to attain a disc diameter of 8·8 mm *0·34 in*, which suggests that large basketstars like *Astrotoma* and *Gorgonocephalus* must take 20 or even 30 years to reach maximum size; but as Fell points out, these giant forms usually live in plankton-rich waters and consequently may have a more rapid growth rate than other ophiuroids.

At least four genera of brittlestars (*Opiura, Amphiophiura, Ophiacantha* and *Ophiosphalma*) live at depths in excess of 4 miles (6432 m *21 120 ft*).

The greatest depth from which an ophiuroid has been recovered is 7210 m *23 359 ft* for an *Ophiosphalma* sp. collected by the Russian research vessel *Vityaz* in the Kurile-Kamchatka Trench, NW Pacific, in *c.* 1962 (Belyaev, 1966).

The Echinoidea are generally small to moderate-sized echinoderms.

The largest of the 800 known species of sea-urchin is *Sperosoma giganteum* of Japan, which is known only from a few specimens. It has a horizontal test (shell) diameter of 320 mm *12·59 in*. The tropical *Diadema setosum* has been credited with measurements up to 457 mm *18 in*, but this figure was obtained *vertically* and included the uppermost spines which may reach 304 mm *12 in* in length in calm waters (the lateral and lower spines are much shorter).

The largest sea-urchin found in British waters is the offshore *Araeosoma fenestratum* which has been measured up to *c.* 280 mm *11·02 in* in horizontal diameter, but the average test diameter is 140–180 mm *5·5–7·08 in*. It is found only to the west of Ireland and Scotland at a depth of at least 500 m *1640 ft*.

The smallest known sea-urchin is *Echinocyamus scaber*, which is found in the waters off New South Wales, Australia. It has a test diameter of only 5·5 mm *0·21 in* (Clarke, 1925). The British *E. pusillus* rarely exceeds 15 mm *0·59 in*.

As with other classes of echinoderms, very little is known about the maximum life potentials of echinoids. It has been claimed that they live indefinitely (*sic*), but this is not borne out by scientific investigation. According to Moore (1935) the oldest *Echinus esculentus* dated by annual rings in the plates was seven years old, but the potential maximum age was *c.* ten years. *Psammechinus miliaris* has been known to live eight years under laboratory conditions (Bull, 1938), and six years has been reported for specimens of *Strongylocentrotus drobachiensis* collected off the coast of Norway (Grief, 1928). In a study of the growth curves of two settlements of *Echinocardium cordatum* it was ascertained that the maximum age reached was *c.* 15 years (Buchanan, 1967). Most other echinoids live less than four years, but some deep-sea species may remain active three to four times as long, depending on the availability of food.

The greatest depth at which a sea-urchin has been recovered is 7250 m *23 786 ft* for an unidentified specimen collected by the Galathea Deep Sea Expedition in the Banda Trench near Indonesia in 1951 (Brunn, 1951).

The most venomous echinoderm in the world is the sea-urchin *Toxopneustes pileolus* of the Indo-Pacific region. Its sting can produce intense pain, followed by a curare-like paralysis and – exceptionally – death (Halstead, 1971). The tropical sea-urchins *Diadem setosum* and *D. antillarum* of the West Indies are also nasty customers; their long, sharp spines can easily impale the unwary and break off in the wound which may take days or even weeks to heal up. (The spine fragments are very difficult to remove and slow to dissolve.)

The holothurians are mostly moderate-sized echinoderms, but the class has wide limitations.

The largest of the 1000 known species of sea-cucumber is the very bulky *Stichopus variegatus* from the Philippines which has been measured up to 1000 mm *39·37 in* in length when fully extended and 210 mm *8·26 in* in diameter (Semper, 1868). Some of the worm-like sea-cucumbers of the genus *Synapta* can stretch themselves out to lengths of 1000 mm *39·37 in* or even 2000 mm *78·74 in*, but they only measure about 12 mm *0·5 in* in diameter. Soule (1974) claims that sea-cucumbers have been known to reach a length of 1820 mm *6 ft* and be 'as thick as a man's thigh', but this claim must be relegated to the world of fantasy.

The largest sea-cucumbers found in British waters are *Cucumaria frondosa* and *Stichopus tremulus*, both of which have been measured up to *c.* 500 mm *19·68 in* (Mortensen, 1927).

The smallest known sea-cucumber is *Psammothuria ganapatii* from southern India, which does not exceed 4 mm *0·15 in* in length (Clark, pers. comm. 28 October 1975).

The smallest British sea-cucumber is *Echinocucumis hispida*, which has a maximum length of 30 mm *1·18 in*.

As with echinoids, longevity records for holothurians are few. Most species probably live at least three years. The Pacific species *Paracaudina chilensis* takes three to four years to reach maximum size (Tao, 1930), and the Japanese sea-cucumber (*Stichopus japonicus*) lives at least five years (Mitsukuri, 1903). One of the longest-lived species may be *Cucumaria elongata*, found in British waters, which reportedly lives ten to twelve years.

The greatest depth from which an echinoderm has been recovered is 10630–10710 m *34 876–35 137 ft* for a *Myriotrochus brunni* collected by the Russian research vessel *Vityaz* in the Mariana Trench, W Pacific, in 1958 (Belyaev, 1970).

The largest of the 650 known species of crinoids as far as arm-span goes is the unstalked Feather-star (*Heliometra glacialis maxima*), which extends from the Sea of Okhotsk, an inlet of the NW Pacific on the coast of Khabarovsk Territory, Soviet Russia, southwards to the Korean Straits. Spans up to 914 mm *36 in* have been claimed for this species (Burton, 1968), but the largest accurately measured specimen had an arm length of 350 mm *13·78 in* and a span of *c.* 700 mm *27·55 in*.

Although feather-stars tend to *decrease* in size in warmer waters, some of the tropical multibrachiate *Comasterids* are also very large, reaching nearly 609 mm *2 ft* in diameter. They are also much heavier than *Heliometra* because they can have as many as 190 arms (cf. ten arms for *Heliometra*).

The largest crinoid found in British waters (if we include the Faroe Channel) is *Heliometra glacialis glacialis* which may have an arm length of at least 200 mm *7·87 in* and a span of *c.* 400 mm *15·74 in* (further north, towards Greenland, this sub-species reaches an expanse of 500 mm *19·68 in*, and some freakish examples *may* match *H. g. maxima* for size).

The largest stalked crinoids are probably some sea-lilies of the genus *Metacrinus* which is widely distributed throughout the Japan-Malay-Australian region. They have a maximum stalk (stem) height of about 609 mm *24 in* plus another *c.* 152 mm *6 in* for the arms. (Some fossil sea-lilies had stalks reaching to a height of more than 21 m *68·9 ft* and must have measured at least 26 m *85·3 ft* to the tips of the arms!)

The strongest contenders for the 'smallest crinoid' title are probably the feather-stars *Antedan parviflora* and *Comissia minuta*, both found in Japanese waters. They have spans of *c.* 40 mm *1·57 in*. The smallest extant sea-lily is *Bathycrinus gracilis* which has a total height of 70–80 mm *2·75–3·14 in*.

Most of the small feather-stars probably have a life-span of only two to three years. Some of the longest-lived species are the tropical comasterids, and Catala (1964) says he has kept a number of them alive in the aquarium at Noumea, New Caledonia, for 'several years'. The Antarctic feather-star (*Promachocrinus kerguelensis*) has an even greater age potential because it does not reach maturity until the tenth year, and Fell – on the basis of the known life-spans in other classes of echinoderms – estimates the maximum age attained by a crinoid at '20 or more years'. Nothing is known about the rate of growth in sea-lilies and the age potential, but the ascending axis probably develops fairly rapidly. 'Consequently', says Fell (1966), 'the great length of the stem in some extinct sea lilies does not necessarily imply a correspondingly lengthy life span.'

The majority of feather-stars – their centre of abundance is the Indo-Pacific region – are found in relatively shallow waters, i.e. *c.* 200 m *656 ft*, but some species of sea-lily are found at much greater depths.

The greatest depth from which a crinoid has been recovered is 9735 m *31 936 ft* for a sea-lily (*?Bathycrinus australis*) dredged up from the Bonin Trench, W Pacific, by the Russian research vessel *Vityaz* in *c.* 1961 (Belyaev, 1969).

Section VII

CRUSTACEANS

(class Crustacea)

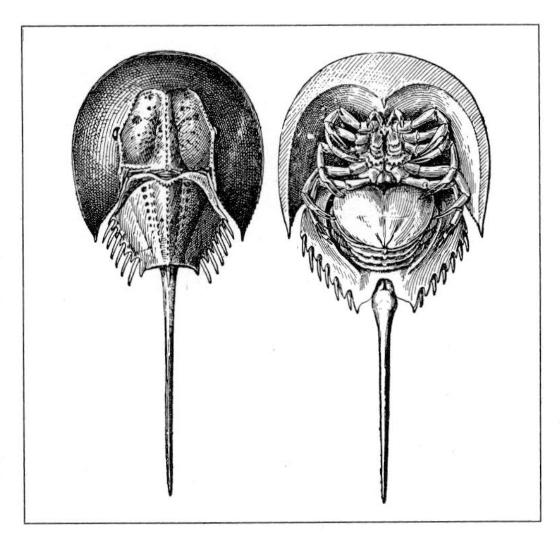

A crustacean is an aquatic invertebrate which breathes through gills. It has a segmented body, paired jointed limbs and a tough outer integument or shell which is pliable at the joints. This external skeleton, which is incapable of growth, is periodically cast off and replaced by a lime-impregnated coat. Another characteristic is the two pairs of antennae in front of the mouth which are used as feelers. The young are produced from eggs shed freely into the water or carried by the female.

The earliest known crustacean was the twelve-legged *Karagassiema* which lived about 650000000 years ago. Its remains have been found in the Sayan Mountains in the USSR.

There are about 26000 living species of crustacean, and the class is divided into eight subclasses. These are the *Cephalocardia* (cephalocarids); the *Branchiopoda* (branchiopods); the *Ostracoda* (mussel or seed shrimps); the *Copepoda* (copepods); the *Mystacocariada* (mystacocarids); the *Branchiura* (fish lice); the *Cirripedia* (barnacles); and the *Malacostraca* (shrimps, prawns, lobsters, crabs and woodlice). The largest subclass is the *Malacostraca*, which contains about 18000 species or 69 per cent of the total number.

The largest known crustacean is the Giant spider crab (*Macrocheira kaempferi*), also called the 'Stilt crab', which is found in the deep waters off the SE coast of Japan from Kamaishi in NE Honshu, southwards to the island of Kyushu.

Very little information has been published about this species, but average adult males have a heart-shaped shell (carapace) measuring *c.* 254 × 304 mm *10 × 12 in* and a biting claw-span of about 1·83–2·13 m *6–7 ft*. Adult females are smaller.

According to R W Ingle, a carcinologist at the British Museum (Natural History), the *largest known specimen* was a male which had a leg span of 3·30 m *10 ft 10 in* and a shell measuring 335 × 305 mm *13·18 × 12 in*, but even bigger examples have been reliably reported. One outsized individual preserved in the Calcutta Museum, Bangladesh, has a claw-span of 3·45 m *11 ft 4 in*, and a measurement of 3·70 m *12 ft 1½ in* has been quoted for another male which weighed 6·35 kg *14 lb*.

The maximum size attained by this bizarre creature has not yet been established with any degree of certainty. One monstrous crab caught in a fisherman's net off Honshu in November 1921 reputedly had a claw-span of 5·79 m *19 ft* and weighed 18·14 kg *40 lb*, but further information is lacking. Holder (1886) says that the first European to set eyes on one of these crabs saw two biting claws leaning against a fisherman's hut which both measured 3·04 m *10 ft* in length and must have belonged to a crustacean with a claw-span of *c.* 6·70 m *22 ft*!

Because of its strange shape the giant spider crab can only move about in the very still waters found at great depths (i.e. down to 305 m *1000 ft*). On dry land it is completely helpless and cannot even raise itself erect.

The largest spider crab found in British waters is the Thornback (*Maia squinada*). Specimens measuring up to 178 mm *7 in* across the shell and weighing 1·81 kg *4 lb* have been recorded.

The heaviest crab in the world is the offshore xanthid *Pseudocarcinus gigas*, found only in the Bass Strait area separating Australia from Tasmania. This edible giant crab has been accurately measured up to 456 mm *18 in* across the carapace, and may reach a weight of 13·60 kg *30 lb* (Griffin, 1970).

Weights up to 7·25 kg *16 lb* have been reported for the King crab (*Paralithodes camtschatica*), which is distributed from the Bering Sea to the Sea of Japan, but the five members of this class are not crustaceans although they are called 'crabs'.

The heaviest crab found in British waters is the Edible or Great crab (*Cancer pagurus*), adult specimens averaging 127–152 mm *5–6 in* across the shell and 0·45–0·90 kg *1–2 lb* in weight. In 1895 one measuring 279 mm *11 in* across the carapace and weighing 6·35 kg *14 lb* was caught off the coast of Cornwall. Another outsized individual caught off Dartmouth, Devon, in October 1952 weighed 3·94 kg *8 lb 12 oz* and one weighing 4·08 kg *9 lb* was taken off the Norfolk coast by Cromer fishermen on 7 September 1958. In June 1972 David Rollinson of Barnsley, Yorkshire, caught a 4·98 kg *11 lb* crab while skin-diving off Brixham, Devon. It had a nipper span of 927 mm *36½ in*.

The heaviest land crab and the largest land crustacean is the Coconut or Robber crab (*Birgus latro*) which lives on tropical Indo-Pacific islands and atolls. Mature examples scale 2·26–2·72 kg *5–6 lb* and have a leg-span of 914 mm *3 ft*, but weights up to 4·08 kg *9 lb* have been reported.

The smallest crabs in the world are the parasitic Pea crabs (family Pinnotheridae) which live in the mantle cavities of bivalved molluscs such as oysters, mussels and scallops. Some species have a shell diameter of only 6·35 mm *0·25 in*, including *Pinnotheres pisum*, which is found in British waters.

Female crabs cover great distances during migration. One individual (*Cancer pagurus*) tagged by the Ministry of Agriculture and Fisheries' laboratory at Burnham on Crouch, Essex, in 1962 travelled 368 km *230 miles* from Whitby, Yorkshire, to Fraserburgh, Aberdeenshire, Scotland, in 18 months, and another one tagged about the same time covered 198 km *124 miles* from Cromer, Norfolk, to Whitby in 21 months.

The speed record is also held by a female which travelled 21 km *13 miles* in 23 days in 1962 (average 910 m *995 yd* a day). Another specimen tagged the same year covered 72 km *45 miles* in 114 days (average 634 m *694 yd* a day). In September 1925 an edible crab with distinctive markings escaped from a covered basket at Sennen Cove, southern Cornwall, and was recaptured two days later just over 3·2 km *2 miles* away! Male crabs are much less venturesome and rarely go further than 8 km *5 miles* in any given direction.

The slowest-moving crab is probably *Neptunus pelagines*. According to Prof Grunel (1931) of the Museum d'histoire Naturelle, Paris, France, one specimen tagged in the Red Sea took 29 years to travel the 162·4 km *101½ miles* to the Mediterranean via the Suez Canal at an average speed of 5·5 km *3½ miles* a year.

Crabs are long-lived animals, as evidenced by the one just mentioned. Another individual caught by French fishermen off Saint-Nazaire on 4 October 1959 had apparently travelled 99 km *62 miles* in 30 years. The words 'Jos. Le Roux' scratched on its shell were identified by Monsieur Joseph Le Roux, who said he had caught the crab in 1929 (further south) and inscribed his name on its shell before throwing the creature back into the sea.

The largest species of lobster is the American or North Atlantic lobster (*Homarus americanus*) and there are several authentic records of specimens weighing over 9·07 kg *20 lb*.

The American carcinologist Dr Francis Herrick (1895) mentions a very large male captured in Penobscot Bay, near Belfast, Maine, on 6 May 1891 which was weighed in the presence of several witnesses and scaled just over 10·43 kg *23 lb*. This lobster (now preserved in the Museum of Adelbert College, Cleveland, Ohio) measures 508 mm *20 in* from the rostrum to the end of the tail-fan, and has a maximum girth of 419 mm *16½ in*. Its crushing-claw is nearly 355 mm *14 in* long, and Herrick says 'it was probably powerful enough to crush a man's arm at the wrist'.

Another male caught earlier at Salem, Massachusetts, in 1850 and now preserved in the Peabody Academy of Science, weighed 11·33 kg *25 lb* when alive, and one weighing 11·11 kg *24 lb 8 oz* was captured at Lubec, Maine, in September 1892. Fragments of another lobster captured with hook and line on the coast of Delaware and stated to have weighed over 11·33 kg *25 lb* are preserved in the Smithsonian Institution in Washington, DC.

In the summer of 1890 a male lobster weighing nearly 11·33 kg *25 lb* was taken in a trawl off Monroe Island, some 8 km *5 miles* east of Rockland, Maine. (The crushing-claw of this specimen measured 406 mm *16 in* in girth.)

Dr Herrick concluded from his diligent researches that although very large lobsters weighing 9·07 kg *20 lb* or more had been caught on occasion, there was no reliable evidence to support claims in excess of 11·33 kg *25 lb*. 'Where lobsters are said to have attained a greater weight', he said, 'measurements of the parts of the skeleton which have been preserved invariably prove that the figures have been exaggerated. I do not maintain that the American lobster does not reach a greater weight than 25 pounds, but that I have been unable, up to the present time, to discover any well-authenticated evidence that this is the case.'

The heaviest American lobster on record (Museum of Science, Boston)

Two years later an enormous lobster was caught off the Atlantic Highlands, New Jersey. It measured 603 mm *23¾ in* from the rostrum to the end of the tail-fan (1·01 m *3 ft 4 in* over-all) and weighed 15·42 kg *34 lb* (Firth, 1939).

This record stood until the autumn of 1934 when Captain Wheeler of the smack *Hustler* caught an enormous lobster weighing 19·25 *42 lb 7 oz* in 100 fathoms (182·8 m *600 ft*) of water off the Virginia Capes, Virginia. A few months later he caught another outsized specimen weighing 17·57 kg *38 lb 12 oz* in the same area. Both these individuals, known as 'Mike' and 'Ike', are now on display in the Museum of Science, Boston, Massachusetts (Gurney, 1950).

During the summer of 1939 an unconfirmed weight of 21·31 kg *47 lb* was quoted for a lobster taken off the coast of New Jersey (Firth, 1939), and another specimen allegedly weighing 21·77 kg *48 lb* was captured off Chatham, New England, in 1949.

The largest crustacean found in British waters is the Common or European lobster (*Homarus gammarus*).

The maximum size attained by this species has been greatly exaggerated in the older literature. Olaus Magnus (1555), for instance, says some of the lobsters found between the Orkneys and the Hebrides were so enormous that they could seize a strong human swimmer and crush him to death in their claws, and Erik Pontoppidan (1735), Bishop of Bergen, claims that one lobster seen by fishermen near Utvaer in the Bay of Erien, Norway, was so huge – it reputedly had a claw-span of over 1·83 m *6 ft* – that no one dared to attack it.

Fortunately the common lobster of today is built along more modest lines (average 0·90–1·36 kg *2–3 lb*) but in earlier times, before it was over-fished, it reached a much larger average size, and probably rivalled *H. americanus* for length.

Frank Buckland (1877) cites a record of a lobster weighing 5·89 kg *13 lb* which was caught at Durgan, south Cornwall, and he says another one weighing 5·44 kg *12 lb* was taken in Saints Bay, Guernsey, Channel Islands, in May 1875.

Other reliable records over 9 lb include: a 5·11 kg *11 lb 4 oz* lobster caught off Guernsey and exhibited in Bond Street, London, in August 1873; one weighing 4·76 kg *10 lb 8 oz* caught off Tenby, Pembrokeshire, South Wales (presented to the Francis Buckland Museum in 1876) and a 14-pounder *6·35 kg* taken in a trammel-net off the coast of southern Cornwall in 1875 (total length 591 mm *1 ft 11¼ in*) (Buckland, 1877).

On 30 July 1842 a gigantic lobster measuring 749 mm *2 ft 5½ in* in total length and having a crushing-claw 355 mm *14 in* long and 406 mm *16 in* in maximum girth was exhibited at Billingsgate Fish Market, London, but as the dimensions of the crushing-claw are proportional to a weight of *c.* 9·07 kg *20 lb*, this crustacean was more probably a member of the American species.

G O Sars (1879) mentions 'an immense specimen' in Bergen Museum, Norway, which was captured in 1850 and weighed an estimated 5·44 kg *12 lb*, but Dr Herrick, after corresponding with Dr Lonnberg of the University of Uppsala, Sweden, came to the conclusion that this lobster (total length 475 mm *18·73 in*) 'probably weighed when alive not over 10 pounds'.

The largest Common lobster on record is generally thought to be a huge example preserved in the Museum of the Academy of Natural Sciences of Philadelphia, Pennsylvania, USA, which has a total length of 825 mm *2 ft 8½ in*, and weighed an estimated 9·5–10 kg *21–22 lb* when alive. Recently, however, this monster was re-examined by another authority and identified as an American lobster.

On 13 August 1935 a lobster weighing 5·55 kg *12 lb 4 oz* was caught by a fisherman in Skelmorlie Bay, Ayr Co., SW Scotland. Another individual measuring 697 mm *2 ft 3½ in* in total length and weighing 6·01 kg *13¼ lb* was taken off Gothenberg, Sweden, in October 1949 and one measuring 1·34 m *4 ft 5 in* from the tip of the feelers to the end of the tail-fan and weighing 5·89 kg *13 lb* was captured by fishermen at Lorient, Brittany, northern France, in February 1952.

In January 1960 a 1·05 m *41½ in* long specimen weighing 5·66 kg *12 lb 8 oz* was netted in Colwell Bay, Isle of Wight, and nine months later one weighing 4·98 kg *11 lb* was taken in Cardigan Bay, W Wales. In July 1962 two skin-divers caught a lobster measuring nearly 914 mm *3 ft* in total length (crushing-claw 254 mm *10 in* long and 304 mm *12 in* in maximum girth), and weighing nearly 4·98 kg *11 lb*, in a wreck off Berry Head, Brixham, Devon.

On 20 June 1964 an enormous crushing-claw was collected in a trawl near Skagen, the northernmost tip of Jutland, at a depth of 40 m *131 ft*. It measured 351 mm *13·81 in* in length and 400 mm *15·74 in* in maximum girth. Dr Wolff (pers. comm. 9 June 1967), Chief Curator of the Universitetets Zoologiske Museum in Copenhagen, Denmark, said: 'When comparing with the list of giant American lobsters . . . the living weight of the whole animal must have been about 13 kg or 30 pounds or perhaps somewhat more as the claws of the American lobster are somewhat larger compared to the body than those of the European lobster.' Dr Wolff's weight estimate of 13 kg *28·66 lb* is probably a bit too high, but the owner of this crushing-claw must have scaled at least 10 kg *22·04 lb*, which would make it the largest Common lobster on record.

In August 1964 a monster lobster with crushing-claws strong enough to sever brass tubing was discovered by members of the Scarborough Sub-Aqua Club in the hold of a sunken wartime minesweeper in Bridlington Bay, Yorkshire. One diver who tried to pull the lobster from the wreck had the brass tubing of his snorkel tube nipped in two by the 914 mm *3 ft* long monster which was estimated to weigh 22·67 kg *50 lb* (*sic*).

Since then, however, there has been no more news, which suggests that this king-sized lobster is still safely ensconced in its underwater den.

On 17 August 1967 a skin-diver caught a lobster weighing 6·57 kg *14 lb 8 oz* off St Ann's Head, Pembrokeshire, W Wales. It is now mounted in the bar of the Amroth Arms, Amroth, Pembrokeshire. This specimen and the one taken off Cornwall in 1875 are the *two largest lobsters recorded in British waters.*

In August 1969 another one measuring nearly 910 mm *3 ft* in total length and weighing 5·29 kg *11 lb 11 oz* was caught by a skin-diver off Oban on the west coast of Scotland. The live specimen was later presented to Flamingo Park Zoo, Yorkshire.

On 4 September 1972 another monster was captured by a skin-diver on a wreck in Ardentrive Bay, Oban. It weighed 5·21 kg *11 lb 8 oz* and had a crushing-claw measuring 292 mm *11½ in* in length (Randall Hinchcliffe, per. comm. 8 January 1973).

In December 1973 an even larger example was reportedly caught off Clacton Pier, Essex, and presented alive to the North Sea World Aquarium on the pier. This lobster was credited with a total length of 887 mm *2 ft 11 in* and a weight of 6·80 kg *15 lb*, but Bloom (pers. comm. 4 February 1974), head of the aquarium, said the crustacean tipped the scales at 4·42 kg *9 lb 12 oz* after two weeks in captivity and probably weighed about 5·21 kg *11 lb 8 oz* when originally taken. In November 1974 a lobster of almost record size (6·12 kg *13 lb 8 oz*) was caught by

A 4·42 kg 9·75 lb *Common lobster caught off Clacton Pier, Essex*

hand off Bournemouth, Hampshire.

The Spiny lobster (*Jasus verreauxi*), also known as the 'Green crayfish', which is found in the waters off the coast of eastern Australia, averages between 4·53 and 6·80 kg *10* and *15 lb* at maturity, but individuals weighing up to 11·33 kg *25 lb* have been reliably reported.

The Crawfish or Spiny lobster (*Palinurus vulgaris*) is also of comparable size, although it has much smaller claws. The largest specimen ever caught in British waters was a female weighing 5·45 kg *12 lb* caught off Plymouth, Devon, in August 1971 and later shipped to France. In June 1974 Mr Randall Hincliffe of Shepley, Yorkshire, caught another huge crawfish weighing 4·77 kg *10 lb 9 oz* (length from rostrum to tip of tail-fan 546 mm *21·5 in*) while skin-diving off the Isle of Mull, western Scotland (pers. comm. 30 March 1976).

Incidentally, one question often asked about lobsters is: Are they dangerous to man? Well, it all depends on your viewpoint, but certainly the skin-diver who had the metal tubing of his snorkel cut in half by one of these creatures would be justified in thinking so, and no doubt Monsieur Roger Magnieu shares the same opinion....

In July 1952 this gentleman walked into a Paris restaurant and ordered a lobster lunch. As is the French custom a basket of these crustaceans was duly presented for his inspection by the owner of the establishment but, when M Magnieu complained that they were *not fresh*, the restaurateur became understandably annoyed. Seizing one of the lobsters he pushed it under the would-be diner's nose and exclaimed: 'Not fresh? Smell that!'

Before M Magnieu could oblige, however, the crustacean suddenly seized him by the nose. It took a good five minutes to prise the stubborn lobster free, and in the process part of the unfortunate man's nose went with it!

In a Paris court M Magnieu was awarded £100 damages for his injury. And the restaurant owner was also fined £3 for not keeping a dangerous animal under reasonable control!

Earlier, in July 1938, lobster fisherman Robert Anthony of Portland, Dorset, was nearly drowned by one of his catch. After emptying his lobster pots, one of the largest specimens clawed the plug from the boat, and the sea poured in through the hole!

The smallest known lobster is the Cape lobster (*Homarus capensis*) of South Africa which usually measures 100–120 mm *3·95–4·72 in* in total length. Frank Buckland (1875) says the 'chicken lobsters' caught off Bognor, Sussex, averaged 14–20 to the pound, but this diminutiveness (the lobsters were in fact *Homarus gammarus*) was due to over-fishing in the area.

One of the most amazing sights in nature can be seen every autumn when thousands of spiny lobsters (*Panulirus argus*) march in parallel single-file columns across the shallow sea floors off the Bahamas and the east coast of Florida, USA.

According to Dr Herrnkind of Florida State University (*Science*, 20 June 1969) the queuing behaviour of these crustaceans is probably a defence mechanism because by migrating in single-file order the vulnerable abdomen of each lobster is protected by the hard thorax of the individual behind. Each lobster manages to keep its position in the queue by clinging on with its front pair of legs to the tail-fan of the animal in front.

The same authority also discovered that migrating lobsters transferred to a pool moved in single-file for 33 days almost without a break!

The largest freshwater crustacean in the world is the crayfish *Astacopsis gouldi*, which is found in small streams in Tasmania. Not a lot of information has been published about this

A 4·78 kg 10 lb 9 oz *Crawfish taken off the Isle of Mull, western Scotland (R Hinchcliffe)*

species, but it is reported to reach a weight of 3·62–4·08 kg *8–9 lb* (Schmitt, 1973). The Murray River 'Lobster' (*Euastacus armatus*) of South Australia also reaches an impressive size, having been credited with a maximum weight of 2·72 kg *6 lb* (François, 1960).

Britain's largest native freshwater invertebrate is the Whiteclaw (*Astacus fluviatilus*). It measures 101–152 mm *4–6 in* in length and may weigh as much as 113 g *4 oz*.

The smallest known crayfish is the freshwater *Tenuibranchiurus* sp. of Queensland, Australia, which does not exceed 25 mm *1 in* in total length.

The smallest known crustaceans are water fleas of the genus *Alonella* which may measure less than 0·25 mm *0·0098 in* in length. They are found in British waters.

The longest-lived of all crustaceans is the American lobster (*Homarus americanus*). According to Dr Herrick (1911) very large specimens may be as much as 50 years old. The king-sized lobster discovered in the wreck of the minesweeper in Bridlington Bay was *believed* to be 80 years old although no explanation was given as to how this figure was arrived at. The Common lobster (*H. gammarus*) has been credited with a maximum life-span of 30 years and the Whiteclaw (*Astacus fluviatilus*) 15–25 years (Comfort, 1964).

The fastest-moving crustacean is probably the American lobster. Dr Herrick states that it can 'shoot backward through the water with astonishing rapidity', and he quotes one observer as saying it can cover 7·62 m *25 ft* in less than a second (more than 27·2 km/h *17 miles/h*).

The common lobster is also a rapid mover. In a letter dated Scarborough 25 October 1768 Travis (cited by Pennant, 1777), reports: 'In the water they can run nimbly upon their legs or small claws and, if alarmed, can spring tail forward to a surprising distance *as swift as a bird can fly*. The fishermen see them pass about 30 feet, and by the swiftness of their motion suppose that they can go much farther.'

The greatest depth at which a crustacean has been seen is 10912 m *35802 ft* for an unidentified red shrimp sighted by Dr Piccard and Lt Walsh, US Navy, from the bathyscaphe *Trieste* in the Challenger Deep (Mariana Trench) in the western Pacific on 24 January 1960 (Piccard, 1960).

The greatest depth from which a crustacean has been recovered is 9790 m *32119 ft* for an isopod *Macrostylis galathae* sledge-trawled by the Danish Galathea Deep Sea Expedition in the Philippine Trench in 1951 (Brunn, 1956). The marine crab *Ethusina abyssicola* has been recovered from a depth of about 4267 m *14000 ft*.

By comparison amphiopods and isopods have been collected in the Ecuadorean Andes at a height of 4053 m *13300 ft*, and in 1875 Agassiz collected nine species of crustacea in Lake Titicaca, the huge freshwater lake in the Peruvian Andes, at 3810 m *12500 ft*.

A 5·21 kg 11 lb 8 oz *Common lobster caught in Ardentrive Bay, western Scotland (R Hinchcliffe)*

Section VIII

ARACHNIDS

(class Arachnida)

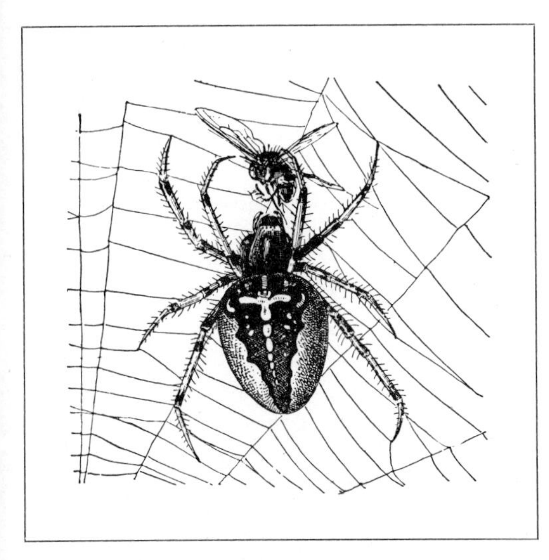

An arachnid is a terrestrial invertebrate which breathes through gill-like structures called 'book-lungs' or by means of a network of air-filled tubes. The body is divided into two main parts: the cephalothorax (head and thorax fused together) bearing four pairs of legs and two pairs of pincer-like appendages, and the abdomen which is limbless. It has a rigid external skeleton which it casts off periodically but no antennae. The function of these organs is served instead by the sensory bristles which cover the body and appendages. The young are usually hatched from eggs after leaving the female, but some are born alive.

The earliest known arachnids were probably scorpions, some species of which lived 400 000 000 years ago. The earliest known spider was *Palaeostenzia crassipes*, which lived about 370 000 000 years ago.

There are about 110 000 living species of arachnid, and the class is divided into ten orders. These are the *Scorpionida* (scorpions); the *Pseudoscorpionida* (pseudoscorpions); the *Solifugae* (camel spiders); the *Palpigradi* (micro-whip scorpions); the *Uropygi* (whip scorpions); the *Amblypygi* (amblypygids); the *Araneae* (spiders); the *Ricinulei* (ricinuleids); the *Opilones* (harvestmen); and the *Acarina* (mites and ticks). The largest order is *Araneae*, which contains about 40 000 species or 36 per cent of the total number.

The largest known arachnid is the 'bird-eating' spider *Theraphosa blondi* of northern South America. A male specimen with a leg-span of 254 mm *10 in* when fully extended and a body length of 89 mm *3½ in* was collected at Montagne, French Guiana, in April 1926. (Now preserved in the American Museum of Natural History, New York.) This individual, however, was probably a freak because females of this species are generally larger.

Wolfgang Bucherl (1971) of the Instituto Bhutantan, São Paulo, Brazil, says a huge female kept in the laboratories there has a body length of 90 mm *3·54 in* and a leg length 'of about 110 cm *4·33 in*'. This latter measurement would indicate an extended leg-span in excess of 270 mm *10·63 in*, and the spider shown in a published photograph certainly justifies this size.

In discussing the large South American 'tarantulas' (Avicularoidea = Mygalomorphae), and in particular *Avicularia (Mygale) avicularia* of Brazil, Henry Bates (1863), the famous English naturalist, writes: 'Some Mygales are of immense size. One day I saw the children belonging to an Indian family who collected for me with one of these monsters secured by a cord round its waist, by which they were leading it about the house as they would a dog.'

In March 1938 a 'tarantula' (probably a specimen of *A. avicularia*) measuring 203 mm *8 in* across the extended legs was found in a consignment of bananas in Exmouth Street Market, London, and bit a fruiterer on the thumb before it was killed. The man was rushed to hospital in great pain, but recovered after treatment.

Some of the wolf spiders (Pisauridae) also have an impressive leg-span, and *Cupiennius sallei* of Central America, which has exceptionally long legs, has been measured up to 254 mm *10 in* across. All the members of this genus, however, are light-bodied creatures and the species just mentioned cannot compete with a large theraphosid in terms of bulk or weight.

Ivan Sanderson (1937) tells of collecting a 'giant hairy spider' in the Assumbo Mountains, British Cameroons, West Africa in 1932 for the British Museum (Natural History) which covered an enamel dish measuring 304 mm *12 in* by 203 mm *8 in* when its legs were fully extended in all directions, but these dimensions were in error. D J Clark (pers. comm. 12 August 1971), an arachnologist at the Museum, said: 'The largest spider sent by Mr Sanderson measures 2½ inches body length, 3 inches if you count the chelicerae (pincers), and has a leg spread of, at most 7 inches. This spider belongs to the genus *Hysterocrates*. The largest African spider is probably *Hysterocrates hercules*, with a body length of 3 inches, 3½ inches if the chelicerae are included, and a leg spread of 8 inches.'

A male specimen of *Grammostola mollicoma* from Brazil preserved in the same museum measures 241 mm *9½ in* across the legs, and the compiler owns another example which measures 222 mm *8¾ in*. Bucherl (1971) contends that this species is 'the largest tarantula I know', and gives the maximum leg-spread as 270 mm *10·63 in*, but he does not quote his source.

The heaviest spider ever recorded was a female of the long-haired species *Lasiodora klugi* collected at Manaos, Brazil, in 1945. It measured 241 mm *9½ in* across the legs and scaled almost 85 g *3 oz*. The huge *Theraphosa blondi* collected in French Guiana weighed nearly 57 gm *2 oz* (Gertsch, 1949).

An enormous 'tarantula' of the species *Dugesiella crinita* collected by Baerg (1958) in Tlahualilo, Mexico, in 1935 had a body length of 85 mm *3·32 in*, and weighed 54·7 g *1·92 oz*.

In 1868 a Mr Rautenbach of Elands River, Natal, South Africa, reportedly killed a 'tarantula' spider as big as a *turkey cock*. He said the monster slew three of his dogs and had three charges of shot fired into it before it succumbed!

Of the 617 known British species of spider, the Cardinal spider (*Tegenaria parientina*) has the greatest leg-span, males sometimes exceeding 127 mm *5 in* (length of body up to 19 mm *0·75 in*). This house spider is found only in southern England, and is so called because Cardinal Wolsey lived in abject fear of them at Hampton Court. In 1974 Mr John Brown of Wokingham, Berkshire, discovered an enormous specimen in the boiler room of his house and managed to take some approximate measurements before it escaped. He told the compiler that it had a leg-span of 134–152 mm *5·3–6 in* and a body length of 20 mm *0·8 in*.

The well-known 'Daddy-Long-Legs' spider (*Pholcus phalangiodes*) rarely exceeds 76 mm *3 in* in leg-spread, but Savory (1928) mentions one outsized male collected in the south of England which measured 152 mm *6 in* across (length of body 0·76 mm *0·031 in*).

The heaviest spider found in Britain is probably the very bulky Orb weaver *Araneus quadratus* (formerly *Araneus reaumuri*). An *averaged-sized* individual collected in October 1943 tipped the scales at 1·174 g *0·041 oz*, and measured 15 mm *0·58 in* in body length (Bristowe, 1947). This weight may be matched by the Six-eyed spider *Segestria florentina* (=*S. perfida*) of southern England, which has been measured up to 23 mm *0·905 in* in body length, and has been referred to as 'the largest British spider' (Locket & Millidge, 1951), and more especially by the very thick-set swamp spiders *Dolomedes fimbriatus* of southern England and *D. planitarius* of East Anglia, which are both almost identical in size (up to 22 mm *0·86 in* body length). The largest British spider seen by Bristowe (1971) was an outsized *Argyroneta aquatica* collected in Kent which had a body length of 28 mm *1·10 in* (normal range 9–13 mm *0·35–0·51 in*), but this water spider was not weighed. Another arachnid worthy of mention is *Drassodes lapidosus*, which usually has a body length of 10–15 mm *0·39–0·59 in*. Recently, however, a giant variety measuring up to 22 mm *0·86 in* was discovered on Skellig St Michael, an islet off the coast of County Kerry, Ireland, (Bristowe, 1971).

The smallest known spider is *Patu marplesi* (family Symphytognathidae) of Western Samoa, south-west Pacific. The type specimen (male), found in moss at an altitude of *c.* 609 m *2000 ft* near Malolelei, Upolu, in January 1956 by T E Woodward of the Queensland Museum, Brisbane, Australia, measures 0·43 mm *0·016 in* in length (abdomen 0·21 mm *0·0082 in*), which means it is half the size of a full-stop on this page (Forster, 1958); (Gray, pers. comm. 6 December 1974).

The smallest recorded British species is the Money spider *Glyphesis cottonae*, which is known only from a swamp near Beaulieu Road Station in the New Forest, Hampshire, and Thurley Heath, Surrey (Locket & Millidge, 1951). Mature examples of both sexes have a body length of 1 mm *0·039 in*. Another money spider (*Saloca diceros*), found among mosses in Dorset and Staffordshire, and the widely distributed Comb-footed spider (*Theonoe minutissima*) are almost equally as diminutive, their body lengths being 1–1·2 mm *0·039–0·047 in* and 1–1·25 mm *0·039–0·049 in* respectively (Bristowe, 1958).

The largest spider webs are the aerial ones spun by the tropical orb weavers of the genus *Nephila*. Several examples found by Captain Sherwill (1850) in the Karrakpur Hills near Monghyr, central Bihar, India, measured 1·52 m *5 ft* in diameter (=4·78 m *15 ft 8½ in* in circumference), and had supporting guy-lines up to 6·09 m *20 ft* in length, and Haeckel (1883) saw 'immense cobwebs one to two metres across' while on an excursion from Bombay. In the lower steppes of central Australia Spencer & Gillen (1912) found the webs made by *Nephila eremiana* so large and strong that they proved quite a hindrance to them as they rode through the scrub. 'The web stretches across from tree to tree for a distance of often twelve to fifteen feet and reaches a height in the middle of fully six feet.'

Some other web-building spiders show a tendency towards social habits, and the Australian species *Amaurobius socialis* constructs communal webs measuring up to 3·65 m *12 ft* in length and 1·21 m *4 ft* in width.

The silk produced by *Nephila* spiders is incredibly strong and possesses great elasticity. K McKeown (1952) reports that it is not unusual for a man to have his hat knocked off his head by one of these snares, and F Ratcliffe (cited by Lane, 1955), describes how he blundered into a huge web 'and almost literally bounced off'.

The smallest webs in the world are spun by spiders like *Glyphesis cottonae*. etc., and measure about 19–20 mm *0·75–0·78 in* in diameter.

The longest-lived of all spiders are the primitive Mygalomorphae ('tarantulas' and allied species). One female theraphosid collected by Baerg (1958) at Mazatlan, Mexico, in 1935 and estimated to be ten to twelve years old at the time, was kept in his laboratory for 16 years, making a total of 26–28 years in all, and he believed that this was the normal life-span for females. Another female (*Eurypelma* sp.) lived in the British Museum (Natural History), London, for 14 years and was thought to have been about 20 at the time of her death (Vesey-Fitzgerald, 1967), and Petrunkevitch (1955) mentions a female theraphosid which lived in the Museum d'Histoire Naturelle, Paris, France, for 25 years.

Most male 'tarantulas' reach maturity in eight to nine years and die a few months later, but Baerg reared specimens of the Arkansas species *Dugesiella hentzi* under laboratory conditions which lived for ten, eleven and even thirteen years.

Some of the funnel-web spiders (family Dipluridae) are also long-lived and Barrett (cited by McKeown, 1952), observed one (*Atrax* sp.) which occupied a cavity in a staghorn fern in his garden at Elsternwick, Victoria, Australia, for 17 years.

The longest-lived British spider is probably the Purse-web spider (*Atypus affinis*), which is found in southern England, parts of Wales and the Channel Islands. Females of this mysterious species spend practically most of their life in a *sealed* silken tube buried in the soil and normally live five and a half to seven years in this tiny underground prison, which also serves as a dining room for any insect unfortunate enough to alight on the 50·8 mm *2 in* long part of the tube sticking above the ground. (When the 'dinner gong' rings, i.e. vibrates, the lady of the house dashes upstairs, punctures the tube with her massive fangs, and then drags the struggling victim through the rapidly tearing slit she has made.) One fully mature female collected by Bristowe (1971) and kept in a cool greenhouse survived for five years, which means she was at least nine years old at the time of her death.

The House spider (*Tegenaria derhami*) is another Methuselah, and two individuals kept by Dr Oliver of Bradford lived for five and seven years respectively (Savory, 1928). Some of the large wolf spiders (Lycosidae and Pisauridae) may also live as long as seven years.

The majority of spiders, however, complete their life-cycle in ten to twelve months. One of the shortest-lived species is the deadly 'Black widow' (*Latrodectus mactans*), which must come as something of a relief to many people! A number of males reared under laboratory conditions lived on an average about 100 days, while females averaged 271 days (Deevey & Deevey, cited by Gertsch, 1949).

The most elusive of all spiders are the rare trapdoor spiders of the genus *Liphistius*, which are found only in SE Asia. Like the Purse-web spider (Atypidae), these arachnids also live in silken tubes buried in the ground, but they are much more difficult to detect because the hinged circular door which secures the entrance to the lair is coated on the outside with plant matter and is beautifully camouflaged.

The most elusive spider in Britain, and the most sought after by collectors, is the Lace-web or Carmine jumping spider (*Eresus niger*). This species is known only from seven specimens (six males and one female) collected in Hampshire, Dorset and Cornwall between 1816 and 1906. The females are a dull black in colour, but the smaller males have a brilliant crimson body with black spots and a series of irridescent white rings round the black fore-legs (see illustration page 175). In the early summer of 1832 another male was seen at Kynance Cove, southern Cornwall, but was not captured (Bristowe, 1971), and in the early 1950s there was a second sighting at Sandown, Isle of Wight (Savory, 1966).

Until very recently only one example was known of *Altella lucida* (genus *Argenna*), a male discovered by F M Campbell in his house at Hoddesdon, Hertfordshire, in 1880 (Locket & Millidge, 1951). In December 1972, however, Dr P Merrett of the Nature Conservancy's research station at Furzebrook, Dorset, discovered a female on nearby Morden Heath, and several more have been collected in the same area since (Locket, Millidge & Merrett, 1974).

The highest speed recorded for a spider on a level surface is 1·73 ft/s (=1·87 km/h *1·17 miles/h*) in the case of a female House spider (*Tegenaria atrica*). This may not seem very fast, but Bristowe (1971) has calculated that this particular individual must have covered a distance equivalent to 330 times her own body length in ten seconds.

The same authority also found that although a mature female *Tegenaria* would run at top speed as soon as she was prodded with a pencil, she could not maintain this 'furious pace' for much longer than 15 seconds before collapsing in a heap. In other words, the greater the size, the less functional the breathing apparatus (book-lungs) becomes.

The swiftest terrestrial invertebrates are the long-legged 'Sun spiders' (*Solpugida*), also known as 'Wind scorpions', which are found in the warmer, drier parts of the world. All of the members of this order have a well-developed tracheal (as opposed to book-lungs), and are active, fast-moving predators.

The fastest solpugids are probably *Galeodes arabs* and *G. araneoides*, both of which have a leg-span of up to 127 mm *5 in*. No speed tests have been carried out on these species, but they may reach as much as 16 km/h *10 miles/h* for short distances. Their range extends from Morocco to India and Turkestan (Cloudsley-Thompson, 1958).

The greatest altitude at which a spider has been found is 6705 m *22 000 ft* for a Jumping spider (family Salticidae) collected by R Hingston, the British naturalist-mountaineer, on Mt Everest in 1924, but immature specimens balloon even higher in their own gossamer. Spiders have also been found deep down in coal-mines and caves.

In February 1949 a seisometer at New York City College, NY, USA, recorded an earthquake in excess of magnitude 9, i.e. the worst earthquake in history, (the current world record is 8·9 magnitude), but engineers later found a spider in the mechanism.

Although all of the 25000 or so species of spider are venomous, only a few are dangerous to man. The title of 'the most venomous spider in the world' is open to question, but the notorious 'Black widow' (*Latrodectus mactans mactans*) of the Americas, Hawaii and the West Indies is usually credited with this dubious honour. Females of this race – male spiders cannot inject a lethal dose of venom into the human body – have a bite capable of killing a human being, but fatalities are comparatively rare. This is because it is timid by nature and bites only when frightened. According to Thorp & Woodson (1945) there were 1291 *reported* cases of 'black widow' bites in the USA during the period 1726–1943 (578 in California), but only 55 proved fatal, and most of these were children or elderly people living in rural areas. (This spider likes to make its home under the seats in outdoor privies.)

On 21 September 1947 Harry Carey, the famous American actor, died of heart complications after being bitten by a black widow spider during the filming of *Red River* in Arizona (Zolotow, 1974). In August 1952 a four-month-old baby was bitten by a member of this race at Logansport, Indiana, USA, and died four days later.

The venom of this spider, and others closely related to it, is neurotoxic in action, which means the toxin attacks the junctions between the nerves, and the muscles they control. The symptoms are excruciating pain, temporary paralysis and profuse perspiration, followed by nervousness and anxiety, which can become so acute that several hospital patients have told doctors they thought they were going mad (Vellard, 1936).

In 1936 D'Amour, Becker & Van Riper carried out a series of experiments to determine the toxicity of this arachnid's venom and found that on a dry weight basis it was 15 times more potent than the venom of the Prairie rattlesnake (*Crotalus terrificus*). This is an interesting point, because at one time the Gosiute Indians of Utah, USA, smeared their arrowheads with a lethal mixture of macerated 'black widow' and rattlesnake venom when hunting big game (Chamberlain & Ivie, 1935).

More recently American scientists have discovered that there is a very marked 'seasonal variation' in the toxicity of *Latrodectus* venom. According to the findings of Keegan, Hedden & Whittlemore (1960) specimens collected near San Antonio, Texas in the autumn and 'milked' in the laboratory had a venom which was ten times as toxic as that taken from others in May. Although the reason for this is not known, the tests did reveal that if you are a Texan you stand a much better chance of survival if you are bitten by one of these creatures in the early summer!

The eggs of the 'black widow' spider are also extremely toxic, and Blair (1934) says that two of them crushed and emulsified in a drop of saline and injected intravenously into an adult mouse (20 g *0·76 oz*) killed the rodent very quickly. A rabbit injected intravenously with a few drops of the same emulsion died within two minutes.

The 'Flax spider' (*Latrodectus m. curacaviensis*) of Argentina has also been held responsible for a number of deaths annually, although Pirosky & Abalos (1963) point out that there are no accurate statistics regarding fatalities. (This race of 'black widow' has often been confused with *L. m. mactans*.)

Another member of the genus *Latrodectus* much in the news is the Jockey or Red-backed spider (*L. hasseltii*), which ranges from Arabia right across southern Asia and the Pacific islands

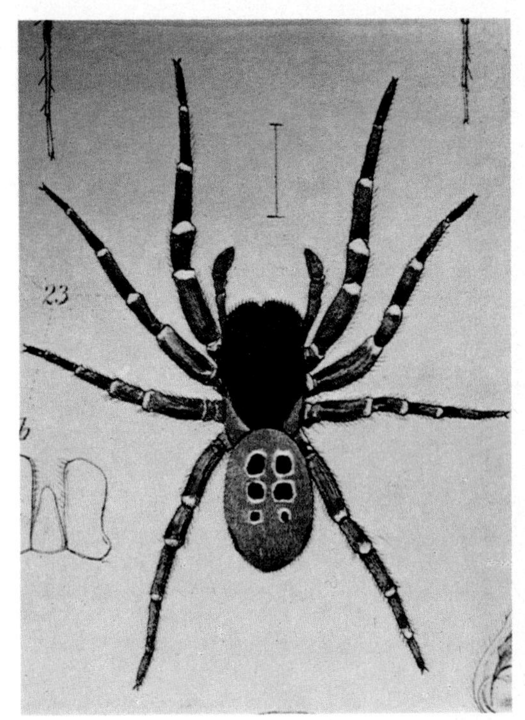

The handsome crimson and black Lace web eresus spider, Britain's most elusive arachnid

to Australia and New Zealand, where it is known as the 'katipo' (night-stinger). According to McKeown (1952) this arachnid has killed *at least* ten people in Australia this century, and there have been five fatalities in New Zealand in the past 60 years. Two of the victims were a girl aged seven and her three-year-old brother who were bitten while asleep after hundreds of these spiders had invaded the country town of Raetahi, South Island, in June 1969 after a minor earthquake.

Latrodectus is also found in all Mediterranean countries, and the typical sub-species *L. mactans tredecimguttatus* (=*L. tredecimguttatus*) was responsible for the virulent epidemics which occurred in Spain in 1833 and 1841, and in Sardinia, Italy, in 1833 and 1839 (Maretic, 1971). In untreated victims the duration of the illness was up to eight days, followed by a long convalescence, but a small number of fatalities were reported. There was also another epidemic in Italy and Yugoslavia during the period 1938–58, but in the 946 cases studied only two people died (Bettini, 1964).

The Button-spider (*Latrodectus indistinctus*) of South Africa is also much feared, and Bucherl (1971) says a number of deaths have been recorded, although Clark (1969) could find only one fatality since 1945 – an eight-year-old girl.

The Funnel-web spider (*Atrax robustus*) of New South Wales, Australia, also has an unsavoury reputation. This species first made the news in 1927 when a two-year-old boy was bitten by one and died within 90 minutes. Since then at least five other deaths have been reported. On 28 December 1970 a 17-year-old pregnant woman was bitten by a funnel-web spider while walking through bushland near Nowra, 152 km *95 miles* south of Sydney. She was on a camping holiday with her husband and suddenly felt something crawling inside her blouse, then a bite on the breast. Shortly afterwards she collapsed and was rushed to hospital, where doctors worked through the night in a vain attempt to save her and the unborn baby.

In December 1971 doctors saved the life of a three-year-old girl in Sydney who had been bitten twice by a funnel-web spider by treating her with the drug athrophine. The child was found with the spider still clinging to her arm.

The Podadora or Bola-spider (*Glyptocranium gasteracanthoides*) of Argentina and the 'Black tarantula' (*Sericopelma communis*) of Panama have also been credited with fatalities.

The Brown recluse spider (*Loxosceles reclusa*) of the central and southern USA and more recently Australia (accidentally introduced) has been said to be even more venomous than the 'black widow', but the majority of people bitten by this arachnid experience only a local reaction. According to Breland (1963) only one death has been attributed to this species, and that was a four-year-old boy, but the records of the Texas State Department of Health list four deaths during the ten-year period 1950–60. Two of them were elderly women, one a six-year-old boy and the other a 46-year-old man (Micks, 1963). The situation is confused, however, by the fact that at least four species of *Loxosceles* (*L. reclusa*, *L. devia*, *L. arizonica* and *L. unicolor*) are known to occur in Texas (Gertsch, 1958), although most of these may have only sub-specific or populational range.

Unlike *Lactrodectus*, the spiders of the sub-family Loxoscelinae have a powerful necrotoxic venom. This means the bite produces an ulcerating wound which often turns gangrenous, and infections are common.

In December 1972 Miss Anna Kashfi, the actress and former wife of Marlon Brando, was bitten by a brown recluse spider while working on a film location. She made a full recovery, but had to undergo skin grafts on her right ankle.

The most dangerous spider in the world is the 'Aranhạ armedeira' (*Phoneutria nigriventer*), the largest true spider of South America, which frequently enters human dwellings and hides in clothing or shoes. Apart from being the most aggressive of all the highly venomous spiders, it also has the largest venom glands (up to 10·4 × 2·70 mm *0·4 × 0·10 in*) of any living spider, including *Theraphosa blondi* (up to 12 × 1·6 mm *0·47 × 0·06 in*), and when approached will bite furiously several times (Bucherl, 1971). Under laboratory conditions it has been discovered that only 0·006 mg *0·00000021 oz* of this spider's neurotoxic venom is needed to kill a 20 g *0·706 oz* mouse when injected into a vein, and 0·0134 mg *0·00000046 oz* when injected under the skin. (cf. 0·110 mg *0·0000038 oz* and 0·200 mg *0·000007 oz* respectively for *Lactrodectus m. mactans* and 0·200 mg *0·000007 oz* and 0·300 mg *0·000010 oz* for *Loxosceles rufipes*). Specimens 'milked' by electric shock at the Instituto Bhutantan, São Paulo, Brazil, yielded an average dry venom weight of 1·25 mg *0·000044 oz* and a maximum of 8 mg *0·00028 oz* (cf. 0·60 mg *0·00021 oz* and 1·3 mg *0·000045 oz* for *L. m. mactans*), the latter amount being enough to kill over 500 adult mice if injected intravenously. The lethal dose for an average adult man (69·85 kg *154 lb*) is not known, but as 0·10 mg *0·0000035 oz* of the less potent *Loxosceles* venom is said to seriously endanger human life, the amount must be extremely small.

Although the Bird-eating spiders (Avicularoidea = Mygalomorphae) of South America and elsewhere can inflict a deep wound with their formidable jaws – the fangs of *Theraphosa blondi* measure 9·80–12·60 mm *0·38–0·49 in* in length – the venom they secrete is generally of a low potency, and has only a local effect in man.

There are exceptions, however, and the most venomous species is the Funnel-web 'tarantula' (*Trechona venosa*) of South America which ranks second only to *P. nigriventer* in terms of neurotoxic potency. Fortunately this creature is sedentary by nature and no human fatalities have been reported, but Vellard says white rats bitten by this arachnid died *within seconds*.

The most dangerous species is probably the aggressive *Euctimsna tibialis* of Australia which has caused at least one death since 1926 (Butler, 1934).

The most feared spiders in the USA after the 'black widow' and the brown recluse are the Great hairy 'tarantulas' (*Aṿicularia*) of the southern states, and some incredible stories have been built up about the 'deadly' nature of its bite.

In the old frontier days it was generally believed that whisky was the only successful cure for a person bitten by one of these monsters, and the fiery liquid was often referred to as 'tarantula juice'. According to Thorp & Woodson (1945) one crafty old Indian became so fond of the remedy that he used to carry a tame 'tarantula' around with him. Finding himself conveniently outside a grocery store, he would suddenly go into a routine of thrashings and screams before staggering into the store, where he would exhibit the spider as proof that he had been bitten. The act must have been very convincing, because one contemporary said he usually ended up being 'gratuitously irrigated with the worthy antidote'.

The venom of *Lasiodora klugi*, the world's heaviest and most formidable looking spider, is only deadly to small animals (the lethal dosages for an adult mouse are 0·640 mg *0·000022 oz* and 1·20 mg *0·000042 oz* respectively); nevertheless, the spiders of this genus can still inflict a very nasty wound with their huge fangs which have been measured up to 8·30 mm *0·32 in* in length.

The largest of the 700 known species of scorpion is *Heterometrus swammerdami* of India, males averaging 180 mm *7·08 in* from the tips of the pincers to the end of the sting. A specimen collected in Madras province, India on 14 September 1869, and now in the possession of the Bombay National History Society, measures 247 mm *9¾ in* in total length (tip of head to point of sting 187 mm *7·37 in*, length of body 82 mm *3¼ in*). Another one found under a rock in the village of Krishnarajapuram, India, by S Burgess during the Second World War and taken to the local military hospital for examination was even larger, measuring 292 mm *11½ in* in total length.

Pandinus imperator of West Africa is also of comparable size. One scorpion preserved in the British Museum (Natural History), London, has a total length of 215 mm *8½ in*. In October 1931 an exceptionally large specimen was received at London Zoo from Ghana. It measured 228 mm *9 in* in total length and was described as the biggest and most perfect example in living memory. *Hadogenes troglodytes* and *H. trichiurus* of South Africa are also very large, males having been measured up to 180 mm *7·08 in* (Bucherl, 1971). A total length of 203 mm *8 in* has been reported for an African rock scorpion (*Pandinus viatoris*), but the average measurement is about 102 mm *4 in*.

The smallest scorpion in the world is *Microbuthus pusillus* from the Red Sea coast, which measures about 13 mm *0·51 in* in total length (Savory, 1964). *M. fagei* from the coast of Mauritania reaches 17–18 mm *0·66–0·70 in*.

Although the scorpion has a very evil reputation, it is largely undeserved, because the venoms of most species produce only a mild to severe local reaction in man (i.e. a sharp burning sensation which may last anything from a few minutes to several hours). The most dangerous scorpions are the buthids. All the members of this family carry very powerful neurotoxic venoms, but they vary greatly in the degree of virulence.

The world's most dangerous scorpion is the large North African species *Androctonus australis*. This scorpion can deliver a massive dose of neurotoxic venom which has been known to kill a man in four hours and a dog in about seven minutes (Millot & Vachon, 1949).

Between 1936 and 1950 *A. australis* was responsible for nearly 80 per cent of the 1300 reported cases of scorpion stings in Algeria, 377 of which proved fatal, but the death-roll would have been much higher if the patients hadn't been treated with an antitoxin produced by the Pasteur Institute in Algiers. In another review of scorpions in North Africa covering a period of 17 years (1946–63) there were 20164 cases of scorpion stings in southern Algeria, but only 386 were fatal, thanks to French medical science. According to Balozet (1971) this species is responsible for 80 per cent of the accidents and 95 per cent of the fatalities, although some other dangerous buthids like *A. aeneas* and *A. amoreuxi* might also be involved (Sheals, 1973).

In September 1938 72 cases of dangerous scorpion sting (mostly *A. australis*) were reported in a *single day* from districts near Cairo, Egypt. On 27 September 1936 nine Pathans sleeping in a guest-house at a village near Peshawar, NW Frontier, Pakistan, were stung by a scorpion (*Androctonus* sp.) and eight of them died within a few minutes. This was reportedly the biggest death-roll from a single scorpion that the Frontier had ever known.

Three scorpions of the genus *Centruroides* (*C. noxius*, *C. suffusus* and *C. limpidus*) found in Mexico are also very dangerous. Data compiled by the State Statistical Dept 1940–9 show that annual deaths from scorpion stings in the Mexican Republic varied between a minimum of 1588 deaths in 1943 up to a maximum of 1944 deaths in 1946. In another survey covering 1957–8 the Institute of Health and Tropical Diseases found that there was a decrease in the number of deaths, the figures being 1495 and

The Fat-tailed scorpion of North Africa, the world's most dangerous scorpion (Peter Ward)

1107 respectively. The total number of fatalities was 20352 (mostly juveniles), while during the same periods 2068 people died from snakebite, 274 spider bite and 1933 from the bites or stings of unidentified venomous animals (Mazzotti & Bravo-Becherelle, 1963).

Two other species, *C. sculpturatus* and *C. gertschi*, restricted to the southern parts of Arizona and New Mexico, USA, also have a nasty reputation and they were responsible for 64 deaths in Arizona between 1929 and 1948 – twice as many fatalities as all other venomous animals combined in that state (Stahnke, 1963).

A number of fatalities have also been attributed to the Brazilian scorpions *Tityus serrulatis* and *T. bahiensis*, both of which have a venom which is even more toxic than that of the Phoneutria spiders. Another member of this genus, *Tityus trinitatis*, has been responsible for deaths in Trinidad. During the five-year period 1929–33, 33 of 698 scorpion sting cases proved fatal (Waterman, 1938).

In most of the above cited cases the sting entered the sole of the naked foot.

During the Second World War a number of soldiers fighting on both sides in the Western Desert, North Africa, died after being stung by scorpions of the genus *Buthacus* (probably the widespread *B. occitanus*).

The stings of South African scorpions (*Buthacus* sp.) are seldom fatal, even among children, although Clarke (1969) says there are at least two fatalities a year in that country. In February 1936 a plague of scorpions descended on Malmesbury, Cape Province, and stung three African children to death, and in November 1959 a two-year-old African girl died in the village of Kenhardtt, Cape Province, ten minutes after being stung by one of these creatures. If the sting enters a vein death comes very quickly, and in one case reported in December 1966 a twelve-year-old girl in Palapye, Botswana, succumbed within five minutes.

Although the huge African scorpions *Pandinus* and *Heterometrus* have the largest stings of any scorpions and can inject a massive dose of venom, the potency is low and human fatalities are extremely rare.

The most venomous scorpion in the world is *Leiurus quinquestriatus* which ranges from the eastern part of North Africa through the Middle East to the shores of the Red Sea. Fortunately the amount of venom it delivers is very small and adult lives are seldom endangered, but fatalities have been reported among young children under the age of five, the mortality rate being as high as 60 per cent (Wilson, 1904).

Persons stung by scorpions carrying a powerful neurotoxic venom quickly develop a tightness in the throat and speech becomes slurred. They then become restless and start perspiring profusely, followed by vomiting and involuntary twitching of the muscles. Eventually breathing becomes laboured, and shortly before death the arms and legs sometimes turn completely blue.

Section IX

INSECTS

(class Insecta)

An insect is primarily a terrestrial invertebrate which breathes through a system of air tubes. The body is divided into three main parts, the head, thorax and abdomen, and is covered by a horny covering or external skeleton which is shed periodically. The head bears a pair of antennae and three pairs of feeding appendages, and there are usually one or two pairs of wings rising from the thorax and three pairs of legs. The capacity for flight is one of the most striking features of adult insects, but not all species have wings. The young usually hatch from eggs after leaving the female, and some are born alive as fully developed larvae.

The earliest known insect was *Rhyniella proecursor*, which lived about 370000000 years ago. Its remains have been found in Aberdeenshire, Scotland.

There are about 1000000 living species of insect, and the class is divided into 29 orders. The largest order is *Coleoptera* (beetles), which contains about 3000000 species or 30 per cent of the total number.

The heaviest insect in the world is the Goliath beetle *Goliathus goliatus* of equatorial Africa. Adult males weigh 70–100 g *2·47–3·53 oz* (females

The huge Goliath beetle compared with a Seven-spot ladybird (Harold Bastin)

The bulky Elephant beetle (*Megasoma*), one of the heaviest insects in the world (C Williams)

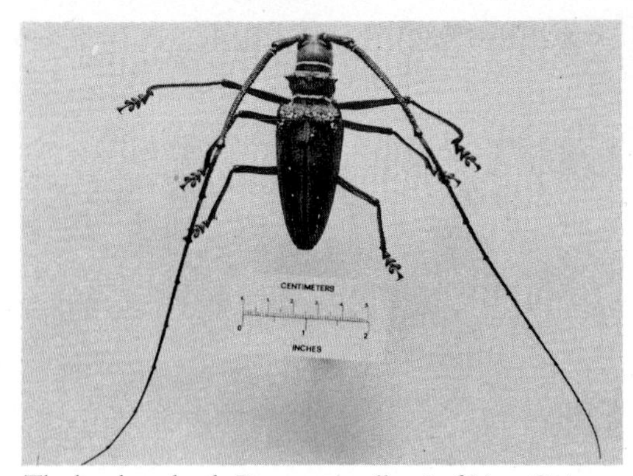

The longhorn beetle Batocera wallacei *of New Guinea has been measured – including antennae – up to 266 mm* 10·5 in *in total length (Eric Lindgren)*

are smaller) and have been reliably measured up to 120 mm *4·72 in* from the tip of the frontal horn to the end of the abdomen. This size is closely matched by *Goliathus regius* of West Africa, and Mr Terry Taylor of Rosemead, California, USA, a leading insect supplier, says he once sold a male which measured 130 mm *5·11 in* in length (Williams, pers. comm. 15 January 1976). Both these species and other goliaths have a massive build-up of heavy chitin forming their thorax and anterior sternum which gives them a distinct weight advantage over other large scarabs of comparable size. The goliath beetles are also the heaviest flying insects, and African children sometimes amuse themselves by tying a piece of string to one of these creatures and flying it in circles. The Elephant beetles (*Megasoma*) of tropical America are also extremely bulky and probably run second to *Goliathus* for weight. The heaviest member of the family is generally thought to be *M. elephas*, which has been measured up to 131 mm *5·15 in*, but *M. chiron* (120+ mm *4·72 in*) has a shorter cephalic horn and is larger bodied. The very rare Longhorn the beetle *Titanus giganteus* of the Amazon basin – natives consider the huge larvae a great delicacy – also exceeds 120 mm *4·72 in* (excluding antennae), and 95 per cent of this is solid beetle. In 1957 Dr Paul Zahl (1959), a naturalist working for the *National Geographic Magazine* in Washington, DC, collected 15 specimens in northern Brazil, four of which measured more than 152 mm *6 in* in length, and a measurement of 190+ mm *7·48 in* has been reliably reported elsewhere for an outsized male. Unfortunately no weight data has been published for this beetle, but as it lacks the body depth and width of *Goliathus* and *Megasoma* it cannot be considered a serious rival.

The female Spiny stick insect (*Extatosoma tirartum*) of Australia has been credited with a weight of 113 g *4 oz* after heavy summer feeding, but this weight is impossibly high for a giant phasmid. The largest female reared by D J Moon, Editor of the *Exotic Entomology Group* (pers. comm. 18 August 1974) was 'an enormous, full-fed, egg-laden beast' which tipped the scales at 27·5 g *0·97 oz*, and he concluded 'Tirartum might conceivably attain a little more than say 30 g, but it would be difficult to imagine one getting up to the size of a goliath beetle.'

Since then, however, Mr Moon (pers. comm. 11 November 1975) has received details of a Malaysian spiny stick insect (*Heperopteryx dilatata*) which he says dwarfs even *Extatosoma tirartum*. He thinks the female of this species may reach 50–60 g *1·76–2·12 oz* after heavy summer feeding, but this is only a calculated guess and the matter will have to remain open until accurate weight data is obtained.

The heaviest insect found in Britain is the Stag beetle (*Lucanus cervus*) which is found in an area stretching from Cornwall to Kent and north just into Suffolk. An adult example of each sex collected by Mr J T Clark (pers. comm. 23 October 1972) weighed 4510 mg *0·159 oz* and 4075 mg *0·143 oz* respectively ('larger adults are known than the ones accessible to me'), and one of his larvae tipped the scales at 10 575 mg *0·373 oz*. In one series of 324 males collected in the Colchester area between 1963 and 1966 by the same authority the body lengths (including mandibles) ranged from 31 mm *1·22 in* to 68 mm *2·67 in*, with a mean value of 49·8 mm *1·96 in*. The record specimen is a male in the British Museum (Natural History) collected in Sheer-

*The Stag beetle, Britain's heaviest insect (*Evening Post*)*

ness, Kent, in 1871 which measures 77·4 mm *3·04 in* in length and probably weighed nearly 9000 mg *0·31 oz* when alive. Another male measuring 76 mm *3 in* was captured at Priest Hill, Caversham, Berkshire, in June 1969 and later presented to Reading Museum.

In October 1939 London Zoo received a large male stag beetle which came with the following letter: 'The defendant was found lying on the pavement in Regent Street, outside the Cafe Royal, at 11.35 pm on the evening of the 4th inst. As he was unable to stand, was causing an obstruction, and inclined to resist, I took him in charge. He did not appear to have any visible means of support and would give no account of himself. A strong tendency to alcoholism was shown by the fact that he refused a meal of drops of ten per cent cane sugar solution, but voraciously attacked the same solution when fortified with two or three drops of sherry. He had probably been overcome at a stag party.'

The longest insect in the world is the tropical stick insect *Pharnacia serratipes*, females of which have been measured up to 330 mm *12·99 in* in length (head and body). A female example of *Phobaeticus fruhstorferi* from Burma preserved in the US National Museum, Washington, DC, measures 300 mm *11·81 in*. A female *Pharnacia maxima* with a head and body length of 270 mm *10·63 in* preserved in the British Museum (Natural History) has a total length of 510 mm *20·07 in* with legs fully extended (Meadows, pers. comm. 10 September 1971).

The longest known beetle (excluding antennae) is the Hercules beetle (*Dynastes hercules*) of Central America and northern South America which has been measured up to 190+ mm *7·48 in*, but half of this length is accounted for by the horn from the thorax. The two largest specimens collected by Beebe (1947) in Avagua National Park, Venezuela, in 1954 measured 74 mm *2·91 in* and 76 mm *2·99 in* excluding the horn and weighed 15 g *0·53 oz* and 16·3 g *0·57 oz* respectively. The longhorn beetle *Batocera wallacei* of New Guinea has been measured up to 266 mm *10½ in*, but 190 mm *7½ in* of this is antennae.

The smallest insects recorded so far are the 'hairy-winged' dwarf beetles of the family Ptiliidae (=Trichopterygidae). One tropical species *Nasonella fungi* measures only 0·25 mm *0·0098 in* (Stanek, 1970), but Evans (1975) says the status of this species is uncertain. The parasitic wasps of the family Myrmaridae are also very tiny and some species like *Camptoptera papaveris* can easily crawl through the eye of a needle.

The most numerous of all insects are the Spring-tails (order Collembola) which are distributed throughout the world including the polar regions. It has been calculated that the top 228 mm *9 in* of soil in one acre of grassland contains 230000000 spring-tails, or more than 5000/ft².

The majority of insects live less than a year, but there are some Methuselahs. One of them is the North American cicada (*Megicica septemdecim*), which is also known as the Seventeen-year locust. Northern broods of this species live underground in a larval state for about 17 years before digging to the surface and living as adults for a few weeks.

Some beetles are also long-lived. In 1846 Sir John Richardson showed members of the British Association a specimen of the Burying or Sexton beetle (*Necrophorus germanicus*) which had been found embedded in concrete and must have been at least 16 years old. The creature was still alive when it was brought to Sir John and lived for another six weeks (Timbs, 1868). A Cellar beetle (*Blaps gigas*) kept by Labitte (1916) lived for 14 years 233 days.

In 1889 the 30 mm *1·18 in* long larva of a Metallic wood-boring beetle (*Burretis splendens*) emerged from the step of a staircase in the library of the University of Copenhagen, Denmark. The stairs, made from foreign pine-wood, were laid in 1860, which meant the beetle was at least 29 years old (Meinert, 1889). A similar case was reported in March 1966 when a 25 mm *1 in* long larva suddenly appeared from the step of a staircase in a house at Thorpe Bay, Essex, after living in the pine-wood for more than 30 years. (The tree in which this beetle was born had been sawn up and shipped to Britain from Canada in 1936.) The specimen, identifed as *B. aurulenta*, was killed and sent to the Forest Products Research Laboratory in Princes Risborough, Buckinghamshire, for analysis. There is also a record of a *B. splendens* living in the wood of a pencil-box for 30 years (Burr, 1954), and unconfirmed claims up to 45 years have been reported for this species.

Certain species of ant (Formicoidea) also have long life-spans. Sir John Lubbock (1894) says he kept a female Small black ant (*Formica fusa*) from December 1874 until August 1888 when it was at least 15 years old, and a queen Westwood's ant (*Stenamma westwoodi*) held captive by Donisthorpe (1936) for 15 years was believed to have been about 18 years old at the time of her death on 2 October 1935.

The longest-lived insects are queen termites (Isoptera), which, although they look like ants, are not closely related. Some have been known to lay eggs for up to 50 years. In 1872 the top of a 4·26 m *14 ft* high termite nest of the Australian species *Nasutitermes triodiae* was broken off to make way for overhead telegraph wires in Queensland, and when it was examined again in 1935 it was still in a good state of repair and had a flourishing population. This means that the king and queen who founded this colony were probably still alive, because secondary reproductions are unknown in this species (Hill, 1942). Klots & Klots (1961) believe that some queen termites may live 100 years or more, but this has not yet been proven.

The shortest-lived of all insects is probably the Common house-fly (*Musca domestica*). In one study of over 8500 specimens males had an average life-span of 17 days and females 29 days (Rockstein, 1959). Some may-flies (Ephemeroidea) survive only a few hours in the adult stage, but they spend two to three years as larvae.

The loudest of all insects is the male cicada (family Cicadidae). The tymbal organs or membranes at the base of the abdomen, which vibrate from 120 to 480 times a second, produce a sound that is detectable more than a quarter of a mile away. According to scientists at Princeton University, New Jersey, USA, the sound produced by thousands of cicadas under a single tree registers between 80 and 100 db at a distance of 18·28 m *60 ft*.

The only British species is the very rare Mountain cicada (*Cicadetta montana*), which is confined to the New Forest area in Hampshire.

The highest gravity force encountered in nature is the 400+ *g* endured by the Click beetle *Athous haemorrhoidalis* when jack-knifing into the air to escape predators. Unlike the flea, however, the British species takes off from a prone position and *without using its legs*! According to Dr Glyn Evans (1972), who has done some research on this 12 mm *0·47 in* long creature at Manchester University, it can catapult itself into the air to a height of at least 300 mm *11¾ in* when theatened by utilising a trigger mechanism similar to the back-breaker type of mousetrap. He also discovered that (1) the parts of the beetle farthest from the central pivot travel at an even greater acceleration, the brain being subjected to a peak deceleration at the end of the movement of 2000 *g*; and (2) that while leaping into the air it both somersaults and rotates on its axis. 'Yet', he says, 'the beetle is prepared to click again and again, so one assumes that it doesn't even get a bad headache. Presumably the secret lies in the very small mass and inertia involved, and the fact that it probably isn't a very intelligent animal to start with.'

The most dangerous insects in the world are the malarial mosquitoes (genus *Anopheles*) which, if we exclude wars and accidents, have probably been responsible directly or indirectly for 50 per cent of all human deaths since the Stone Age (Weimer, 1961). At the present time these tiny creatures still kill at least 1000000 people a year in Africa and SE Asia, and an estimated 600000000 are constantly exposed to the disease. Four species of malarial mosquito are found in Britain, but they can only carry malaria if they come into contact with someone suffering from the disease. The North Kent marshes have long been breeding grounds for this demon, and when soldiers carrying the disease returned to the area from abroad after the First World War it was quickly spread by the mosquitoes. At one stage half the population of the Isle of Grain were affected, and as recently as the 1940s malaria cases were still being noted. In 1974 workers building the new Grain power station at the edge of the marshes complained of being bitten by mosquitoes, and a specialist firm was called in to deal with the menace. There are also a small number of cases of people dying in this country from mosquito dermatitis after being bitten on the face or neck. In each instance the person was

The Desert locust, the most destructive insect in the world (Crown Copyright)

highly allergic to the salivary secretion of the insect.

The most potentially dangerous insect in the world is the Common house-fly (*Musca domestica*). It can transmit 30 diseases and parasitic worms to man, including cholera, typhoid, dysentery, bubonic plague, leprosy, cerebro-spinal meningitis, diphtheria, scarlet fever, smallpox and infantile paralysis via 'vomit drop' (ingested food which has been allowed to flow back to the outer end of the proboscis).

If we exclude disease carriers, then the most dangerous insects are bees (Apidae) and wasps (Vespidae) in that order. Between them they kill more people each year than the estimated 40000 who die from snakebite.

In September 1964 a young Rhodesian set up an unenviable world record when he was stung 2243 times by a swarm of Wild bees (*Apis adonsonii*). The attack took place on a river bank, and the youth was forced to take refuge in the water where he remained for four hours with only his head showing. When he was eventually rescued his scalp, face, neck, trunk and arms were black with stings. More than 2000 were removed from his eyelids, lips, tongue and mouth alone. To the astonishment of everyone he made a complete recovery.

In actual fact the sting of this African bee is no more potent than that of the ordinary honey-bee, but unlike their relatives these winged killers attack in great numbers and with ferocious intensity. On one occasion a scientist juggled a leather ball outside a hive and saw it stung 92 times in five seconds before he himself was chased for half a mile by the angry insects. In 1956 thousands of African bees were imported into Brazil from Tanzania to improve honey production, but the following year a number of queens and their swarms escaped. They mated with local honey-bees, and soon afterwards the new vicious hybrids began moving north at a rate of 320 km *200 miles* a year. Since then they have stung to death at least 150 people – the man who imported them has had his life threatened several times by families of the bees' victims – and all attempts to halt them have been

A close-up shot of a Desert locust (Crown Copyright)

unsuccessful so far. At their present rate of progress they should reach Texas between 1988 and 1994, but by then it is hoped that their aggressive behaviour will have been considerably diluted by further hybridisation with millions of docile bees set down in their migratory path.

In July 1970 an 80-year-old woman died after being attacked by a colony of bees she kept in her garden at Selsdon, Surrey. More than 300 stings were removed from her body.

The most destructive insect in the world is the Desert locust (*Schistocera gregaria*), the locust of the Bible, whose habitat is the dry and semi-arid regions of Africa, the Middle East, Pakistan and India.

The desert locust is also mentioned by Paulus Orosius (*fl.* 415 BC), the Spanish historian and theologian, who wrote: 'In the year of the world 3800 such infinite myriads of locusts were blown from the coast of Africa into the sea and drowned that, being cast upon the shore, they emitted a stench greater than could have been produced by the carcases of one hundred thousand men, and caused a general pestilence.'

This species of locust, which in fact is a short-horn grasshopper, can eat its own weight in food a day, and during long migratory flights of anything up to 3200 km *2000 miles* a large swarm will consume 3054 tonne *3000 ton* of greenstuff a day and bring famine to whole communities.

The greatest swarm of desert locusts ever recorded was one covering an estimated area of 3200 km^2 *2000 miles*^{2} seen by Fletcher in South Africa in 1784. He says this swarm was blown out to sea by a strong wind, and when the tide washed back their bodies they formed a bank 1·21 m *4 ft* high along the beach for 80 km *50 miles*. In 1889 another swarm also covering an area of 2000 *miles*^{2} and containing an estimated 250000000000 insects weighing about 508000 tonne *500000 ton* was observed crossing the Red Sea.

Because of the practical difficulties of measurement very little reliable information has been published on the flight speeds of insects, but their general small size gives them the false impression of being very much faster than they really are. One of the first men to carry out experiments in this field was Demoll (1918), the German entomologist. He released a number of different species in a room lit by a single window and timed them by stopwatch as they flew directly from the dark side of the room to the light. It was found that the swiftest flier was a Hawk-moth (family Sphingidae) which travelled at a rate of 15 m/s 53·6 km/h *33·5 miles/h*. A Horse-fly (*Tabanus bovinus*) and a Dragon-fly (family Agrion) were not far behind with a speed of 14 m/s 50 km/h *31·25 miles/h*. The other insects tested were much slower. A Common house-fly (*Musca domestica*) achieved a velocity of 2·3 m/s 8·19 km/h *5·12 miles/h*, a Honey-bee (*Apis mellifera*) 3·7 m/s 11·6 km/h *7·26 miles/h* and a Bumble-bee (*Bombus lapidarius*) 5 m/s 17·85 km/h *11·16 miles/h*.

The most accurate lists of insect flying speeds published so far, however, are those of Magnan (1934), the French entomologist, who obtained his statistics by two methods. In one series of experiments he tethered the insect to a thread wound round a small drum mounted on ball-bearings, the revolutions of which were recorded on a kymograph as the creature was in flight. In another series he timed the insect with a chronometer (aided with cinephotography) as it flew between two markers set against a grid at a measured distance. Of the 32 species of insect tested by Magnan, the fastest was the Dragon-fly (*Anax parthenope*) which travelled at a rate of 8 m/s 28·57 km/h *17·86 miles/h*. A Hornet (*Vaspa crabro*) recorded a velocity of 6 m/s 21·42 km/h *13·39 miles/h* and the fastest of the five species of flies tested was a Horse-fly (*Tabanus bovinus*) which travelled at 4 m/s 14·28 km/h *8·93 miles/h*. These speeds are considerably lower than those given by Demoll, but the French investigator pointed out that the measurements he obtained did not represent the full maximum velocity of the insects as they fly in nature. In the final analysis he felt that certain insects could reach speeds of 10 m/s 35·72 km/h *22·33 miles/h*. From these observations, it would appear that hawk-moths, horse-flies and certain species of dragon-fly are the fastest-moving insects, all of which can exceed an air speed of 32 km/h *20 miles/h* in level flight over short distances.

Dr R J Tillyard (1917), the Australian entomologist, says he once timed a large dragon-fly of the species *Austrophlebia costalis* at 98 km/h *61·3 miles/h* along a stretch of stream 73–82 m *80–90 yd* but this was ground speed. According to the calculations of Hocking (1953) a dragon-fly this size could not possibly fly faster than 57·6 km/h *36 miles/h* even in short bursts, over a level course, and the absolute maximum air speed over a sustained distance would be about 38·4 km/h *24 miles/h*. He added, however: 'This does not necessarily mean that Tillyard's observation is not reliable; a following wind, downhill flight and the short distance (850 × body length) taken together could account for the difference in the figures. It is well known that muscle may develop power much greater than normal for very brief periods. It is worth noting that even the great fossil species *Meganeura monyi*, with its 29 inch wing-span would on a similar basis only have been capable of 43 mph; indeed it probably had to fly at close to this speed in order to remain airborne.'

In April 1926 Dr Charles Townsend, an American zoologist, startled the scientific world by announcing that the Deer bot-fly *Cephenemyia pratti* was the fastest-moving living creature. Writing in the *Journal of the New York Entomological Society*, he said: 'On 12000 foot summits in New Mexico I have seen pass me at an incredible velocity what were certainly the males of Cephenemyia. I could barely distinguish that some had passed – only a brownish blur in the air of about the right size for these flies and without a sense of form. As close as I can estimate, their speed must have approximated 400 yards per second.' (Equivalent to 1308 km/h *818 miles/h*!)

Astonishing as it may seem, a great number of people took this claim seriously, despite the fact that no instruments were used to measure the speed, and the figure was widely quoted for a number of years. It was eventually shot down in flames by Dr Irving Langmuir (1938), the Nobel Prize winner for chemistry in 1932, who carried out some calculations and experiments on a model of a deer bot-fly in his laboratory. He said the fly would have to develop the equivalent of 1·5 hp and consume one and a half times its own weight in food per second to acquire the energy that would be needed to reach a velocity of 1280 km/h *800 miles/h*, and even if this was possible the fly would still be crushed by the air pressure and incinerated by the friction. He also revealed that a whirling piece of solder the same size as the fly was 'barely visible' at 41·6 km/h *26 miles/h*, 'a very faint line' at 69 km/h *43 miles/h* and 'wholly invisible' at 102 km/h *64 miles/h*. He deduced from these tests that 'a speed of 25 miles per hour is a reasonable one for the deer fly, while 800 miles is utterly impossible'.

Despite these very careful calculations, however, Langmuir's figure of 40 km/h *25 miles/h* is probably too low. This is because he carried out his tests in *normal* air conditions and used a model of a 10 mm *0·39 in* deer bot-fly (*Chrysops*) instead of a 15 mm *0·59 in Cephenemyia* – there was some confusion over identity. But even if we make due allowances for these differences, it is still doubtful whether this high-altitude fly ever exceeds 56–64 km/h *35–40 miles/h* in level flight, even in the rarified air of its normal habitat.

The frequency of wing-beat among insects varies enormously and is closely linked to bodily size.

The highest wing-beat frequency so far recorded for any insect under natural conditions is 1046 c/s (=62760 a minute) by a tiny midge of the genus *Forcipomyia*. This measurement was based on an aural estimate by Dr Olavi Sotavalta (1947), the remarkable Finnish entomologist, who has a sense of absolute pitch accurate to within 2·5 per cent of a cycle. Later he told Frank Lane (1955) that another specimen which had been truncated and exposed to a temperature of 37 °C 98·6 °F produced a flight tone equivalent to a frequency of 2218 c/s (=133080 a minute). Measurements calculated from pitch-of-sound experiments, however, have been questioned by some scientists who contend that such values are 50 per cent too high. The highest figure obtained by Magnan (1934), who used high-speed cinematography and a tuning fork in his tests, was 250 c/s (=15000 a minute) for the Honey-bee (*Apis mellifica*).

The lowest frequency of wing-beat reported for any insect is 5 c/s (=300 a minute) for a Swallowtail butterfly (*Papio machaon*)

(Sotavalta, 1947). Most butterflies beat their wings at a rate of 460–636 a minute.

The largest ant in the world is the Ponerine ant *Dinoponera gigantea* of Brazil, workers of which have been measured up to 33 mm *1·31 in* in length. Workers of the rare Bull ant *Myrmecia brevinoda* (= *M. gigas*) of Queensland and northern New South Wales, Australia, have been measured up to 37 mm *1·44 in* in length, but this species has much longer mandibles than *Dinoponera gigantea* and loses out in terms of bulk. Both these sizes, however, are exceeded by the queens of the genera *Dorylus* and *Anomma* of Africa after hypertrophy, and Step (1924) says the huge wingless queens of the species *Dorylus helvolus* of South Africa are sometimes nearly 51 mm *2 in* in length.

The largest ant found in Britain (27 species) is the Wood ant (*Formica rufa*), males reaching 9 mm *0·35 in* and queens 11 mm *0·43 in*. The Blood-red robber ant (*F. sanguinea*) and the Meadow ant (*F. pratensis*) are also of comparable size, and a measurement of 11·5 mm *0·45 in* has been reported for a queen of the latter species (Donisthorpe, 1927).

The smallest ant in the world is the worker minor of *Oligomyrmex bruni* of Sri Lanka, which measures 0·8–0·9 mm *0·031–0·035 in* (Yarrow, pers. comm. 21 November 1971).

The smallest ant found in Britain is *Solenopsis fugax*, workers measuring 1·5–3 mm *0·059–0·18 in*.

The most dangerous ant in the world is the Black bulldog ant (*Myrmecia forficata*) of the coastal regions of Australia and Tasmania which uses its sting and jaws simultaneously when attacking. In November 1936 a 50-year-old man died at Mount Macedon, Victoria, after being 'bitten' by one of these ants. Another fatality occurred in September 1963 when a woman was bitten on the foot by a black bulldog ant in her suburban garden at Launceston, Tasmania, and died 15 minutes later.

The Fire ants (Myrmicinae) of Argentina and *now* the southern USA also have a very unpleasant reputation, and Clarke (1969) says at least one death has been attributed to them.

The Army ants (*Eciton*) of South America and the Driver ants (Dorylinae) of Africa also have a nasty reputation – but for an entirely different reason. When searching for food they move off in highly organised columns numbering anything up to 150000 individuals and will devour any animal that is too slow to get out of the way (even anteaters flee when they see a horde of army ants on the march). As far as humans and other large mammals are concerned, however, the fearsome reputation of these ants is not really justified, although a tethered horse, an injured man in a hammock or a baby in a cot would be doomed. (On 2 March 1922 Prof Lefreoy startled his audience at the Royal Institute in London by citing an authentic case of a baby which had been eaten alive by driver ants.) In December 1973 a column of army ants a mile long and half a mile wide reportedly marched on the town of Goiania, central Brazil, and devoured several people including the *chief of police* before being driven back into the jungle by 60 militiamen armed with flame-throwers but this story probably owed much to journalistic licence! Albert Schweitzer, the world-famous mission doctor and Nobel Prize winner, once watched an army of driver ants invade the hen house attached to his hospital on the Ogowe River, Gabon, French Equatorial Africa, and was amazed by their military precision. They marched six abreast in four parallel columns, and as they neared their objective warrior ants raced into position on either side of the main force. Dr Schweitzer, his wife and servants counter-attacked with pails of water mixed with disinfectant, but the ants got in a few choice bites before they were driven back by the fumes from the chemical. On another occasion a horde of driver ants ate three dead goats in three days (Clarke, 1969).

In 1958 the bite of an ant (species not identified) indirectly caused the death of a 42-year-old woman at Kenninghall, Norfolk. The unfortunate victim had been pulling bulbs in her garden when suddenly she felt a sting in her ankle. Shortly afterwards she became ill and died three days later. Apparently the woman had rubbed the affected area with her hands, and the earth sticking to the skin had contained a tetanus germ.

One of the most bizarre uses of ants is that made by the Tamil natives of Sri Lanka, who take them as smelling salts! They rub a number of red ants violently between their palms and then inhale the strong ammonia-like fumes which rise from the mass of crushed and bruised insects. This is said to instantly relieve a severe cold in the head – if the sufferer has no objection to a few of the more active ants burying their mandibles in various parts of his person.

A huge mound built by African termites

The master builders of the insect world are termites (Isoptera) which are commonly but erroneously called 'white ants' because they are superficially ant-like and live like ants. Some of the mounds of these creatures found in northern Australia measure up to 6·09 m *20 ft* in height and nearly 30·5 m *100 ft* in diameter at the base, and Howse (1970) has calculated that if termites were man-sized their largest citadels would be four times as high as the Empire State Building and measure 8 km *5 miles* in diameter. The African termite *Macrotermes bellicosus* builds even taller mounds (height up to 12·80 m *42 ft* have been reported in the Congo), but the bases are usually less than 3·04 m *10 ft* in diameter (Harris, 1961). These termite cities are as well run as a human city, and the insects even install 'central heating' by bringing in fermenting vegetation which gives off warmth.

Apart from their architectural ability, termites are best known as destroyers of wood which they digest with the aid of intestinal protozoa, but they don't just stop there! Their tiny bodies can also distil an acid which eats through lead, manufacture a liquid capable of disintegrating glass, and spread a substance on metal which rusts it and enables the insects to bore through. In 1949 termites quietly and quickly penetrated the walls of the Vatican in Rome and got to the library where they damaged valuable books and manuscripts. Five years later they ate through the concrete floor of the almost completed Legislative Council Chamber at Darwin, Northern Territory, Australia, and then attacked the doors and other woodwork. There is also a record of a man coming home one night and finding four of his framed pictures missing. All that was left were the glass coverings which the termites had cemented to the walls.

In 1953 a woman living in Esher, Surrey, noticed that there were a lot of ants wandering around her house, but she didn't pay much attention because she thought they were harmless. She soon changed her mind, however, when she found that they were eating her window frames and furniture and, when insecticides failed to do the trick, she called in pest experts. They were also unsuccessful, and in 1957 the desperate woman called in two scientific boffins, who eventually traced the colony to the sun lounge which was panelled with maple. Everything had to come down, of course, until finally a hammer smashed through the plaster revealed the citadel. It was 1·5 m *5 ft* high and 425 mm *18 in* across and literally crawling with inhabitants. A powerful insecticide soon destroyed them and the woman was told by the scientists that she would have no more trouble. They were wrong, however, because three weeks later the uninvited guests returned in strength, and the next few months were a nightmare. Then, suddenly, the ants left, leaving behind over £5000 worth of damage. This is the only known case of termites at work in Britain.

The largest termite in the world is the *Macrotermes bellicosus* of Africa which reaches 127 mm *5 in* in length (Harris, 1961).

The smallest known termite is *Afrosubulitermes* sp., which measures about 3·5 mm *0·13 in* in length.

The largest bush-cricket in the world is the rare *Phyllopora grandis* of New Guinea. The largest of several female examples in the collection of the Papua New Guinea Department of Agriculture, Stock and Fisheries has a half-span of over 127 mm *5 in* and would certainly measure very nearly 266 mm *10½ in* from wing tip to wing tip if fully spread. Another specimen (unspread) has a forewing length of very nearly 127 mm *5 in* (Fenner, pers.

Phyllopora grandis, *the world's largest bush-cricket (Dept. of Information and Extension Services, Papua)*

comm. 15 February 1972). *Pseudophyllanax imperialis* of New Caledonia, SW Pacific, is also of comparable size. A female preserved in the British Museum (Natural History) has a body length of 105 mm *4·13 in* including the ovipositor, or 125 mm *4·02 in* with wings folded. The wing-span is 240 mm *9·44 in* (Meadows, pers. comm. 10 September 1971).

The largest bush-cricket found in Britain is *Tettigonia viridissima*. Adult males measure 28–33 mm *1·1–1·3 in* body length and adult females 32–35 mm *1·26–1·37 in*. In August 1953 an enormous female measuring 77 mm *3·03 in* in body length was caught in a sand-pit at Grays, Essex, and later presented to London Zoo.

The largest true grasshopper (14 species) found in Britain is *Mecostethus grossus*, females of which measure up to 38 mm *1·53 in* in body length.

The smallest grasshoppers are members of the family Eumasticidae, where the males are often under 10 mm *0·39 in* in body length, and one specimen preserved in the British Museum (Natural History) has a body length of only 6 mm *0·23 in*.

The smallest grasshopper found in Britain is *Myrmeleotettix maculatus*, which has a maximum body length of 16 mm *0·63 in*.

The largest dragon-fly in the world is *Tetracanthagyna plagiata* of NE Borneo which is known only from the type specimen preserved in the British Museum (Natural History). This specimen has a wing-span of 194 mm *7·63 in* and an over-all length of 108 mm $4\frac{1}{4}$ *in* (Ward, pers. comm. 7 September 1971). This size is closely matched by the damsel-fly *Megaloprepus caerulatus* of Central and South America which has been measured up to 190 mm *7·48 in* across the wings and 127 mm *5 in* in over-all length. The largest specimen in the British Museum (Natural History) has a wing-span of 188 mm *7·40 in* and an over-all length of 118 mm *4·64 in*.

The largest dragon-fly found in Britain (43 species) is the Golden-ringed dragon-fly (*Cordulegaster boltoni*). In July 1974 Master Stephen Coates (pers. comm. 21 April 1976) of Dunfermline, Scotland, found an outsized golden-ringed dragon-fly floating in the sea at Amroth, South Wales, while on holiday. The specimen measured 86 mm *3·38 in* in length and 105 mm *4·13 in* from wing tip to wing tip.

The smallest dragon-fly in the world is *Agriocnemis naia* of Burma. A specimen preserved in the British Museum (Natural History) has a wing-span of 17·6 mm *0·69 in* and an over-all length of 18 mm *0·7 in*.

The smallest dragon-fly found in Britain is the Scarce ischnura (*Ischnura pumilio*) which has an over-all length of 25 mm *0·98 in*.

The largest known fly is the Robber-fly *Mydas heros*, which is found in tropical South America. It has a body length of up to 60 mm *2·36 in* and measures about the same in wing-expanse (Oldroyd, 1964). Some of the daddy-long-leg species of crane-fly of the tropics have a wing-span of nearly 100 mm *3·93 in* and a length with legs extended of about 200 mm *7·87 in*, but these insects have very little bulk. A specimen of *Holorusia brobdignagius* preserved in the British Museum (Natural History) measures 228 mm *8·97 in* from the tips of the front legs to the tips of the hind legs (Cogan, pers. comm. 13 October 1971).

Various writers have tried to estimate the possible number of descendants a pair of common house-flies might have in a summer if all the individual flies survived. According to the calculations of Wardle (1929) there would be 5 598 720 000 000 progeny at the end of the season if each female laid 120 eggs, but this figure falls far short of the maximum potential (one female laid 2387 eggs!). Fortunately the mortality rate of these insects is enormously high.

The largest known flea (1830 species) is *Hystricopsylla schefferi schefferi*, which was described from a single specimen taken from the nest of a Mountain beaver (*Aplodontia rufa*) at Puyallup, Washington, USA, by T H Scheffer. In one series of five males and 30 females, the largest male measured 6 mm *0·23 in* in length and the largest female 8 mm *0·31 in*.

The largest British flea is the Mole flea (*Hystricopsylla talpae*) which measures 5–6 mm *0·19–0·23 in* in length (Rothschild & Clay, 1952). This flea is also found on other animals, and a specimen measuring 6 mm *0·23 in* was once collected from a Pigmy shrew (*Sorex minutus*)! In human terms this is equivalent to a man carrying an adult rat around with him (Lehane, 1969).

The most dangerous flea in the world is the Oriental rat flea (*Xenopsylla cheopis*) carrier of the dreaded bubonic plague bacterium *Pastuerella pestis*. In the 14th century this blood-sucking parasite caused the deaths of 25 000 000 people in Europe or approximately one-quarter of the total population, and in the great Indian plague of 1896–1917 over 10 000 000 people died.

Apart from their danger as carriers of disease, fleas are best known for their jumping abilities. In 1910 M B Mitzmain of the US Public Health Service carried out some tests on these insects and found that the jumper *par excellence* among fleas was the Common flea (*Pulex irritans*). One energetic performer allowed to leap at will executed a high jump of at least 196 mm *7¾ in* and a long jump of 330 mm *13 in* but these measurements were exceptional. In jumping 130 times its own height a flea subjects itself to a force of 200 *g*, the energy coming from the elasticity of resilin protein situated above the jumping legs. (A human would black out at 14 *g*.)

Although the entire life-cycles of some fleas can be as short as two weeks in hot, dry weather, life can be prolonged considerably in cool, moist conditions. Bacot (1914) found that *unfed* adults of the Common rat flea (*Ceratophyllus fasciatus*) lived for 22 months, those of the Common flea (*Pulex irritans*) for 19 months and the Dog flea (*Ctenocephalides canis*) for 18 months in cool temperatures, but these fleas would have lived much longer if they had been supplied regularly with blood.

The oldest flea on record was a well-nourished Russian bird flea (*Ceratophyllus* sp.) which lived for 1487 days or 4 years 27 days.

The Queen Alexandra birdwing, the largest known butterfly (Clarke & Hyde)

The most popular 'host' for fleas is the Red squirrel (*Sciurus vulgaris*). More than 13 000 Squirrel fleas (*Ceratophyllus acutus*) were removed from one individual.

The largest wasps in the world are the spider-hunting wasps of the genus *Pepsis* found in South America. *P. atrata* of Amazonia has a body length of 63 mm *2½ in* (excluding the 13 mm *0·5 in* stinger) and a wing-span of 101 mm *4 in* (Zahl, 1959), but *P. frivaldskii* is even larger. A specimen preserved in the British Museum has a wing-expanse of 114 mm *4½ in* (Yarrow, pers. comm. 21 October 1971).

The largest wasp found in Britain is the wood-boring Greater horntail (*Urocerus gigas*), females of which have been measured up to 40 mm *1·57 in* in length (excluding the ovipositor). If, however, the *U. gigas* is considered to be a saw-fly, then the largest British wasp is the Hornet (*Vespa crabro*), queens of which measure 30 mm *1·17 in* head to tip of abdomen.

The smallest known wasps are the parasitic wasps of the family Myrmaridae (see page 181).

The largest known butterfly is the Queen Alexandra birdwing (*Ornithoptera alexandrae*) of New Guinea. Females (males are smaller) of this rare tree-dwelling species average 203–228 mm *8–9 in* in wing-expanse, but D'Abrera (1975) says specimens have been collected which exceeded 280 mm *11·02 in*. A huge female preserved in the British Museum (Natural History) has a fore-wing measurement of 135 mm *5·31 in*, which is equivalent to a wing-span of *c.* 280 mm *11·02 in*. This birdwing is also the heaviest known butterfly. The Queen Victoria birdwing (*Ornithoptera victoriae*) of the Solomon Islands, SW Pacific and the African giant swallowtail (*Papilio*

The Milkweed – Britain's largest butterfly (British Museum, (Natural History))

antimachus) are also exceptionally large, females of both species measuring up to 254–279 mm *10–11 in* across the wings. Butterflies of these dimensions are virtually impossible to catch with a net because they frequent the tops of high trees and fly very rapidly. Instead collectors have to resort to 'shooting' them with sporting guns loaded with dust or water.

The largest butterfly found in the British Isles is the Monarch butterfly (*Danaus plexippuss*), also called the Milkweed or Black-veined brown butterfly, a rare vagrant which breeds in the southern USA and Central America. Adult females average 38 mm *1½ in* in body length and have a wing-expanse of 89 mm *3½ in*. On 2 October 1937 a specimen with a wing-span of 109 mm *4·3 in* was caught in a garden at Lydney, Gloucestershire, and later presented to Gloucester Museum (Darknell, pers. comm. 9 August 1973). The first monarch butterfly ever recorded in the British Isles was caught at Neath, South Wales, on 6 September 1876 by Sir John Llewelyn. Since then more than 200 examples have been observed or captured.

The largest native butterfly (65 species) is the Swallowtail (*Papilo machaon*), females of which have a wing-expanse of 70–100 mm *2·75–3·93 in* and weigh 500–600 mg *0·017–0·021 oz*. This species is now confined to a small area of the Norfolk Broads.

The world's smallest known butterfly is the Dwarf blue (*Brephidium barberae*) of South Africa. It has a wing-span of 14 mm *0·55 in* and weighs less than 10 mg *0·00034 oz*.

The smallest butterfly found in the British Isles is the Small blue (*Cupido minimus*) which has a wing-span of 19–25 mm *0·75–1 in*.

Some butterflies have extraordinary powers of flight and migrate great distances. The champion is the Painted lady (*Vanessa cardui*). In the early spring hundreds of thousands of these seemingly fragile insects leave their homes in North Africa and Asia Minor to fly across the Mediterranean and Europe to southern England where they arrive in late May or early June after a journey of 3200–4800 km *2000–3000 miles*. Some take a further trip to Scotland, and individual stragglers have even been seen in the extreme north of Iceland within a few degrees of the Arctic Circle, which means they must have flown nearly 6400 km *4000 miles*!

Monarch butterflies also cover enormous distances during migration. This species leaves the Canadian border in September and arrives in Florida, Mexico or California two months later after a journey of 3200 km *2000 miles*. Then, after spending the winter in a state of semi-hibernation, they fly back to the Canadian border the following March. They have also been found in the Hawaiian Islands 3200 km *2000 miles* from the American continent, where they were unknown until the milkweed, their favourite food, was first grown. The occasional flights made by vagrant specimens across the Atlantic are not nearly so arduous because they are assisted by strong westerly winds, and in calm conditions they can settle on the sea and take off again without difficulty. In January 1966 the BBC Natural History Unit at Bristol released 50 marked monarch butterflies (received from Toronto University) near the city in an experiment to determine how far they would travel. One individual flying at 27 km/h *17 miles/h* reached Alvechurch 108 km *80 miles* away in less than five hours.

Like migrating locusts, butterflies often travel in huge swarms. One mass migratory flight of painted ladies observed moving on a 64 km *40 mile* wide front in the southern USA reportedly took three days to pass a given point, and Williams (1958) says he witnessed a single flight of the Pale form butterfly (*Catopsilia florella*) in East Africa in 1928–9 which continued steadily for more than *three months*!

The majority of butterflies fly 1·52–2·43 m *5–8 ft* above the ground during migration, but some species travel at great heights. According to Williams dead Cabbage whites (*Pieris brassicae*) have been found in the Alps above the 3657 m *12000 ft* mark and butterflies of the genera *Cosmosalyrus*, *Lymanopoda* and *Pedaliodes* have been seen crossing the Andes at heights up to 4700 m *15419 ft* (Klots & Klots, 1958).

The greatest height reliably reported for migrating butterflies is 5791 m *19000 ft* for a few 'small tortoiseshells' (probably *Aglais urticae*) seen flying over the Zemu Glacier (Sikkim) in the eastern Himalayas. This altitude is probably also reached by the Queen of Spain fritillary (*Argynnis lathonia*) which has been sighted at nearly 6000 m *19650 ft* in the Himalayas.

The rarest of all butterflies (and the most valuable in cash terms) is the birdwing *Ornithoptera allottei*, which is found only on Bougainville in the Solomon Islands. A male specimen from the collection of G Rousseau Decelle, the French collector, was sold for £750 at an auction in Paris on 24 October 1966 (Howarth, 1967). In November 1955 a collector paid DM1500 (then about £125) at the annual Insect Trading Day in Frankfurt for the only known example in Europe of *Teinopalpus imperialis*, a butterfly found in the mountains of SE China. Abnormally coloured butterflies also fetch high prices. In 1921 Lord Rothschild paid £75 for a black swallowtail caught by a fisherman on the Norfolk Broads. At a London auction in April 1941 a collector paid £49 for an entirely white Marbled white (*Melanargia galathea*), and an all-black example was sold for £41. In 1958 a black Cabbage white (*Pieris brassicae*) was sold in London for £44, and in 1962 the British Museum (Natural History) paid £50 for a Chalk hill blue (*Lysandra coridon*) which was pure white.

The rarest and most endangered native British butterfly is the Large blue (*Maculinea arion*), which is now confined to three or four tiny localities in north Cornwall which are permanently guarded to stop unscrupulous collectors from moving in. The total population is now so small (estimated at only 35–70 in 1975) that the British Butterfly Conservation Society believe that it would need only poor weather during the flight season – large blues will not seek mates in wet conditions – for the species to become extinct.

Mr Robert Goodden, vice-chairman of the society, has been trying to breed large blues at his butterfly farm at Over Compton, Sherborne, Dorset (headquarters of the BBCS) 'in the hope that if it does ever die out we can try to produce fresh stocks for restocking', but so far he has been unsuccessful. One of the main problems is the large blue's strange eating habits. As a caterpillar it lives only on wild thyme which it needs through its first two skin changes. It is then 'adopted' by Red ants (*Myrmica rubra*) and taken into their underground chamber where they 'milk' the creature of its thyme honey which is stored in two glands near its tail. In return the caterpillar is fed with ant grubs and spends the winter hibernating in the chamber. In the spring it turns into a chrysalis and eventually leaves the ant hole as a butterfly.

The Large tortoiseshell (*Nymphalis polychloris*), which is now confined mainly to Essex and the Kent/Sussex border, is another seriously threatened species. According to Goodden (pers. comm. 15 February 1973) there have been no reports of this butterfly breeding in this country for several years – 'which might be explained by the fact that the larvae tend to live right at the tops of elm trees' – and he thinks that some of the specimens under observation may have been migrants from the Continent. Mention should also be made of the Black hairstreak (*Strymonidia pruni*), found only in Huntingdonshire and adjacent counties, although one researcher who has been studying this species for several years has recently found more localities than had previously been thought possible. The swallowtail of the Norfolk Broads is not in any immediate danger, but the BBCS say 'a watch needs to be kept . . . so that any decline, and even the reason for it, may be noticed in time'.

The most fanatical butterfly collector of all time was James Joicey of Witley, Surrey, who died in March 1932 aged 61. For over 30 years this man, the son of a millionaire, spent up to £10000 annually in sending agents all over the

The Large blue, the most seriously endangered of all British native butterflies (Robert C Goodden)

world in search of rare specimens, and by 1913 his collection had become so important (later to become the third largest private collection in existence) that a special building was erected in the grounds of his magnificent home at Witley where a curator and seven assistants were in constant attendance. He was also an extremely generous benefactor of the British Museum (Natural History), and during his lifetime he presented to the Trustees between 200000 and 300000 specimens. In 1927 Joicey went bankrupt, and it was related in court that because of his 'special hobby' he found it impossible to live on £20000 a year. It had always been the intention of this fanatical lepidopterist to bequeath his collection of 1500000 butterflies and moths estimated to be worth nearly £1000000 to the nation, but the involved conditions of his financial affairs at the time of his death made this impossible. Joicey's proudest boast was that he had added 500 new species to the order.

By the end of 1975 Mr John Glick (b. 1917) of Los Angeles, California, USA, had personally netted 57100 of the 60000 known species of butterfly, and discovered four new species. His most valuable capture to date is a female example of the Butterfly of paradise (*Troides paradisea*) from New Guinea which commands prices in excess of £100.

The largest moth in the world is the bulky Hercules emperor moth (*Coscinoscera hercules*) of tropical Australia and New Guinea. Females of this species have been reliably measured up to 267 mm *10½ in* across the broad wings, giving a total surface area of nearly 2·54 cm² *100 in²*. According to Lord (1948) one outsized specimen found near the post office at the coastal town of Innisfail, Queensland, Australia, measured 355 mm *14 in*, but the Australian Museum in Sydney have no knowledge of this giant. The largest examples in their collection measure 228–254 mm *9–10 in*, and Mr C N Smithers, Principal Curator (pers. comm. 23 November 1973) said that as this species has long extensions to the wings it would be possible by manipulation to obtain a large expanse. This statement was later picked up by David J Moon, Editor of the *AES Exotic Entomology Group*, who said (pers. comm. 9 December 1973): 'I was most interested in Mr Smithers' comments re manipulation of the wings to increase the span; it is customary to give wing-spread measurements for the forewings only . . . but there is no anatomical reason to prevent the hindwings from being set at right-angles to the insect's thorax – except that one set in this manner wouldn't exactly look aesthetically pleasing! If this *was* how the 355 mm Hercules was measured, then it would indeed produce a fractionally greater figure than for the forewings . . . Another interpretation is that stretching of the basal wing muscles during the "relaxing" process (which often precedes "setting") could increase the wing-span substantially, although there is little justification for supporting this view. It is possible to pull the wings out a little way, say a couple of mm, but too much strain results in the muscles tearing away from the thorax to which they are attached . . . and in

A Great owlet moth with a wing expanse of 360 mm 14·17 in *(Oberthur Collection)*

any case obvious distortion would be the outcome. . . .'

Another candidate for the title is the little-known Great owlet moth (*Thysania agrippina*) of Central America and northern South America. This species has the greatest wing-span of all the Lepidoptera, but it lacks the bulk of hercules moths and the wings are much narrower. The largest specimen in the British Museum (Natural History) – collected in Chiriqui province, Panama – has a wing-expanse of 304 mm *11·96 in* (Nye, pers. comm. 19 March 1975), and another one preserved in the Dorman Museum, Middlesbrough, measures 281 mm *11·06 in* (Thornton, pers. comm. 5 January 1975). M Charles Oberthur (1841–1924), the famous French millionaire lepidopterist, once purchased a *T. agrippina* which measured an astonishing 360 mm *14·17 in* across the wings, but the present whereabouts of this monster is not known. The Oberthur Collection of 1 140000 butterflies and moths was sold in 1927 and 70 per cent of the specimens were acquired by the British Museum (Natural History), but the superlative female was not among them. The Atlas moth (*Attacus atlas*) of SE Asia is also exceptionally large – a female in the Dorman Museum has a wing-span of 264 mm *10·39 in* – and the species runs second to *Coscinoscera hercules* in terms of bulk.

The largest moth found in Britain is the very rare Death's-head hawk-moth (*Acherontia atropos*) which breeds in Africa and occasionally reaches these shores during migratory flights in the late summer. Females of this species have a body length of *c.* 60 mm *2·36 in*, a wing-expanse of up to 133 mm *5¼ in* and weigh about 1·6 g *0·056 oz*.

The largest native moth is the Privet hawk-moth (*Sphinx ligustri*). Adult females of this species have a body length of *c.* 50 mm *1·96 in*, a wing-span of over 110 mm *4·33 in* and weigh about 1·3 g *0·045 oz*.

The smallest of the 140000 known species of Lepidoptera is the moth *Stigmella ridiculosa* which has a wing-span of only 2 mm *0·08 in*. It is found in the Canary Islands.

Up until very recently the smallest British moth was believed to be *Nepticula microtheriella*, which has a wing-span of 3–4 mm *0·11–0·15 in* and a body length of 2 mm *0·078 in* (Meyrick, 1928). On 28 March 1974, however, Mr A M Emmett, a leading authority on the Nepticulidae, put the question to the test by exhibiting a series of this moth against a series of *Johanssonia*

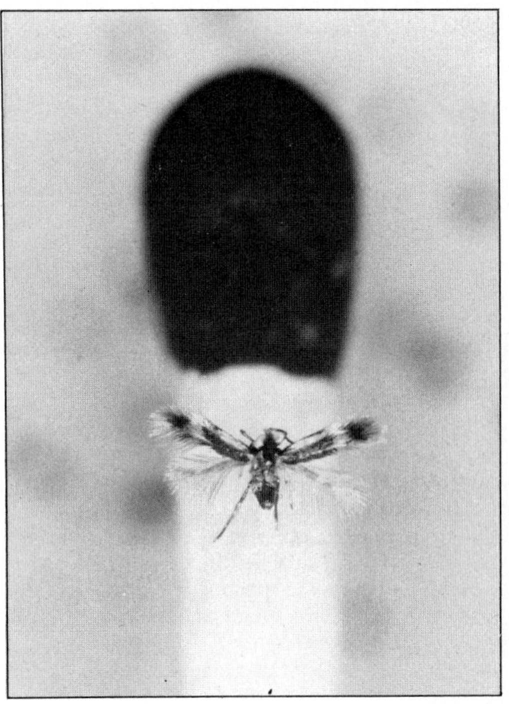

Britain's smallest moth compared with the head of a match (M Emmett)

acetosae, another diminutive species, at a meeting of the British Entomological and Natural History Society. Afterwards he said (pers. comm. 2 March 1974): 'To prevent the influence of preconception in the voting the specimens were unlabelled except as Series A and Series B. Between 40 and 50 members were present, including professional entomologists from the British Museum (Natural History), and they voted unanimously that Series B (*acetosae*) was the smallest species. H T Stainton named both these species in 1854, and at that time stated that *microtheriella* was the smallest moth known to science, a statement which has been repeated uncritically ever since. While the largest specimens of *acetosae* are larger than the smallest specimens of *microtheriella*, there is no doubt at all that, when seen in series, *acetosae* is, on the average, the smaller species. In 1854 there was only a single specimen of *acetosae* available for study.'

Although moths are not normally considered dangerous to man, a larger number of caterpillars have poisonous hairs or spines which produce varying degrees of irritation and swelling on contact with skin. One of the most

The rare Death's head hawk moth, Britain's largest insect in terms of wing-span

feared species in the southern USA is the abundant Puss caterpillar (*Megalopyge opercularis*). Children 'stung' by larvae of this moth often develop a high fever and nervous symptoms (Riley & Johannsen, 1932), and the poison secreted by a closely related species, *M. lanata*, is stated to be *potentially lethal*, although no human fatalities have been reported. The poison of the Emperor moth caterpillar *Lonomia cynira* of Venezuela is said to cause a fatal type of haemophilia in humans, but further information is lacking.

Another dangerous moth, though for quite different reasons, is the recently discovered 'Vampire' moth (*Calyptra eustrigata*). Dr Hans Banziger (1968), a Swiss entomologist working in Malaysia, discovered that this large moth uses its strong barbed proboscis to extract blood from large mammals like water-buffalo, deer and tapir, and he says its nasty habit of regurgitating some of the blood it has drunk makes it a potential carrier of disease. In the laboratory this moth has been induced to feed on man – 'it felt like being stabbed with a hot needle' – but so far this performance has not been witnessed under natural conditions.

The most acute sense of smell exhibited in nature is that of the male Lesser emperor moth (*Eudia pavonia*), which is widely distributed throughout Europe. In 1961 some German scientists released a number of marked males from a moving train at regular intervals in an experiment to determine the maximum distance the love scent could be picked up through their olfactory phermone-receptor systems, and they found that some of the insects could track the perfume to its source from the almost unbelievable range of 11 km *6·8 miles* (Droscher, 1964). Another female put in a gauze cage attracted 127 males from 3·2 km *2 miles* away in the space of only three hours. This scent has been identified as one of the higher alcohols ($C_{16}H_{30}O$), of which the female carries less than 0·10 mg *0·0000035 oz*, and even then only a very few molecules of this phermone are released into the air at any given time. Yet the male moths still manage to pick up these inconceivably minute traces in the 40000 sensory nerve cells of their feather-like antennae.

Section X

CENTIPEDES & MILLIPEDES

(classes Chilopoda and Diplopoda)

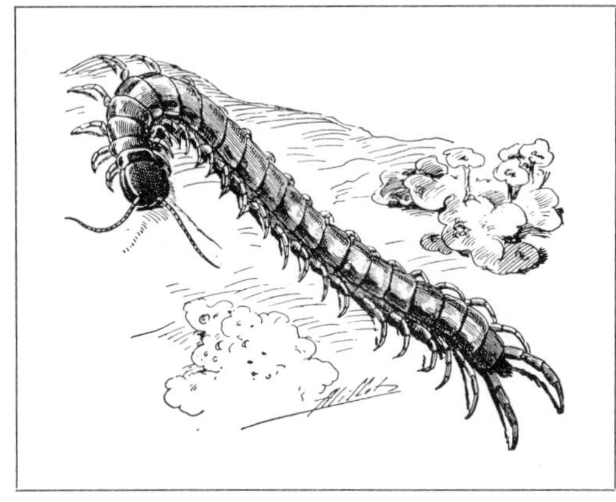

At one time centipedes and millipedes were grouped together in the class Myriapoda, but they are now placed in separate classes.

A true centipede is basically a carnivorous land-dwelling invertebrate (a few are marine) which breathes by means of tracheae or air tubes distributed throughout the body. It has a distinct head bearing one pair of antennae and three pairs of jaws and an elongated body composed of many segments, each of which bears one pair of jointed walking legs. The appendages of the first body segment are modified into poison fangs called maxillipeds. During growth the animal frequently sheds its external skeleton or horny skin. Young are hatched from eggs after leaving the female. The earliest known centipedes lived about 350000000 years ago, and today there are about 3000 living species which are divided into four orders. These are: *Geophilomorpha*; *Lithobiomorpha*; *Scolopendromorpha* and *Scutigeromorpha*.

The millipede is a herbivorous, land-dwelling invertebrate which also breathes by means of tracheae. It has a distinct head bearing one pair of antennae like the centipede, but each segment of the elongated body has two pairs of legs instead of one. It also differs from the centipede in having two rows of stink glands down each side of the body which secrete an evil-smelling substance capable of repelling insect enemies. The young, however, are born the same way. The earliest known millipedes lived about 400000000 years ago and today there are about 8000 living species which are divided into ten orders. These are: *Polyxenida, Glomerida, Glomeridesmida, Nematophora, Stemmiulida, Polydesmida, Julida, Spirobolida, Spirostreptida* and *Cambalida*.

The longest known species of centipede is the 46-legged giant scolopender *Scolopendra gigantea* of the rain forests of Central and South America, adults measuring 250–265 mm *9·84–10·43 in* in length when fully extended and 25 mm *1 in* in breadth (Bucherl, 1971). Measurements of 304 mm *12 in* and even 381 mm *15 in* have been reported for individuals collected in Guayana, but these figures need confirming. Much less reliable is the statement made by the Spaniard Ulloa, Christopher Columbus's gold assayer, that he saw centipedes on the north coast of South America which were 914 mm *36 in* long and 152 mm *6 in* in breadth (*sic*), and the claim by Baron Alexander von Humboldt (1769–1859) that some of the centipedes he encountered in Venezuela measured 457 mm *18 in* in length.

The widely distributed *Scolopendra morsitans* usually measures 80–120 mm *3·14–4·72 in* when fully extended, but some giant forms are found in India and Malaysia. On one occasion H S Wood (1935), the British naturalist, saw a 254 mm *10 in* long electric blue centipede with bright coral-red fangs in the Kubbo-Kale Valley, India, and described it as 'the most terrible thing I have

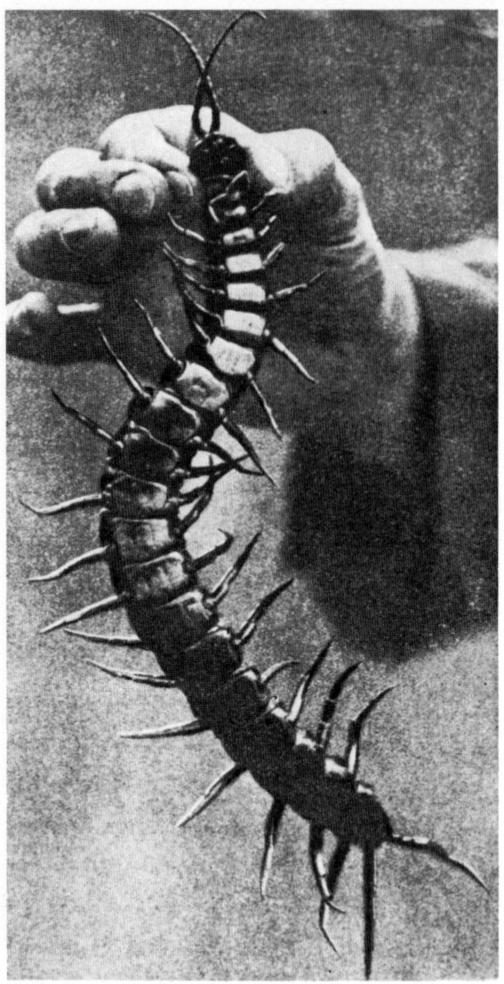

A giant scolopender from South America (Photo-Reportage Ltd)

ever seen in my tramps through the forest'. Another very large variant is found on the Andaman Islands in the eastern part of the Bay of Bengal, specimens having been measured up to 330 mm *13 in* in length when fully extended and 38 mm *1½ in* in breadth (Bayley-de Castro, 1921), and an unconfirmed measurement of 317 mm *12½ in* has been quoted for another outsized example collected in West Africa. *Scolopendra subspinipes*, another widely distributed species, also reaches an impressive size (average length 215 mm *8·46 in*), and a measurement of 254 mm *10 in* has been reported for one individual collected in Brazil (Cloudsley-Thompson, 1958).

Outside the genus *Scolopendra* the only other centipede worthy of mention in terms of size is the worm-like *Orya barbarica* (order Geophilomorpha) of North Africa which has been measured up to 177 mm *7 in*.

The longest centipede found in Britain is *Haplophilus subterraneus*, which has between 69 and 89 pairs of legs and measures up to 75 mm *2·95 in* in length and 1·4 mm *0·005 in* in breadth (Eason, 1964). On 1 November 1973 Mr Ian Howgate of St Albans, Herts, reportedly saw a specimen measuring at least 114 mm *4½ in* in length, but the centipede was not captured.

The sub-species *Nesoporogaster souletina brevior*, which may have been introduced from abroad, is also of comparable size, having been measured up to 70 mm *2¾ in*. It is restricted to a single wood in Cornwall.

The largest scolopender found in Britain is *Cryptops anomalans* of southern England which has been measured up to 40 mm *1·57 in* in length and 2–3 mm *0·078–0·11 in* in breadth.

The shortest known centipede is probably an unidentified species which measures only 5 mm *0·19 in* in length.

The shortest centipede found in Britain is the widely distributed *Lithobius duboscqui*, which measures 5·5–9·5 mm *0·21–0·37 in* in length and 0·7–1·1 mm *0·02–0·04 in* in breadth (Eason, 1964).

Although the word 'centipede' means 'hundred-legs', only some of the Geophilomorpha justify this name.

The centipede with the greatest number of legs is *Himantarum gabrielis* of southern Europe which has 171–177 pairs when adult.

Half of the known species of centipedes only have 15 pairs of legs when adult, but they all start life with about six pairs, including *H. gabrielis*.

The fastest-moving centipede is probably *Scutiger coleoptrata* of southern Europe which can travel at a rate of 50 cm *19·68 in* a second (7·15 km/h *4·47 miles/h*) over short distances (Manton, 1952). It can thus outpace a fast-walking pedestrian.

Centipedes are reasonably long-lived animals. The Methuselahs of the Chilopoda are the giant scolopenders, which do not reach maturity until they are four years old, and probably live at least ten years. The longest-lived British species is *Lithobius forficatus* which has a maximum life-span of five to six years. The geophilid centipede *Pachymerium ferrugineum* also lives at least four years (Palmen & Rantala, 1954).

Although all centipedes are provided with poison glands, for paralysing or killing prey, the bite is not inflicted by the jaws but by the curved, horny claws of the front pair of legs which serve as fangs. Most centipedes are relatively harmless to man, bites occurring very rarely and without serious consequences, but some species can administer a damaging bite. The most dangerous centipedes are the large tropical and subtropical scolopenders which can secrete a large amount of venom. One of the most feared is a variety of *Scolopendra subspinipes* found in the Solomon Islands which reportedly has a bite that defies description. According to one writer the pain is so intense that victims have been known to plunge their bitten hand into boiling water as a counter-irritant – the attempted cure being more disastrous in its effect than the original injury.

An unidentified Malayan species (probably a variant of *S. morsitans*) is said to be particularly dangerous, the symptoms being more severe than those from the bite of indigenous vipers, and victims are sometimes laid up for as long as three months (Keegan *et al.*, 1963).

Some persons, particularly small children, are especially sensitive to the toxins of centipedes capable of puncturing the human skin, and a number of deaths have been *claimed*. Wood (1886) and Chalmers (1919) both mention several fatal human accidents, and Faust (1928) describes several fatalities in India, but Bucherl (1971) has queried the validity of these reports. Remington (1950) could find only one authentic case of a person dying after being bitten by a centipede, and that was a seven-year-old Philippine child who succumbed 29 hours after being bitten on the head (Pineda, 1923).

During his stay in the Philippines, Remington himself was bitten under the arm by one of these centipedes which crawled into his bed, and he said the pain at first was almost unbearable. The following morning he found the region round the arm-pit was severely swollen and very painful to the touch, and this condition persisted for nearly three weeks. At the time of writing no serum exists against Scolopendromorph bites.

Centipede pseudoparasitism is another hazard to man, and a number of cases of animals living inside the nasal sinus or the alimentary canal of man, after being accidentally introduced, have been reported. For instance, Tartaglia (1961) mentions a case of intestinal pseudoparasitism lasting nine hours after a Yugoslav woman had accidentally swallowed a centipede (identified as *S. cingulata*). Symptoms included vomiting, cold perspiration and irregular heartbeat.

In another case reported in the *British Dental Journal* (October 1972) a London dentist removed a woman's aching tooth – and out strolled a centipede! 'It had probably crawled in when she was asleep with her mouth open', explained the dentist. 'It may have been there for as long as two days. This is very rare, although I do remember some years ago a report of a patient who had a tomato seed behind a tooth which started to grow into a small plant!'

The longest known species of millipede are *Graphidostreptus gigas* of Africa and *Scaphistostreptus seychellarum* of the Seychelles in the Indian Ocean, both of which have been measured up to 280 mm *11·02 in* in length and 20 mm *0·78 in* in breadth (Kaestner, 1968).

The longest millipede found in Britain (44 species) is *Cylindroiulus londinensis* which measures up to 50 mm *1·96 in*. The common Black millipede (*Tachypodoiulus niger*) is also of comparable size, males and females having been measured up to 45 mm *1·77 in* and 49 mm *1·92 in* respectively (Cloudsley-Thompson & Sankey, 1961).

The shortest millipede in the world is the British species *Polyxenus lagurus*, which measures 2–3 mm *0·07–0·11 in* in length and 0·5 mm *0·019 in* in breadth. Another British species, *Macrosternodesmus palicola*, measures 3·5 mm *0·137 in* in length.

Although the word 'millipede' means 'thousand legged', very few species have more than 200. The record appears to be held by a South African millipede described by Schubart (1966) which has 355 pairs (710 legs). Another species (*Siphonophora panamensis*) discovered by H Loomis (1964) in Panama has 175 segments, but not every segment has two pairs of legs. Millipedes start life with even fewer legs than centipedes, the usual number being three pairs.

The rarest species of millipede found in Britain is *Isobates littoralis*, which lives on the sea-shore between high and low tides. It is known only from three specimens collected from the Isle of Man, Lancashire and Llandudno, Caernarvonshire respectively. *Cylindroiulus parisiorum* is known only from four specimens, all of them collected in churchyards (Blower, pers. comm. 6 November 1971).

No speeds have been published for millipedes, but they move at a much slower rate than centipedes.

Most species of millipede live one to two years, but some like *C. londinensis* may live up to seven years.

Millipedes are not venomous because they do not possess fangs, but some of the large tropical species can squirt a very strong hydrocyanic acid secretion through the pores on the body surface which is unpalatable to other animals.

One of the nastiest members of the Diplopoda is *Rhinocricus lethifer* of Haiti which can discharge this repugnatorial fluid several inches from the body. Loomis (cited by Sheals & Rice, 1973) found this to his cost when he picked up a large specimen and was hit in the face and left eye from a distance of about 457 mm *18 in*: 'The pain was instantaneous, intense and of a burning and smarting nature. It persisted for several hours despite immediate bathing with ice-water. Swelling of the eyelid and cheek progressed rapidly and soon the eye was closed. The following morning the pain was gone but the eyelid again was swollen shut. On the day following the attack the skin over the affected area had turned dark brown and was raised into blisters where the concentration of the secretion had been greatest. The blisters persisted for nearly a week after which the discoloured skin peeled off without leaving any scars.'

The millipede has been known to cause permanent blindness in chickens and other small animals in Haiti (Halstead & Ryckman, 1949).

Mention should also be made of an African millipede (probably *Spirobolus* sp.) described by Burtt (cited by Sheals & Rice, 1973) who – rather unwisely – left a living specimen in his hip pocket for nearly an hour: 'A soreness developed in the skin below the pocket, and soon afterwards it was found that the skin over an area of about nine square inches had become black. No blisters were raised, and four days later the blackened skin peeled off leaving a raw wound. The scar of the injury was clearly defined more than a year later.'

A large species of *Spirostreptus* found in the Sunda Isles, Malay Archipelago, is also much feared by the natives, and a species of *Julus* found on Ambon, an island in the central Moluccas, is reported to have an exceptionally venomous secretion (Kopstein, 1932).

The sex life of the millipede is also interesting, and this subject was discussed by an international group of experts at a conference held at Manchester University in April 1972. There it was revealed that a millipede with 120 legs uses only 16 of them to woo his mate into submission . . . that it can make love for one and a half hours at a time, three times a day . . . and a millipede's eyesight is so poor that the suitor cannot see his loved one, and she cannot see his advances (Romeo tries to impress her with good vibrations instead).

In addition to these findings Dr Ulrich Haacker of Hamburg University, West Germany, reported that the male millipedes of the British family Chordeumidae bang their heads on the ground at a rate of five times a second to attract a mate, and that one of the South African pill millipedes (Sphaerotheriidae) has a screeching love call which can be heard by the human ear 4·57 m *15 ft* away.

Section XI

SEGMENTED WORMS

(phylum Annelida)

A segmented worm is a soft-bodied invertebrate which lives on land or in water and breathes through the skin or by means of gills. Its body is divided into equal parts and the muscular part of each segment is covered by a thin transparent skin from which bundles of horny bristles protrude. The main function of these bristles is to aid locomotion, but they are also used to anchor the worm firmly in its burrow. Young are hatched from eggs deposited on land or shed in water.

The earliest known segmented worms lived about 500000000 years ago and were marine.

There are about 6800 living species of segmented worms and the phylum is divided into three classes. These are: the *Oligochaeta* (earthworms); the *Polychaeta* (bristleworms); and the *Hirudinea* (leeches). The largest class is *Polychaeta*, which contains about 4000 species or nearly 32 per cent of the total number.

The longest known species of giant earthworm is *Microchaetus rappi* (= *M. microchaetus*) of South Africa. An average-sized specimen measures 1360 m *4 ft 6 in* in length (650 mm *25½ in* when contracted), but much larger examples have been reliably reported. In 1936 a giant earthworm measuring an incredible 6·70 m *22 ft* in length when *naturally* extended and 20 mm *0·78 in* in diameter was collected in the Transvaal (van Heerden, 1937), and when P O Ljungstrom and Dr A J Reinecke of the Institute for Zoological Research at Potchesfstroom, South Africa, visited the Eastern Cape in November 1969 they were told by a resident of Debe Nek, a few miles from King Williams Town, that earthworms had recently been seen reaching over the national road (width 6 m *19 ft 8½ in*).

According to Ljungstrom and Reinecke (1969) the only enemies of this giant worm are the Hamerkop bird (*Scopus umbretta*) and the Night adder (*Causus rhombeatus*), the latter being the only known viperid that eats earthworms.

The only other giant earthworm that reaches anything like the size of *Microchaetus rappi* is the better known *Megascolides australis*, first discovered in Brandy Creek, southern Gippsland, Victoria, Australia, in 1868. An average-sized example measures 1·22 m *4 ft* in length (1·11 m *2 ft* when contracted), and nearly 2·13 m *7 ft* when *naturally* extended.

The longest accurately measured *Megascolides* on record was one collected before 1930 in southern Gippsland which measured 2·18 m *7 ft 2 in* in length and over 3·96 m *13 ft* when naturally extended. Barrett (1931) says another huge individual measured by his friend I C Cook on his farm at Holbrook, southern Gippsland, in the 1920s was 1·83 m *6 ft* long, and 3·35 m *11 ft* when naturally extended. Lengths up to 4·57 m *15 ft* have been claimed for earthworms collected near Loch, but Barrett thinks these measurements were probably obtained by 'stretching' the worm to its utmost limit – 'Megascolides is liable to break under such barbarous treatment, despite a generous allowance of rings and its surprising elasticity.'

In 1952 excavations at West Burleigh, Queensland, uncovered another giant worm belonging to the species *Digaster longimani* which measured 910 mm *3 ft* in total length, 22 mm *0·87 in* in diameter and weighed 254 g *9 oz*. Five years later, in May 1957, a scientific expedition from the Australian Museum, Sydney, collected another 44 examples in Toonumbar State Forest, southern New South Wales. The largest worm measured 1651 mm *5 ft 5 in* and over 25 mm *1 in* in thickness when suspended alive by the tail, but it contracted to 1650 mm *3 ft 6 in* on being preserved in alcohol (Pope, 1958).

Other giant earthworms found round the world include *Rhinodrilus fafner* and *Glossoscolex giganteus*, both of Brazil, which have been measured up to 2100 mm *6 ft 10¾ in* and 1260 mm *4 ft 1½ in* respectively when naturally extended, *Drawida grandis* of India which reaches 1080 mm *3 ft 6½ in* and *Spenceriella gigantea* of New Zealand which has been measured up to 1300 mm *4 ft 6 in* (Stephenson, 1930).

In May 1961 Mrs Marte Latham, an American explorer and animal-collector, discovered a new genus of giant earthworm near Popayan in the Columbian Andes at a height of 4267 m *14000 ft*. The following month she presented a *live* specimen measuring 1574 m *5 ft 6 in* when naturally extended to London Zoo. Unfortunately 'Gertrude' (also known as 'Willie') died ten days later on 21 July, and is now preserved in the British Museum (Natural History).

The heaviest giant earthworm must be *Microchaetus rappi*. Unfortunately no weight data has been published for this species, but the 6·70 m *22 ft* example already mentioned could hardly have weighed less than 1359 g *3 lb*. The runners-up are *Megascolides australis*, *Rhinodrilus fafner* and *Digaster longimani*, all of which have been credited with weights slightly in excess of 453 g *1 lb*.

The longest earthworm found in Britain is *Lumbricus terrestris*. The normal range is 90–300 mm *3·54–11·81 in*, but this species has been reliably measured up to 350 mm *13·78 in* when naturally extended. Measurements up to 508 mm *20 in* have been claimed, but in each case the body was probably macerated first.

The longest segmented worm found in Britain is the King ragworm (*Nereis virens*) which is found on the British foreshore. The average length of this marine bristleworm is in excess of 300 mm *11·81 in*, but larger specimens have been recorded. On the 19 October 1975 a king ragworm measuring 969 mm *38 in* in length was dug up by Mr James Sawyer at Hawkley Bay, Northumberland (pers. comm. 5 December 1975).

The smallest known Oligochaete worm is *Chaetogaster annandalei*, which measures less than 0·5 mm *0·019 in* in length. Two other worms in this class, *Aeolosoma kashyapi* and *Chaetogaster langi*, are 1 mm *0·039 in* long (Stephenson, 1930).

It has recently been discovered that earthworms are not so stupid or insensitive to what is happening round them as some people might think. In fact, they can sense danger and even warn other worms. According to scientists working in the Psychological Department at Milwaukee University, Wisconsin, USA, when an earthworm is disturbed (e.g. by a gardener's fork) it releases a kind of oily perspiration along its body. This lubricant not only helps the worm to travel faster through tunnels, but it also acts as an alarm signal for other worms. The team reached their conclusions after giving earthworms a mild electric shock on a sheet of glass. The startled worms immediately released the fluid which turned white on drying. Other worms put near the substance hurriedly moved away – much more so when they were placed on clean glass. It was also found that the secreted substance retains its potency for several weeks because it is not readily soluble in water.

The 'richest' earthworm on record was an artistic specimen named 'Willie' owned by Miss Lena Mueller of San Francisco, California, USA. His particular calling was painting, and during the two-year period 1963–5 he turned out nearly 200 paintings which sold for anything up to £35 a time. Instead of painting by numbers little Willie painted by wiggles. His owner would dip him in the paint, drop him on to the canvas – and away he would go. When one colour was completed the Rembrandt of the Oliogochaeta would be dipped into another. It is rumoured that Willie eventually ran out of inspiration and was put back in the soil by his grateful mistress, but the compiler has not been able to confirm this.

Some species of segmented worm are relatively long-lived. According to Wilson (1949) the Peacock bristleworm (*Sabella pavonina*) is not sexually mature until it reaches ten years of age, and a figure of ten and a quarter years has been quoted for the earthworm *Allolobophora longa*. Nothing is known about the longevity of giant earthworms like *Megascolides australis* in the wild state, but they probably live at least eight years.

The longest-lived segmented worms are leeches (Hirudinea). Most species go on growing for five years, and they can live up to 20 years (Graf, 1968).

Section XII

MOLLUSCS

(phylum Mollusca)

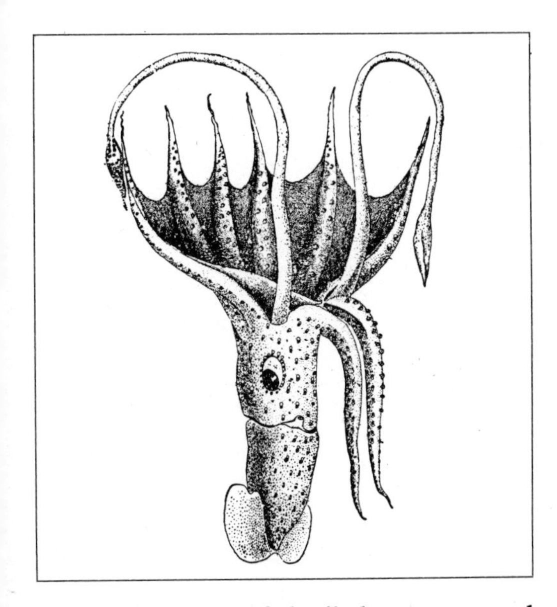

A mollusc is a soft-bodied unsegmented invertebrate which lives on land or in water, and breathes by means of a mantle cavity folded to form a lung or by gills. The body, which does not have a standard shape, is divided into four sections: a well-developed head with tentacles (missing in bivalves); a muscular foot or 'arms' which serve for locomotion or are modified to perform other functions; a rounded visceral mass in which the internal organs are housed; and a protective shell or mantle which grows with the body and is not shed periodically. It also has a highly complex nervous and circulatory system, and young are usually hatched from eggs.

The earliest known mollusc was *Neophilina galathea*, which lived about 500000000 years ago. Its remains were first discovered off Costa Rica in 1952.

There are about 45000 living species of mollusc, and the phylum is divided into six classes. These are: the *Monoplacophora* (monoplacophs); the *Chitons* (chitons); the *Gastropoda* (slugs, snails and limpets); the *Scaphopoda* (tusk shells); the *Lamellibranchia* (bivalve molluscs); and the *Cephalopoda* (squids, cuttlefish and octopods). The largest class is *Gastropoda*, which contains about 35000 species or 77 per cent of the total number.

The largest living invertebrate is the Giant squid (*Architeuthis* sp.), the many-armed 'Kraken' of Scandinavian legend, which allegedly dragged small fishing vessels down to their doom.

Although some of the great museums of Europe had physical evidence of this fabulous sea monster, in the shape of preserved fragments of arms or bodily remains from as early as the 16th century, the scientific world remained dubious of the existence of the giant squid until right up to the middle of the 19th century.

Then, in December 1853, a gigantic cephalopod was washed up on Aalbaek Beach in Jutland, Denmark. Unfortunately the body of this specimen, which 'represented a full cartload', was cut up for fish bait before it could be secured for scientific examination, but the black horny parrot-like jaws measuring 114 × 82 mm *4½ × 3¼ in* and the attached muscles came into the hands of Prof Japetus Steenstrup, the eminent Danish zoologist, who described the specimen under the name *Architeuthis monachus* in 1857.

In 1856 the mutilated remains of another giant squid found floating in the sea near the Bahama Islands by a Captain Hygom the previous autumn were brought back to Copenhagen and described by Steenstrup under the name *Architeuthis dux*. This animal had a 1·89 m *6 ft 2½ in* long head and mantle and arms the same length. (The two long tentacles or tentacular arms were missing.)

Four years later a giant squid (*A. monachus*) was stranded between Hillswick and Scalloway on the west coast of Shetland. This one had a head and mantle length of 2·13 m *7 ft* and measured 7·01 m *23 ft* from the tip of the caudal fin to the end of the longest tentacle.

The scientific doubters were knocked back even further on their heels in 1861 when the crew of the French steam despatch vessel *Alecton* tried unsuccessfully to capture a giant squid weighing an estimated 2000 kg *4409 lb* some 192 km *120 miles* north-west of Tenerife in the Canary Islands. After a three-hour-long battle the sailors managed to pass a slip noose along the body until it was held fast over the junction of the caudal fin, but when they attempted to hoist the squirming creature aboard, the rope cut through the base of the fin like a knife through butter and the rest of the brick-red monster crashed back into the sea. The head and mantle of this specimen was estimated to have measured 4·57–5·48 m *15–18 ft* in length and its eight arms 1·52–1·83 m *5–6 ft* (the two tentacles were missing). A drawing of the huge cephalopod sketched by one of the ship's officers was later sent to the French Academy of Sciences in Paris with a full report and the animal was subsequently named *Architeuthis bouyeri* (Cross & Fischer, 1862).

If any doubts still lingered in the minds of men after this dramatic incident that there was such a creature as the giant squid, they were finally put to rest in 1873 when two herring fishermen and a twelve-year-old boy were attacked by a huge individual off the coast of Newfoundland. The men, Theophilus Piccot and Daniel Squires, and Piccot's son Tom, were fishing in a dory off Portugal Cove, Conception Bay, about 14 km *9 miles* from St John's, on 26 October when they saw a dark mass floating in the water.

Thinking it was the debris of a wreck they decided to investigate, and one of the men prodded the matter with a boating hook. Suddenly the 'mass' opened out like a gigantic umbrella to reveal two huge green eyes and a parrot-like beak 'as big as a six-gallon keg' (*sic*), and the fishermen saw to their horror that they had accidentally disturbed a kraken.

The monstrous creature immediately launched an attack and hit the gunwhale with its horny beak. Then it wrapped one arm and a tentacle round the flimsy craft and threatened to drag it, occupants and all, below the surface. The two men were terrified and sat there virtually paralysed as water began pouring into the boat, but fortunately young Tom was made of sterner stuff. With great presence of mind for a person of such tender years he snatched up a small axe and cut through the two appendages; whereupon the injured cephalopod ejected a tremendous amount of ink into the surrounding water and slowly sunk from view.

When the badly shaken trio returned to shore the severed arm was thrown to the ground and eaten by dogs. But the long, thin tentacle was kept by Tom Piccot, who later sold it to the Rev. Moses Harvey, a local amateur naturalist interested in kraken lore, after whom the giant squid (*Architeuthis harveyi*) was subsequently named.

According to Harvey (1874), who had the pale pink specimen preserved in strong brine, the tentacle measured 5·79 m *19 ft* in length and 88 mm *3·5 in* in maximum circumference, but the fishermen told him that 1·83 m *6 ft* had already been destroyed and there had been a further 1·83–3·04 m *6–10 ft* still attached to the creature.

From this data – and the diameter of the largest sucker – it was possible to estimate the size of this giant squid: head and mantle length 3·65 m *12 ft*, tentacles 9·75 m *32 ft*, total length 13·41 m *44 ft*.

There was also a stranding in Coomb's Cove, Fortune Bay. This one had a total length of 15·84 m *52 ft* (head and mantle 3·04 m *10 ft*, tentacles 12·80 m *42 ft*). Another very large example washed up at West St Modent in the Strait of Belle Isle, Labrador, also measured 15·84 m *52 ft*, (tentacles 11·27 m *37 ft*), and must have been much heavier (Verrill, 1879–81).

The giant squid was now *legitimate*, and in the next decade there were numerous sightings and strandings in Newfoundland waters. In October 1875 alone more than 30 dead specimens were found on the Grand Banks, most of them measuring over 9·14 m *30 ft*.

In the 1900s and 1930s there were further sightings and strandings in the same waters, and during the period 1963–7 ten giant squids were brought to the Marine Science Research Laboratory at the Memorial University of Newfoundland, St John's, for scientific examination. The largest example (*Architeuthis dux*), stranded at Conche on White Bay on 24 October 1964, measured 9·60 m *31 ft 6 in* in total length (head and mantle 3·20 m *10 ft 6 in*, tentacles 6·40 m *21 ft*) and weighed 150 kg *331 lb*, and none were under 5·48 m *18 ft* (Aldrich, 1967).

The largest giant squid on record ran aground in Thimble Tickle Bay, Newfoundland, in 1878. The capture of this monster (*Architeuthis princeps*) was described by the Rev. Harvey in a letter to the *Boston Traveller* (30 January 1879):

'On the 2nd day of November last, Stephen Sherring, a fisherman, residing in Thimble Tickle . . . was out in a boat with two other men; not far from the shore they observed some bulky object, and, supposing it might be part of a wreck, they rowed towards it, and, to their horror, found themselves close to a huge fish, having large glassy eyes, which was making desperate efforts to escape, and churning the water into foam by the motion of its immense arms and tail. It was aground and the tide was ebbing. From the funnel at the back of its head it was ejecting large volumes of water, this being its method of moving backward, the force of the stream, by the reaction of the surrounding medium, driving it in the required direction. At times the water from the siphon was black as ink.

'Finding the monster partially disabled, the fishermen plucked up courage and ventured near enough to throw the grapnel of their boat, the sharp flukes of which, having barbed points, sunk into the soft body. To the grapnel they had attached a stout rope which they had carried ashore and tied to a tree, so as to prevent the fish from going out with the tide. It was a happy thought, for the devil-fish found himself effectually moored to the shore. His struggles were terrific as he flung his ten arms about in dying agony. The fishermen took care to keep a respectful distance from the long tentacles, which ever and anon darted out like great tongues from the central mass. At length it became exhausted, and as the water receded it expired.

'The fishermen, alas! knowing no better, proceeded to convert it into dog's meat. It was a splendid specimen – the largest yet taken – the body measuring 20 ft from the beak to the extremity of the tail. . . . The circumference of the body is not stated, but one of the arms measured 35 ft. This must have been a tentacle.'

Very little is known about the weights attained by exceptionally large giant squids and some incredible estimates have been made. Dr Bernard Heuvelmans (1968), for instance – comparing the shape of a giant squid with its arms (excluding tentacles) held together to that of a cigar – has calculated that the Thimble Tickle cephalopod may have weighed as much as 27 tonne/*ton* provided it was 'very thickset' i.e. maximum bodily circumference *c*. 3·65–4·57 m *12–15 ft*, but this weight is *impossibly high* for an animal of this bodily size. What Heuvelmans is saying, in effect, is that an invertebrate with a head and mantle length of 6·09 m *20 ft* and a maximum bodily girth of *c*. 4·57 m *15 ft* is *twice as heavy* as a Whale shark (*Rhineodon typus*) with a body length of 11·58 m *38 ft* and a maximum girth of 6·09 m *20 ft*, *and five times as heavy* as an adult bull African elephant (*Loxodonta africana africana*). This, of course, is absurd.

In actual fact if we apply the Belgian zoologist's formula to the 9·60 m *31 ft 6 in* Conche giant squid already mentioned the weight comes out at 274 kg *603 lb* – an error by almost a factor of 2!

Earlier MacGinitie & MacGinitie (1949), basing their calculations on a maximum bodily girth of 3·65 m *12 ft*, had arrived at a weight of 29 tonne/*ton*, or 30 tonne/*ton* including the arms and tentacles (*sic*) for this individual, but in 1968 they admitted that this figure was exaggerated and corrected it to a more reasonable 7·63 tonne/*ton 16827 lb*. Even this poundage, however, is still excessive.

According to Dr Igor Akimushkin (1965), the Russian teuthologist, a 12 m *39 ft 3 in* long giant squid will weigh 1 tonne/*ton* if the head, mantle and arms combined make up half the total length. Since there is a cubic relationship between the linear dimensions of *Architeuthis* and its volume or weight, this means the Thimble Tickle enormity must have scaled about 2·84 tonne/*ton* (i.e. the weight of a large bull hippopotamus), although 2 tonne/*ton* is probably a more realistic figure.

Giant squids are also found in other parts of the world, but the lengths they attain outside the North Atlantic are generally less spectacular. This has prompted one or two teuthologists to query the accuracy of some of the measurements given by Verrill (1879), but although a small number of them may have been taken roughly, i.e. paced out, there is no real reason to suppose that any of the lengths were deliberately exaggerated.

On 16 April 1930 an *Architeuthis* was washed ashore on the Miura Peninsula, Japan. Its head and mantle measured 3·59 m *11 ft 9¾ in* in length and its tentacles 3·57 m *11 ft 9 in*, making a total length of 7·16 m *23 ft 6¾ in* (Tomilin, 1967).

On 2 November 1874 a French expedition to the uninhabited island of St Paul in the southern Indian Ocean found 'a great calamary' (*Architeuthis sanctipauli*) cast up on the northern shore which had a head and mantle length of 2·13 m *7 ft* and tentacles measuring 4·87 m *16 ft*, making a total length of 7·01 m *23 ft* (Verlain, 1877).

There are also several records from New Zealand. In September 1870 a giant squid with 1·67 m *5 ft 6 in* long arms (the tentacles were missing) and a head and body measuring 3·17 m *10 ft 5 in* was washed ashore at Waimarama, North Island. On 23 May 1879 three boys found another specimen (*Architeuthis stockii*) on the beach in Lyall Bay, Cook Strait, which had a head and mantle length of 3·38 m *11 ft 1 in* and arms measuring 1·30 m *4 ft 3 in*. (The mutilated tentacles of this squid measured 1·87 m *6 ft 2 in* in length, but they were probably over 3·65 m *12 ft* long when intact – Kirk, 1882.) Another specimen (*Architeuthis kirki*) discovered *alive* among the rocks at Cape Campbell on 30 June 1886 measured 8·78 m *28 ft 10 in* in total length (head and mantle 3·04 m *10 ft*, tentacles 5·74 m *18 ft 10 in*) (Robson, 1887).

In October 1887 a giant squid with exceptionally long tentacles was washed up in Lyall Bay, Cook Strait. This bizarre creature was described by T W Kirk of the Dominion Museum, Wellington, who wrote:

'Early last month, October 1887, Mr Smith, a local fisherman, brought to the Museum the beak and buccal mass of a cuttle which had that morning been found lying on the "Big Beach" [Lyall Bay], and he assured us that the creature measured 62 ft in total length. I that afternoon proceeded to the spot and made a careful examination, took notes, measurements, and also obtained a sketch. Measurements showed that, although Mr Smith was over the mark in giving the total length as 62 ft (probably not having a measure with him, he only stepped the distance), those figures were not so very far out; for, although the body was in all ways smaller than any of the hitherto-described New Zealand species, the enormous development of the very slight tentacular arms brought the total length up to 55 ft 2 in, or more than half as long again as the largest species yet recorded from these seas.'

It should be pointed out here that the tentacles of *Architeuthis* are highly elastic and can be extended or retracted at will by the animal; so it is quite feasible that the length of 18·89 m *62 ft* paced out by the fisherman may have been correct at the time he found the squid. This probably also explains the discrepancy in Kirk's figures, because in the table of measurements for this individual he gives the total length as 17·37 m *57 ft* (head and mantle 2·36 m *7 ft 9 in*, tentacles 15·01 m *49 ft 3 in*).

Architeuthis longimanus, as it was named, is the *longest* giant squid so far recorded, although nothing like the heaviest. (This specimen probably weighed not much more than 136 kg *300 lb*.)

None of the remains of the New Zealand strandings already mentioned have survived apart from a tentacle club and two pairs of beaks preserved in the Dominion Museum. 'Unfortunately', writes Dr Dell (1970), 'the labels have deteriorated so badly that no detail can be read. The club should belong to either the original specimens of *A. kirki* or *A. longimanus* since the clubs were missing in both *verrilli* and *stocki*.'

This means that *A. longimanus* is only known from Kirk's description – and there are certain anomalies in that. In the 1950s what *may* have been a second specimen of *A. longimanus* was removed from the stomach of a sperm whale captured by the Russian whaling flotilla *Stovetskaya Ukraina* in the Indian Ocean. This squid was credited with a length of 19 m *62 ft 4 in* (Berzin, 1972), but a published photograph does not bear out this extreme measurement – at least not in terms of proportionate bulk. The animal shown has a head and mantle length of *c.* 2·74 m *9 ft* and arms about 2·13–2·43 m *7–8 ft* in length, which should indicate an over-all length of *c.* 10·66 m *35 ft*; but the mantle is very slender and the squid must have weighed less than 226·8 kg *500 lb*. However, if the over-all measurement is genuine and not a misprint, then the tentacles would have measured about 15·84 m *52 ft* in length (cf. 15·01 m *49 ft 3 in* for *A. longimanus*).

In June 1955 some members of the Zoology Dept at Wellington University announced that they were shortly going on a giant squid-baiting expedition in the Cook Strait – the 48 km *30 mile* wide channel separating North and South Islands. Prof I R Richardson, leader of the 'bring 'em back alive' project, said he and his team had built several 6·09 m *20 ft* long steel, spring-back 'mousetrap' devices and planned to suspend them from buoys down to a depth of 1066 m *3500 ft* in a newly discovered undersea canyon in the Hikurangi Trench only a few miles from the shore. It was hoped that *Architeuthis* would be lured by the shark meat in the trap, seize the bait – and become fastened by a spring-loaded grapnel.

It certainly looked good on paper, but sad to relate although these snares were used in the Cook Strait on several occasions no cephalopods of any shape or size were captured. Dr R K Dell (pers. comm. 17 January 1973), Director of the Dominion Museum, said later: 'I have always regarded the "mousetrap" as a publicity gimmick. . . . It always seemed extremely doubtful if such a piece of equipment could really capture such an active creature as a squid, much less a giant *Architeuthis*.'

Although Norway is closely associated with kraken lore, surprisingly few giant squids have been stranded on her coasts in the past 100 years. The largest recorded specimen appears to have been one found on a beach at Kyrksaeterora (formerly Heven) in 1896 which measured about 11·27 m *37 ft* in total length (Heuvelmans, 1968). Another one stranded at Ranheim on Trondheim Fjord in October 1964 measured just under 9·14 m *30 ft* in length and was estimated to weigh between 200 and 300 kg *441 and 661 lb* (Sivertsen, cited by Lane, 1957).

If we exclude the alleged 19 m *62 ft 4 in* long giant squid already mentioned (?*Architeuthis longimanus*), **the largest giant squid recorded this century** was a 14·32 m *47 ft* long specimen captured by a US Coast Guard vessel near the Tongue of the Ocean on the Great Bahamas Bank in 1966 after being involved in a fight with a sperm whale. The carcass was later deposited with the Institute of Marine Sciences at the University of Miami, Florida. Dr Gilbert L Voss (pers. comm. 29 October 1971), Chairman of the Division of Biology at Miami University, said: 'Unfortunately much of this specimen was destroyed through the actions of a graduate student who attempted to preserve a couple of shark heads in the same formalin tank, with the result that most of the contents spoiled. One arm is on deposit in the Natural History Museum at Vienna, and the head is still in our possession.'

In July 1968 two doctors fishing off the resort of Luanco, on the NW coast of Spain, found a dead giant squid and towed it to shore where it was weighed and measured before being cut up for fish bait. It scaled 256 kg *564·38 lb* and measured 9·5 m *31 ft 2 in* in total length.

On 4 July 1972 the Portuguese fishing trawler *Elisabeth* caught an *Architeuthis* near the Flemish Cap Bank which had a mantle length of 1·60 m *5 ft 3 in* and a total measurement of 8·20 m *27 ft 5 in*. It weighed 207 kg *456·35 lb*. This specimen is now on display in the Aquario Vasco da Gama, Lisbon (Vasconcelos, pers. comm. 16 June 1973).

Very few giant squids have been found on British shores. The largest recorded stranded specimen was one (*Architeuthis monachus*) found at the head of Whalefirth Voe, Shetland, on 2 October 1949 which had a 1·21 m *4 ft* long head and mantle and a total length of 7·31 m *24 ft* (Stephen, 1950). In August 1971 a giant squid measuring 6·75 m *22 ft 2 in* in total length was caught in Scottish waters (McCall, pers. comm. 10 October 1971).

A 9·14 m 30 ft *giant squid which ran aground at Ranheim, Norway in 1964 (Camera Press)*

Incidentally, Dr F A Aldrich (1967) is of the opinion that only one kind of giant squid inhabits the North Atlantic – *Architeuthis dux*. He says *A. princeps*, *A. harveyi* and *A. clarkei* are the same species at different stages of decomposition, which suggests that the systematics of the North Atlantic forms will probably be revised soon.

At the moment nobody really knows what the absolute size limit is for a giant squid, but it would be foolish to suppose that the one which ran aground in Thimble Tickle Bay represents the ultimate in terms of length and bulk. Captain A Kean claims he found a huge individual stranded in Flowers Cove, Newfoundland, which measured 21·64 m *72 ft* in total length (Frost, 1934). Another one washed up on the same coast in *c.* 1882 was credited with a length of 26·82 m *88 ft* (head and mantle 9·14 m *30 ft*, tentacles 17·67 m *58 ft*), and Murray (1874) says two giant squids stranded on the coast of Labrador before 1870 measured 24·38 m *80 ft* and 27·43 m *90 ft* respectively.

Measurements of 27·43 m *90 ft*, 39·63 m *130 ft* and even 60·96 m *200 ft* have been conjectured for giant squids from the size of sucker marks found on the skins of captured sperm whales, but it is dangerous to place too much weight on this evidence. Verrill (1879) says the largest suckers on the arms of a 9·75 m *32 ft* long specimen measured 31 mm *1¼ in* in diameter, and those on a 52-footer *15·84 m* about 51 mm *2 in*. Daniel (1925), however, examined sucker marks on the head of one cachalot which measured 89 mm *3½ in* across, and others measuring up to 127 mm *5 in* in diameter have been found on the skins of sperm whales captured in the North Atlantic.

The general consensus of opinion is that exceptionally large sucker marks, i.e. over 51 mm *2 in* in diameter, are old scars that have increased in size as the sperm whale grew. Thus, if a cachalot grows by a factor of 4× after being marked by a squid, and the original tooth ring was 25 mm *1 in* across, it will end up with a 106 mm *4 in* wide scar.

In an unpublished letter to Frank Lane, author of the definitive *Kingdom of the Octopus*, dated 4 May 1967, Prof E Bullock of the Memorial University of Newfoundland, St John's, gave details of some calculations he had made regarding the weight of a hypothetical squid with a tooth ring measuring 304 mm *12 in* across. The answer came out at 259 tonne *255 ton* provided the animal had the same proportions as the 9·60 m *31 ft 6 in* Conche specimen already mentioned, or an even more nightmarish 1038 tonne *1020 ton* if the squid was 'thick set'! In other words, there ain't no such animal.

Another possibility is that other giant squids exist which have much larger suckers in proportion to arm length than *Architeuthis*; for instance, an exaggerated form of *Stenoteuthis caroli* (the largest suckers on a 2·15 m *7 ft 1½ in* female stranded on the beach at Withernsea, Yorkshire, after a gale in February 1925 measured 25 m *1 in* in diameter).

It is more likely, however, that the animal responsible for most of the abnormally large sucker marks is not a squid at all but the blood-sucking Sea lamprey (*Petromyzon marinus*). This species, which can grow to a length of more than 0·91 m *3 ft* is closely associated with the sperm whale and feeds by pressing the circular edge of its mouth hard against the skin until it is punctured.

Huge fragments of squid's arms (not tentacles) said to have been recovered or vomited up from the stomachs of sperm whales have also been cited as *evidence* for the existence of really monstrous specimens. Heuvelmans (1968) speaks of arms measuring 8·22 m *27 ft*, 10·66 m *35 ft* and even 13·71 m *45 ft* in length and up to 0·76 m *2 ft 6 in* thick, but as none of these appendages have ever found their way into a museum or research institute, such measurements must be considered spurious.

Although the North Atlantic appears to have a monopoly of exceptionally large squids, some of the cephalopods found in the rich feeding grounds of the Antarctic Ocean also reach impressive sizes. The largest form so far discovered is the gelatinous cranchid *Mesonychoteuthis hamiltoni*. According to Dr Malcolm Clarke (1966), formerly of the National Institute of Oceanography, Wormley, Surrey, one of these 'passive, balloon-like drifters' removed from the stomach of a sperm whale had a length of 3·5 m *11 ft 6 in* excluding the tentacles which are comparatively short, but he says even larger individuals were collected by the S.S. *Southern Harvester* in the Bellingshausen Sea in 1955–6, including one which had a mantle length of *c.* 3 m *9 ft 9½ in*.

The only known enemy of the giant squid in its natural habitat is the adult bull sperm whale which sometimes swallows its victims whole (female cachalots feed on smaller squids), although killer whales and large sharks will attack disabled specimens on the surface. A cachalot harpooned off São Lourenço, Madeira, on 12 June 1952 vomited a giant squid which still showed signs of life. This one measured 10·36 m *34 ft* in total length and weighed 149·6 kg *330 lb* (Rees & Maul, 1956). On 4 July 1955 an intact *Architeuthis* was found in the stomach of a 14·32 m *47 ft* long bull brought into the whaling station at Fayal Island in the Azores. It measured 10·48 m *34 ft 5 in* in total length and weighed 183·7 kg *405 lb* (Clark, 1955). In 1956 the remains of another squid estimated to have measured 12 m *39 ft 4½ in* in total length when alive were discovered in the food cavity of a 15·8 m *51 ft 9½ in* sperm whale captured by the Russian whaling flotilla *Slava* in the Atlantic off the southern Orkneys (Korabel'nikov, 1959). Another one taken *alive* from the stomach of a large bull caught by a Russian whaling flotilla in the North Pacific on 31 December 1964 weighed 204·5 kg *450 lb* (total length 12 m *39 ft 4½ in*).

Young Basking sharks (*Cetorhinus maximus*) measuring up to 3 m *9 ft 9½ in* in length have also been recovered from this cetacean's stomach, which suggests that a weight of 204–227 kg *450–500 lb* is probably about the most an adult bull can get down its gullet in one go.

The giant squid (*Architeuthis* sp.) has the largest eye of any living or extinct ani-

mal. In a 12 m *39 ft 4½ in* specimen the ocular diameter was 180 mm *7·08 in*, and in a 55-footer *16·76 m* the measurement was *c.* 400 mm *15·74 in* (cf. 304 mm *12 in* for a 33⅓ long-playing record!). In the latter case the eyeball must have weighed several pounds, but most of this bulk would have been fluid. By comparison, the largest blue whales have an ocular diameter of 100–120 mm *3·93–4·71 in*, the southern elephant seal *c.* 70 mm *2¾ in* and man 24 mm *0·94 in*.

The smallest squid so far recorded is *Parateuthis tunicata*, which is known only from two specimens collected at depths of 3000 m *9842 ft* and 3425 m *11 235 ft* respectively, in the Antarctic Ocean by the German South Polar Expedition 1901–3. The larger of the pair had a head and mantle length of only 7·87 mm *0·309 in*, and a total length of 12·7 mm *0·50 in* (Thiele, 1921).

The smallest squid found in British waters is *Alloteuthis media* which has a maximum total length of 177 mm *7 in*.

The most dangerous squid is *Dosidicus gigas* (=*Ommastrephes gigas*) of the Humboldt Current off Peru which may reach a total length of 3·65 m *12 ft* and a weight of 158·7 kg *350 lb*. This extremely aggressive cannibalistic animal is greatly feared by native fishermen, and Lane (1957) believes that anyone unfortunate enough to fall overboard in the vicinity of these demons would be torn to pieces in less than half a minute. (The chitinous jaws of this species are so powerful that they can bite through the extra-tough wire leaders used by fishermen to catch tunny.) So far, however, no human fatalities have been reported, although American big-game fishermen now regularly angle for this cephalopod.

Nature gave the squid a jet propulsion unit long before man ever thought of using similar methods for his own transport, and Bartsch (1917) claims that decapods 'inch for inch' . . . will compete in swimming power with any other creature that lives in the sea'. Certainly some of the smaller surface-dwelling varieties like *Stenoteuthis bartrami* and *Onychoteuthis banksii* are among the swiftest creatures in the sea, and it has been calculated from their flight trajectory – there is a record of a specimen of *O. banksii* landing on a ship's deck 7 m *23 ft* above sea level (Rees 1949) – that squids of these genera leave the water at speeds up to 55 km/h *34·35 miles/h* when escaping from enemies like tunny (Akimushkin, 1965).

Fast-swimming squids are very popular with neurological researchers because they have the largest nerve fibres (axons) of any known creature. In the case of *Dosidicus gigas* they have been measured up to 18 mm *0·7 in* in diameter (Akimushkin, 1965), which means they are about 500 times thicker than human nerves (Young, 1944).

In January 1961 a gold medallion was found in the stomach of a European common squid (*Loligo vulgaris*) caught at San Sebastian, Spain. It had been lost two years previously by a woman on a beach at Barcelona – 2500 sea miles away!

Very little research has been done on the longevity of squids, but most species probably have a natural life-span of less than four years. Akimushkin thinks it must take 'several decades' for giant squids like the one caught in Thimble Tickle Bay to reach such dimensions, but in marine animals there is no real relationship between size and age.

On 25 March 1941 the British troopship *Britannia* was attacked and sunk by a German raider in the Atlantic about 2240 km *1400 miles* west of Freetown. Eleven survivors managed to cling to a tiny raft and took it in turns to sit on the floating platform. One moonlit night a large squid (?*Dosidicus gigas*) seized one of the sailors and dragged him screaming below the surface as his horrified companions looked on helplessly (Lane, 1957). **This is the only known record of a man being attacked and killed in the water by a large squid.**

The record 53·8 kg 118 lb 10 oz *Common Pacific octopus with its captor (Donald E. Hagen)*

The largest of the 150 recognised species of octopod is *Paroctopus apollyon* (=*Octopus punctatus*) of the North Pacific which regularly exceeds 4 m *13 ft 2½ in* in radial spread and 18 kg *39·68 lb* in weight. One huge example trapped in a fisherman's net in Monterey Bay, California, USA, had a radial spread of over 6·09 m *20 ft* and scaled 49·89 kg *110 lb* (MacGinitie & MacGinitie, 1949).

According to one authority (Akimushkin, 1965) this species attains a 'maximum size' of 5 m *16 ft 4¾ in*, but he does not indicate whether this measurement refers to length (tip of head to end of tentacles) or radial spread. If, as the compiler suspects, the figure refers to length this is equivalent to a tentacular span of *c.* 9 m *29 ft 6¼ in*. Heuvelmans (1968) goes even further and credits this octopus with a radial spread of up to 9·75 m *32 ft* ('some say 38 ft') and a weight of 124·73 kg *275 lb*, but neither he nor Akimushkin back up their statements with any sort of proof.

In May 1944 a Canadian diver named Ralph Wood was attacked by a very large *P. apollyon* while exploring the wreck of a tug which had sunk in the Juan de Fuca Straits, British Columbia. After a desperate struggle he managed to sever the arms grasping his left leg with a knife, and was then hauled to the surface by the anxious crew of the diving barge working overhead. The mutilated appendages were said to have measured from 1·21 m *4 ft* to 1·83 m *6 ft* in length, and scientists at the University of British Columbia estimated that the villain of the piece must have had arms measuring from 3·04 to 4·57 m *10 to 15 ft* in length when intact and a body measuring 1·52 m *5 ft* in diameter (length?).

The largest octopus (*P. apollyon*) on record was a specimen weighing 53·80 kg *118 lb 10 oz* which was 'wrestled' to the surface single-handed by a skin-diver named Donald E Hagen in Lower Hoods Canal, Puget Sound, Washington, USA, on 18 February 1973. Mr Hagen (pers. comm. 9 August 1974), who lives in Camas, Washington, said he first spotted the octopus lying on a rock reef in 18·28 m *60 ft* of water. He immediately plunged down and tried to loosen the invertebrate from its resting place with his *bare hands*, but the task proved too much for him and eventually he was forced to surface because 'my arms had weakened and breathing was becoming difficult'. Fortunately the octopus then moved on to some sand and rubble and Hagen decided to make another attempt. This time, watched by his diving partner Robert Click, he managed to wrench the cephalopod off the bottom without too much trouble and succeeded in towing the squirming animal some 60·96 m *200 ft* – 'it seemed like one of the longest swims I have ever made' – to the beach.

This mollusc measured 3·86 m *12 ft 8½ in* over-all – several lengths were actually obtained ranging from 3·55 m *11 ft 8 in* to 4·08 m *13 ft 3 in* – and had a *relaxed* tentacular span of *c.* 7·01 m *23 ft*. (The length of the arms in this species accounts for about 78 per cent of the total length.)

The closely related *P. asper*, another octopus of the North Pacific region, has also been credited with impressive measurements. Akimushkin states that this species matches *P. apollyon* for size, i.e. maximum length 5 m *16 ft 4¾ in*=radial spread *c.* 9 m *29 ft 6¼ in*, but a *large* specimen described by Joubin (1897) from Avacha Bay, on the SE coast of the Kamchatka Peninsula, northern Siberia, USSR, measured only 1232 mm *4 ft 0½ in* in length=radial spread *c.* 2290 mm *7 ft 6 in*.

Octopus dofleini, found off the coasts of Japan and Korea, is probably a more worthy rival in terms of size. This species has been measured up to 3 m *9 ft 10 in* in over-all length (Sasaki, 1929), which gives it a maximum radial spread of *c.* 5·80 m *19 ft*.

Some of the octopods found on the coral reefs off Port de Papeari, Tahiti, also reach a large size. Wilmon Menard (cited by Lane, 1957) was present on the Rimaroa atoll when one spanning 5·48 m *18 ft* was killed by the native population with clubs and spears and this was not considered a record specimen.

In 1874 Dr William H Dall, the Curator of Molluscs at the US National Museum in Washington, DC, speared an octopus of the North Pacific variety *Paroctopus hongkongensis* in Illiuliuk Harbour, Unalaska Island, Alaska, which had a radial spread of 9·75 m *32 ft*, but the body of this cephalopod was diminutive by comparison – 305 × 152 mm *12 × 6 in* – and the creature probably weighed less than 9·07 kg *20 lb* (parts of this octopus are preserved in the US National Museum).

At the end of November 1896 the remains of a large marine animal were found by two boys on a beach 19·2 km *12 miles* south of St Augustine, Florida, USA. At first the fleshy mass was thought to be part of a large whale, but after a careful examination Dr DeWitt Webb, president of the local scientific society, pronounced it to be gigantic octopus of a type unknown to science. On 3 January 1897 the *New York Herald* devoted considerable space to the story, the following being an extract:

The 5-tonne/ton cephalopod Octopus giganteus, *once believed to be a myth but now known to exist*

'It had evidently been dead some days and was much mutilated. Its head was nearly destroyed and only the stumps of two arms were visible. Its gigantic proportions, however, were astounding. The body, as it lies somewhat embedded in the sand, is 18 ft long and about 7 ft wide, while it rises $3\frac{1}{2}$ ft above the sand. This indicates that when living its diameter must have been at least $5\frac{1}{2}$ ft. The weight of the body and head would have been at least four or five tons [US short tons of 2000 lb *907 kg*]. If the eight arms held the proportions usually seen in smaller species of the octopus, they would have been at least 75–100 ft in length and about 18 in in diameter at the base.

'The form of the body and its proportions show that it is an eight-armed cuttlefish, or octopus, and not a giant ten-armed squid like the devil fishes of other regions. No such gigantic octopus has been heretofore discovered.'

Full details of the monster, together with photographs taken by Dr Webb, were sent to Prof A E Verrill of Yale University, the noted authority on giant squids and other cephalopods, who later wrote in the *American Journal of Science*: 'These photographs show that it is an eight-armed cephalopod, and probably a true octopus of colossal size. Its body is pear-shaped. . . . The head is scarcely recognisable, owing to mutilation and decay. Dr Webb writes that a few days after the photographs were taken . . . excavations were made in the sand and the stump of an arm was found, still attached, 36 ft long and 10 in in diameter, where it was broken off. . . . This probably represents less than half of their original length. . . . The length, given as 18 ft, includes the mutilated head region. . . . The parts cast ashore probably weighed at least 6 or 7 tons, and this is doubtless less than half of its total mass when living . . . this species is evidently distinct from all known forms, and I therefore propose to name it *Octopus giganteus*.'

Soon afterwards, however, Verrill had second thoughts about the true identity of this animal and made a retraction in a later issue of the same journal. He said the tissue samples Dr Webb had sent him were not from a cephalopod, and suggested that the great mass might be part of the head of a creature like a sperm whale, although he admitted that it was decidedly unlike the head of an ordinary cachalot.

Interest in the mysterious carcass rapidly waned after that, and within a few years the St Augustine monster had been forgotten. The story, however, didn't quite end there. . . .

In 1970 Joseph Gennaro, Jr, Associate Professor of Biology at New York University, carried out some microscopic tests on one of the tissue samples of '*Octopus giganteus*' preserved in the Dept of Molluscs at the US National Museum and his findings, along with a detailed article on the strange creature by Forest G Wood, former Curator of the Marineland Research Laboratory at St Augustine, were published in *Natural History* (March 1971), the journal of the American Museum of Natural History.

Prof Gennaro concluded: 'Viewing section after section of the St Augustine sample, we decided at once, and beyond any doubt, that the sample was not whale blubber . . . the connective tissue pattern was . . . similar, if not identical with, that in my octopus sample. The evidence appears unmistakable that the St Augustine sea monster was in fact an octopus.'

Since then Prof Gennaro has carried out further biochemical tests on the sample and these 'indicate rather conclusively that the creature was not a decapod'.

The half-buried remains of the giant octopus washed up on the beach at St Augustine, Florida, USA (Argosy *magazine)*

In an unpublished letter to Mr F G Wood, now working for the US Naval Undersea Research and Development Center at San Diego, California, dated 18 May 1971, Mr John C Martin, USN Ret'd of San Diego, gave details of another giant octopus he had seen in the same area during the Second World War. He said: 'In 1941 I was a coxswain in the first division aboard the USS *Chicopee* AO-41. My section was on duty in the second dog watch manning the 3 inch gun on the forecastle and keeping a very good look-out for periscopes and any other part of the enemy. The ship had departed Baton Rouge, Louisiana, with a cargo of aviation gasoline and fuel oil for Portland, Maine.

'It was in the last of March or April that the ship was steaming off the coast of Florida in the general area of Fort Lauderdale and Saint Augustine. Dead ahead of our course appeared something on the surface of the water that could not be readily described. The closer we approached it looked like a huge pile of brown kelp seaweed. As it hove into view there was no doubt as to its identity. The coils of its arms were looped up like huge coils of manila rope. However, the coils were over 36 in in circumference. This last deduction was compared to the girth of my waist at that time. There was no mystery as to why we were able to see this monster animal after we had time to reflect on the circumstances preceding its occurrence. There had been preceding our arrival at this segment of our journey two destroyers. They were rearranging the ecology on the floor of Torpedo Junction like an underwater Fourth of July!'

Mr Wood (pers. comm. 4 February 1975) said he later learned from this correspondent that when the alarming apparition was sighted there was enough light for good visibility. He also ascertained that the creature measured an estimated 9·14 m *30 ft* in diameter and that 'its arms seemed about equal length; coiled but moving slowly'. The giant polyp was apparently observed by all the members of Martin's gun crew and details were given to the watch officer, but the incredible encounter was not recorded in the ship's log.

A giant octopus was also probably responsible for the sinking of the 152 tonne *150 ton* schooner *Pearl* in the Bay of Bengal, Indian Ocean, although most writers who have cared to pass an opinion on this story seem to think the creature was a gigantic kraken. The tragedy occurred about an hour before sunset on 10 May 1874 and was witnessed by a number of travellers aboard the passing steamer *Strathowen*, who first noticed a large brownish mass like a bank of seaweed floating in the water less than a mile from the becalmed schooner. Suddenly a rifle shot rang out from the pretty little vessel and the 'mass' immediately erupted into life and headed for the ship. Seconds later it struck the schooner with a tremendous thud and the vessel visibly reeled before righting herself. The next moment the nightmarish animal was on board, squeezed in between the two masts, and as the passengers looked on aghast the schooner was slowly pulled over on to her beam-ends where she remained for a short while before going down.

The steamer quickly put out boats and picked up five men swimming in the water, including the master of the *Pearl*, James Floyd, who said later that one of his crew had been crushed between the mast and one of the 'thick as a barrel' arms of the devil fish, and another had been sucked down by the ship.

It is interesting to note that the colour of this marine enormity was described as 'brownish' – the same as that of the outsized cephalopod seen by Martin off Florida. Other pieces of evidence supporting the giant octopus theory – apart from the fact that the creature had 'monstrous arms like trees' and a body 'as thick as the schooner and half as long' – lie in the statement by James Floyd that (1) the monster was able to *swarm* aboard the schooner (this feat would be impossible for a giant squid because large decapods are practically immobile out of water); and (2) he could see 'a huge oblong mass moving by jerks just under the surface of the water, and an enormous train following'. (All octopods swim jerkily because of the intermittent jets from their syphon or funnel, and they also travel backwards, i.e. with the bulbous part of the body to the fore with the arms trailing out behind; the giant squid, on the other hand, propels itself through the water in the forward position with the partly retracted tentacles and arms out front.)

Frank Lane contends that 'the great disparity in weight and lack of support are probably the strongest arguments against the authenticity of this report', but when he wrote these words he didn't know of the existence of a giant octopus weighing 10 or even possibly 20 tonne/*ton*!

How this colossus – whose existence is still not generally recognised – manages to maintain its enormous bulk must remain something of a mystery. Most of the members of this order are essentially carnivorous and feed on crustaceans and other molluscs, but it is difficult to imagine a cephalopod of this magnitude subsisting on such meagre fare. Robson (1932) believes that most benthic octopuses live on food which does not take any effort to capture such as rotting flesh and plankton, but in the case of *O. giganteus* it would have to be something much more substantial (bottom-feeding sharks?).

The 'sea monster' washed up on a desolate beach on the West Tasmanian coast in 1960 may also have been a giant octopus. The huge carcass, measuring 6·09 × 5·48 m *20 × 18 ft* and weighing an estimated 8 tonne/*ton*, lacked any recognisable feature and the flesh was described as *extremely tough*. (Prof Verrill also remarked upon the 'extreme firmness and toughness' of the tissue samples sent to him of the St Augustine monster.)

Scientists from the Commonwealth Scientific and Industrial Research Organisation (CSIRO) who later examined the remains said they could find no trace of vertebral structure and admitted that they were completely baffled as to what the animal could be.

The largest octopus found in British waters is the Common octopus (*Octopus vulgaris*), which is confined mainly to the English Channel. A specimen with a radial spread of just over 1·83 m *6 ft* and weighing 3·30 kg *7 lb 8 oz* was spear-gunned by a member of Brighton Swimming Club near the Palace Pier, Brighton, Sussex, in September 1960. Another one with a tentacular span of nearly 1·52 m *5 ft* was captured in a lobster pot at Freshwater Bay, Isle of Wight, about the same time and on 23 November 1960 an octopus with a radial spread of 1·60 m *5 ft 3 in* was caught in a fisherman's net off Margate, Kent and handed over to Whitstable's marine research laboratory.

This species reaches a much greater size in the warmer waters of the Mediterranean. The largest males examined by Packard (1961) at the famous Zoological Station at Naples weighed between 8 and 10 kg *17·63* and *22·04 lb* and the largest female was 6·3 kg *13·88 lb*. Verany (1851) speaks of one which had a radial spread of 3 m *9 ft 10 in* and scaled 11 kg *24·24 lb*, and other examples weighing up to 25 kg *55·11 lb* have been reliably reported.

The record 3·30 kg 7·5 lb *Common octopus harpooned off the Palace Pier, Brighton in September 1960*

The smallest known octopus is *Octopus arborescens* of Sri Lanka which has a radial spread of less than 508 mm *2 in*.

Although a lot of blood-curdling stories have been written about octopuses, they are normally shy, inoffensive creatures and rarely attack man deliberately. They are curious, however, and will sometimes investigate the arm or leg of a diver simply because it is moving, but if the person is experienced and remains motionless, the octopus soon loses interest. The diver is only in danger if he struggles and the mollusc is firmly anchored to a rock, because it then gets excited and holds on tightly to the limb it has seized. (American skin-divers have discovered that if they offer a plug of strong-smelling tobacco to an octopus it will immediately release its limpet-like hold on a rock.)

In April 1935 a large octopus (*Paroctopus apollyon*) seized a fisherman named Frank Contrin while he was wading waist-high in the surf at a point south of Golden Gate – the entrance from San Francisco Bay, California, to the Pacific Ocean. The arms of the monster were wrapped round the man's body, legs and left arm. Fortunately a friend, Harry Simmons, who was scraping shellfish off the rocks 30·48 m *100 ft* away with a 33 cm *12 in* butcher's knife heard his companion's cries and rushed to his assistance as he was being slowly dragged out of his depth. As fast as Simmons severed one arm another seemed to replace it, but after a desperate struggle he managed to mortally wound the creature with a knife thrust between the eyes. Contrin, his body covered with sucker marks, was then helped ashore in an exhausted condition, and a few minutes later the octopus was washed up dead. It had a radial spread of 4·57 m *15 ft* and weighed 19·50 kg *43 lb*.

All octopuses have poison glands which secrete a venom for capturing or killing prey.

The most venomous cephalopods in the world are the blue-ringed octopods *Hapalochlaena lunulata* and *H. maculosa* of the Indo-Pacific region. The fast-acting neurotoxic venom carried by these small molluscs (radial spread 101–152 mm *4–6 in*) is so potent that scientists at the Commonwealth Serum Laboratories in Melbourne, Victoria, Australia, say the amount ejected through the horny beak in one bite is sufficient to kill seven people.

Curiously enough persons bitten by these demons do not feel any initial pain. The first symptoms are usually a dryness of the mouth and difficulty in swallowing, followed by vomiting, loss of co-ordination, failing eyesight and a paralysis which spreads to all parts of the body.

In September 1954 a 21-year-old skin-diver was bitten on the shoulder by a blue-ringed octopus off East Point, near Darwin, Northern Territory, Australia, and collapsed shortly afterwards. He was immediately rushed to hospital where he was given emergency treatment and then placed in an iron lung, but died less than two hours after being bitten (Lane, 1957). The octopus was almost certainly *H. lunulata*=*Octopus rugosus* (McMichael, 1964).

In 1968 there were two more fatalities in Australian waters, including a 23-year-old soldier who was bitten on the hand while paddling in a rock pool near Sydney, New South Wales, and died 90 minutes later. In both cases the octopus was believed to have been *H. maculosa*.

On the credit side, however, it is interesting to note that at least four people are known to have survived a bite from *H. maculosa* despite the potency of its venom. Two of them were discharged from hospital after 24 hours' treatment, and the others made a gradual recovery over a period of several days.

The greatest depth at which an octopus has been recovered is 8100 m *26574 ft* for a specimen (?*Octopus* sp.) collected in a trawl by the Russian research vessel *Vityaz* at Station 162 in the Kuril Kamchatka Trench in *c.* 1950, but the animal was unfortunately lost before it could be positively identified (Zenkevitch *et al.*, 1955).

The largest of all existing bivalve molluscs is the Giant clam (*Tridacna derasa*) which is found on the Indo-Pacific coral reefs and off the E coast of Africa. Weights up to 454·4 kg *1000 lb* have been claimed for this species, but there are very few records of giant clams reaching even half that poundage.

The largest giant clam ever recorded is one preserved in the American Museum of Natural History which measures 1·090 × 0·73 m *45 × 29 in* and weighs 263·4 kg *579·5 lb*. It was collected from the Great Barrier Reef off the NE coast of Queensland, Australia, in 1917 and probably weighed about 272·7 kg *600 lb* when alive. (The soft parts weigh up to 9·09 kg *20 lb*.) Another huge specimen collected at Tapanoeli (Tapanula) on the NW coast of Sumatra before 1817 and now preserved at Arno's Vale, County Down, Northern Island, measures 1·37 m *4 ft 6 in* in length and weighs 230·4 kg *507 lb* (Rosewater, 1966), and the Australian Museum in Sydney, New South Wales, has a giant clam weighing exactly 227·2 kg *500 lb*.

Although the giant clam is popularly believed to be a man-killer, and there have been a number of stories of pearl divers and others drowning after getting their foot or leg caught in the steel-trap jaws of one of these creatures (the serrated edges fit so tightly that they can grip a piece of wire), not one single case has ever been authenticated. Caras (1964) says the huge purple-green mantle of this clam is so conspicuous that only a fool or a very unobservant person would tread on it. He also points out that even if somebody did accidentally put his foot between the giant teeth he would still have plenty of time to withdraw it because the valves do not *snap* shut as many people imagine. (The speed is governed by the size of the clam . . . small specimens closing much more rapidly than large ones.)

According to Dr R T Abbott of the Philadelphia Academy of Natural Sciences (cited by Breland, 1963), the first attempt by one of the Dayak divers to recover the famous 'Pearl of Laotze' (see page 214) ended in tragedy when the giant clam closed on the man's arm and drowned him, but this report has never been proven and is probably a glamorous extension of the story. Caras solicited the opinions of 13 specialists (six of them Australian) regarding the danger or otherwise of the giant clam to man, and the general consensus of opinion was that evidence incriminating this species was 'circumstantial and largely hearsay'. Certainly none of them knew of a human fatality or injury due to this mollusc.

This doesn't mean though that *T. derasa* is not dangerous. On the contrary, if a large specimen did trap a limb and the man was unarmed and unable to sever the great adductor muscle which controls the huge valves, then he would surely drown.

In 1963 Dr Joseph Rosewater, Curator of Molluscs at the US Museum of Natural History, Smithsonian Institution, Washington, DC, received a personal communication from a Malaysian named Johnny Johnson, who attributed the loss of one of his legs to a giant clam, but further details are lacking.

Monsieur Vaillant, the French naturalist, once put the strength of the giant clam to the test in 1883 by fastening an exceptionally large specimen (254·5 kg *560 lb*) to a post and then hooking buckets of water to the lip of one of the valves until the weight forced it open. By this method he discovered that the clam yielded to a pressure of 891 kg *1960 lb*, or three and a half times its own weight, which means it would be impossible for a trapped man to wrench open the two valves of a *Tridacna* this size with his bare hands.

The largest bivalve mollusc found in British waters is the Fan mussel (*Pinna fragilis*) which is most abundant off the southern coast of England. One specimen found at Torbay, Devon, measured 370 mm *14·56 in* in length and 200 mm *7·87 in* in breadth at the hind end which was buried in the sea-bed.

The smallest bivalve mollusc found in British waters is the Coin-shell (*Neolepton sykesi*), which measures less than 1·6 mm *0·062 in* in length. This species is only known from a few examples collected off Guernsey, Channel Islands, in 1894 (Tebble, 1966).

The value of a sea-shell does not necessarily depend on its rarity or its prevalence. Some rare shells are inexpensive because there is no demand for them, while on the other hand certain common shells fetch high prices because they are not readily accessible. In addition 'live' shells (with the live creature inside) fetch higher prices than 'dead' examples, and exceptionally large specimens in good condition are more valuable than ones of average size. In theory the most valuable shells in the world should be some of the unique specimens collected in deep-sea trawls, but these shells are always dull and unattractive and hold very little interest for the collector.

The most highly prized of all molluscan shells in the hands of conchologists is the White-tooth cowrie (*Cypraea leucodon*), which is found in the Philippines. Up to 1960 this shell was known only from the type specimen preserved in the British Museum (Natural History), London, which was included in the Broderip Shell Collection purchased by the Trustees of the British Museum in 1837 for £1575 – but that year a second specimen was 'rediscovered' in the Shell Collection of the Boston Society of Natural History and is now preserved in the Museum of Comparative Zoology at Harvard University, Cambridge, Massachusetts, USA. In 1965 a third example was found in the stomach of a fish caught in the Sulu Sea, a large inter-island sea of the Philippine Islands (Dance, 1971).

The rarest highly prized shell is *Conus dusaveli*, which is known only from the type specimen recovered from the stomach of a fish caught off Mauritius in the Indian Ocean.

The highest price ever paid for a sea-shell is £1350, which was the sum given at a Sotheby & Co., London, auction for one of the four known examples of *Conus bengalensis*. The 101 mm *4 in* long shell was trawled by fishermen off NW Thailand in December 1970. It has been described as 'one of the most beautiful cone shells in the world'.

The Precious wentletrap, a 76 mm *3 in* long white shell resembling a spirally twisted trumpet

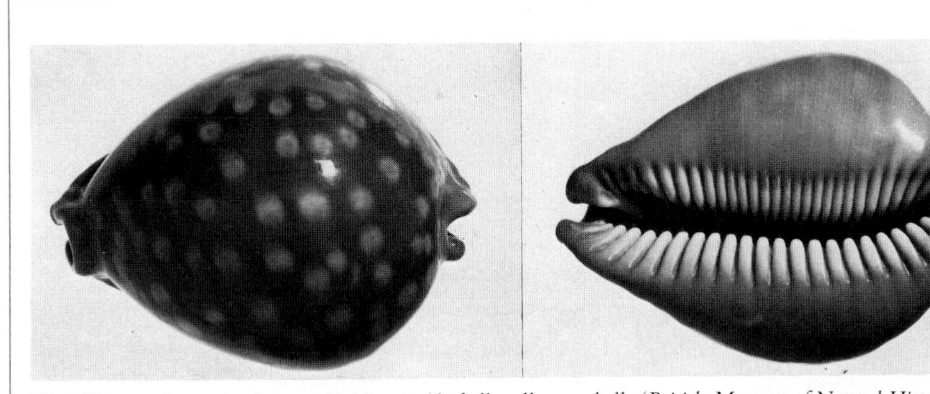

The White-tooth cowrie, the most highly prized of all molluscan shells (British Museum of Natural History)

and open from top to bottom, was once so scarce that nearly £1000 was reputedly paid for a specimen. Today, however, they are collected in quantity and now sell for only a few pounds.

Pearl is a dense, lustrous concretion that is formed in various molluscs by deposition of thin concentric layers of nacre about a foreign particle (e.g. a minute parasitic worm) within or outside the mantle. Biologically speaking all shelled molluscs are capable of producing a pearl of sorts but the specimens fit for use in jewellery come chiefly from the large tropical pearl oysters (family Pinctada) and the Freshwater mussels (family Quadrula).

The largest known natural pearl is a milk-white opaque nacreous mass known as the 'Pearl of Laotze', owned by Mr Wilburn Dowell Cobb of San Francisco, California, USA, which measures 240 mm *9½ in* in length by 101–139 mm *4–5½ in* in diameter and weighs 98437 grains or 6·37 kg *14 lb 1 oz* (one pearl grain=50 milligrams).

According to the owner's romantic account of how he came to own this huge pearl, which is shaped like a human brain and has the same convolutions and furrows, the specimen was found by Dayak divers in a giant clam (*Tridacna derasa*) at Boligay Cove, in the Philippine island of Palawan on 7 May 1934. They presented the pearl to Panglima Pisi, lieutenant to a Moro chief, who, in return, gave them a sack of rice which was considered generous payment. Later the same year Cobb, a collector of early Chinese pottery, saw the misshapen freak and tried to buy it, but Panglima refused to sell because he looked upon the pearl as a talisman. Two years afterwards the American revisited southern Palawan and went to see Panglima again. On his arrival he found the man's only son critically ill with malaria, and by applying his small stock of medical knowledge he was able to effect a cure. The grateful father then offered him the pearl as a gift if he would return for it in a year – for Panglima had vowed to pray by it for twelve months in thanksgiving for the recovery of his son. Cobb accepted the offer and took possession of the pearl in July 1937. Shortly afterwards the specimen was examined by Dr Hilario A Roxas of the Bureau of Science of the Philippine Commonwealth Government in Manila who found that part of one of the valves of the *Tridacna* fitted the pearl 'almost like a glove'. He suggested that a detached piece of brain coral may have accidentally got between the valves of the mollusc. The clam, unable to eject the coral, then tried to minimise the irritation caused by covering the unwelcome guest with mother-of-pearl.

In April 1969 the 'Pearl of Laotze' was put up for sale in London, the price tag being £1458333, but there were no takers and one appraiser put its value at less than one-twentieth of this amount. In July 1971 the pearl, which is now kept in a San Francisco bank vault, was reportedly worth $4080000 (*c.* £1700000).

If we exclude this monster, then the largest pearl in the world is probably the poorly formed 'Hope Pearl' of 1800 grains (116·3 g *4·11 oz*). It is fractionally over 76 mm *3 in* in length and has a circumference at its globular end of 114 mm *4½ in*. Henry Thomas Hope, the 19th-century Amsterdam and London banker, reputedly owned a baroque pearl weighing 1860 grains (120·2 g *4·25 oz*), but the compiler has not yet been able to confirm this figure.

The largest known pearl of regular shape is an example known as 'La Peregrina' (The Wanderer), which was discovered by a Negro slave in the Gulf of Panama in the early 16th century. It was taken to Europe, and Philip II of Spain gave it to Princess Mary Tudor (later Mary I) of England when he married her in 1554. Later it came into the possession of the Bonaparte family. Napoleon's brother Joseph, King of Spain (1808–13), is said to have taken La Peregrina when he abdicated, and the Duke of Abercorn's family acquired it from Napoleon III (1808–73), who spent his last years in exile at Chislehurst, Kent. In January 1969 the pearl was sold in New York to Richard Burton, the Welsh film actor, for £15420. He wanted it as a present for his wife, Elizabeth Taylor.

The largest known freshwater pearl is the famous 'Little Willie' which measures 12 mm *0·5 in* in length and weighs 41·45 grains (2·68 g *0·095 oz*). The pearl was found in a freshwater mussel (*Margaritana margaritifera*), collected by William Abernethy, a professional pearl wader, in the River Tay, Perthshire, Scotland, in August 1967. (Pearl waders scan the river beds through glass-bottomed drums.) He sold the pearl to A & G Cairncross, jewellers, of Perth for an undisclosed sum. (The pearl has been valued at £10000.) In September the same year another giant pearl weighing 28 grains (1·81 g *0·064 oz*) was also taken from the River Tay. This one was found by Donald McGregor of Rattray, Blairgowrie, who sold it to the Perth jewellery firm of R W Proudfoot Ltd. Freshwater pearls lack lustre, but can still be attractive.

Pearls vary in colour from a dull white, pink, apricot, gold and rose through to green, purple and black. The most sought after pearls are rose or greenish black in colour, but blue-black ones are nearly as costly and true gold pearls also fetch extremely high prices. (In 1947 an oyster was found with one pink and one black pearl inside it.)

The longest-lived of all molluscs is probably the deep-sea clam *Tindaria callistiformis* which takes an estimated 100 years to reach a length of 8 mm *0·31 in* (which must be the slowest rate of growth in the animal kingdom!). This figure was obtained by a team of scientists from Yale University, Connecticut, USA, after they had dredged up a number of these clams from a depth of 3800 m *12467 ft* in the North Atlantic and measured the amount of the radioactive element radium 228 in their shells at different stages of growth. The clams were put into four different size categories and the average radium content of the shell at the time of capture was calculated for each class by analysis of individual shells. Then the relationship between radium content, shell mass and age was plotted for a range of hypothetical ages. The greatest age recorded using this method was 98 years for a clam in the largest size category and this tied in nicely with the number of rings, i.e. *c.* 100 found on its shell. (It has been suggested that each band represents one year's growth – Turekian, 1975.)

The largest known species of snail is the seaweed-eating Sea hare (*Tethys californicus*), also known as the 'Sea slug', which is found in the shallow waters off the coast of California, USA. The average weight is 3·18–3·63 kg *7–8 lb*, but one enormous individual tipped the scales at a massive 7·15 kg *15 lb 13 oz* (MacGinitie & MacGinitie, 1949). The only other marine snails which remotely approach this sort of weight are the Queen conch (*Strombus gigas*) and the Horse conch (*Fasciolaria gigantea*) of the Florida Keys and the West Indies, and the Trumpet or Baler conch (*Syrinx auranus*) of Australia, all of which weigh up to 2·27 kg *5 lb* and have a shell measuring 304–609 mm *12–24 in* in length.

The largest marine snail found in British waters is the Sea hare *Aplysia limacina*, which has been measured up to 380 mm *14·96 in* in length (McMillan, 1968).

The largest known land snail is the Giant African snail (*Achatina achatina*). This species was originally found only in East Africa and Madagascar, but at the beginning of the 19th century it turned up in Mauritius, the Seychelles and Réunion in the Indian Ocean. Since then it has spread to India, the Far East and the United States. The Japanese also took it to all the Pacific Islands they occupied during the Second World War for use as food, and several thousand were accidentally taken to California with American army surplus material after the war. Today this nightmarish destroyer of almost any growing thing is a major pest, and although countless millions of them are wiped out annually by fire and poison, they reproduce at such a staggering rate – it has been worked out that one snail can theoretically produce 11000000 descendants in five years – that their numbers remain virtually unaffected.

The average adult specimen measures 203 mm *8 in* in length when fully extended (shell length *c.* 127 mm *5 in*) and weighs about 227 g *8 oz*, but much larger examples have been recorded.

The heaviest African giant snail on record is one owned by Mr Chris Hudson of Hove, Sussex which measures 342 mm *1 ft 1½ in* in total length (nose to tail) and weighs 200 g *1 lb 7 oz*. He collected the snail in Sierra Leone, West Africa in June 1976.

The largest land snail found in Britain (about 85 species) is the Roman or Edible snail (*Helix pomatia*) of southern England which measures up to 101 mm *4 in* when fully extended (shell 45 × 45 mm *1·77 × 1·77 in*) and weighs up to 85 g *3 oz*. The Common garden snail (*H. aspersa*) has a shell measuring 34 × 34 mm *1·37 × 1·37 in* (Ellis, 1969).

The smallest land snail found in Britain is *Punctum pygmaeum*, which has a shell measuring only 0·023–0·035 × 0·047–0·059 in.

The average life-span of large land snails like *Helix pomatia* and *H. aspersa* is two to three years. Ellis, however, says the common snail has been known to live for ten years (a male specimen brought back to London by a woman after a holiday in Yugoslavia in 1963 was still flourishing in 1972), and a figure of eight years has been reported for the Roman snail. There is also a record of a European garden snail, *Copaea nemoralis*, being kept in captivity at the Royal Ontario Museum of Zoology, Toronto, Canada, from 28 April 1938 to the middle of December 1944 (Dymond, 1947).

According to John Slee (*Garden News*, 4 August 1972) the longevity record for a garden snail is held by a *Helix desertorum* from Egypt which was presented to the British Museum

(Natural History), London, as a dead specimen in 1846. 'It was displayed on a table until 1950', he writes, 'when one of the curators noticed some movement in it. It was sprinkled with warm water, began to feed, and lived happily for a year. In 1951 it became torpid and died in 1952.' Unfortunately this rather nice story was exposed by Mrs M A Edwards (née Elsnore) of the Dept of Zoology (Mollusca), British Museum (Natural History), who stated: 'As you suggest, this is in fact an exaggeration and cannot be taken as true fact' (pers. comm. December 1972).

The giant African snail has a maximum life-span in captivity of *c.* ten years, but probably lives less than six years in the wild state. Of the marine gastropoda the Red abalone (*Haliotis rufescens*) and the Japanese limpet (*Acmaea dorsuosa*) are known to live to 13 years and 17 years respectively, and observations made by MacGinitie & MacGinitie (1949) on the growth rate of another limpet, *Lottia gigantea*, indicate a life-span of at least 15 years.

The most fertile of all known gastropods is the Sea hare *Tethys californicus*. MacGinitie & MacGinitie (1949) give a record of a 2·60 kg *5 lb 12 oz* female which laid 478 000 000 eggs in 4 months 1 week (41 000 eggs per minute), but they say larger females lay considerably more. At the other end of the scale, some species of gastropod lay less than 1000 eggs a year, and there is a record of a Black slug (*Arion ater*) laying only 477 eggs in 480 days (Bartsch, 1934).

The most dangerous snails in the world are the tiny aquatic species of Africa, the Far East and South America which carry the flukes responsible for the terrible wasting disease schistosomiasis (formerly called bilharzia).

Three species of the genus *Schistosoma* – *S. haematobium*, *S. mansoni* and *S. japonicum* – habitually live in man, and in some countries schistosomiasis causes more sickness and death than any other single disease. In many parts of Africa and tropical America it is ranked among the most dangerous of human diseases, along with malaria, hookworm and trypanosomiasis.

At the present time there are over 100 000 000 people round the world with the malady, traces of which have been found in ancient Egyptian mummies, and it is the only major tropical disease actually on the increase. This is mainly due to the spread of irrigation and hydro-electric schemes which create new breeding-grounds for the water-snails.

The most venomous gastropods are the very attractive Cone shells (genus *Conus*) of the Indo-Pacific region. These molluscs all possess a highly developed neurotoxic venom apparatus consisting of a single poison gland plus duct and a retractable proboscis which contains a number of harpoon-like rasping teeth.

The most dangerous known species is the very rare *Conus geographus* which, according to McMichael (1971) has been *definitely* responsible for at least four (actually five) fatal stings – all of them amateur shell-collectors. Three of the victims were natives of New Caledonia Territory, SW Pacific, one of them a nine-year-old girl. In the case of Charles Garbutt, 27, who was stung on Hayman Island, off the coast of Queensland, Australia, on 27 June 1935, the first symptom was a numbness which spread to the lips. Twenty minutes later his eyesight started to fail and then his legs became paralysed. After 60 minutes he lapsed into a coma and died four hours later (Flecker, 1936).

Conus omaria also has a very bad reputation. In November 1963 a nine-year-old native girl died on Tanga Island, New Guinea, after being stung by a cone shell (probably *C. omaria*), and on 27 August 1964 another native girl aged eight nearly succumbed on Manus Island, New Guinea, after being stung by another *Conus* which was later positively identified as *C. omaria*. In the latter case the symptoms were slurred speech, palsy and laboured breathing and her life was only saved by artificial respiration (Petrauskas, 1955).

Unfortunately very little is known about the virulent nature of the venom of *C. geographus* or the quantity a single individual can produce, but it is reportedly more powerful than that of the Asiatic cobra (*Naja naja*) which, in turn, has a venom only one-twentieth as potent as that of the tiger snake (see page 128).

Section XIII

ROUNDWORMS

(class Nematoda)

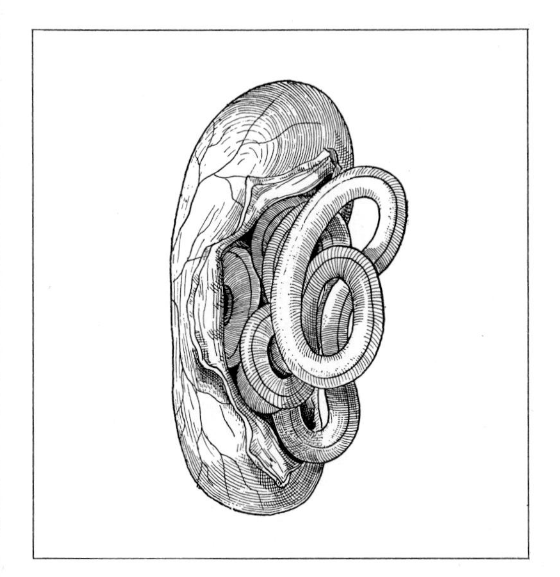

A true roundworm or nematode is a partially or wholly parasitic soft-bodied invertebrate which lives on land or in water and breathes through the skin or by means of gills. It has a cylindrical, elongated body covered by a thick layer of horny skin, and during growth this skin or cuticle is shed four times. Young are hatched from eggs deposited on land or shed in water.

The earliest known roundworms lived about 370 000 000 years ago.

There are about 10 000 known species of roundworms, and the class is divided into seventeen orders. These are: the *Enoploidea*; the *Dorylaimoidea*; the *Mermithoidea*; the *Chromadoroidea*; the *Araeolaimoidea*; the *Monhysteroidea*; the *Desmoscolecoidea*; the *Rhabditoidea*; the *Rhabdiasoidea*; the *Oxyuroidea*; the *Ascaroidea*; the *Stronglidea*; the *Spiruroidea*; the *Dracunculoidea*; the *Filaroidea*; the *Trichuroidea*; and the *Dioctophymoidea*.

The largest known roundworm is *Placentonema gigantissima*, a marine nematode found in the Pacific. Adult females measure 6·75–8·40 m *22 ft 1 in–27 ft 6 in* in length, and adult males 2·04–3·75 m *6 ft 10 in–12 ft 3 in*.

The largest non-marine roundworm is the Giant kidney worm (*Dioctophyma renale*), a species found in the kidneys and liver of dogs and other mammals. (At least nine cases of human infection are known.) Females have been measured up to 1000 mm *39·37 in* in length and 12 mm *0·47 in* in diameter, with a maximum weight of 43·3 g *1·53 oz* (Von Brand, 1957; Cox, 1967). Males are much smaller, measuring 140–200 mm *5·51–7·87 in* and 4–6 mm *0·15–0·23 in* in diameter (Craig & Faust, 1945).

The longest non-marine roundworm is the very slender Guinea worm (*Dracunculus medinensis*), females measuring up to 1200 mm *47·24 in* in length and 0·9–1·7 mm *0·03–0·06 in* in diameter. Adult males are tiny by comparison, measuring 20–29 mm *0·7–1·1 in*.

The smallest known parasitic roundworm is *Ollulanthus tricuspis*, which lives in the stomach walls of cats. Adult females measure only 1 mm *0·039 in* in length (Rauther, 1925).

Roundworms are long-lived creatures. The filarial worms *Wuchereria bancrofti* and *Loa loa* in man have been known to live 17 and 15 years respectively (Coutelen, 1935), and Taylor (1933) says the hookworms *Ancylostoma duodenale* and *Necator americanus* may live 15–16 years. There is also a record of a nematode (*Anguina tritici*) living in a dried state in laboratory storage for 28 years before it came back to life, and 39 years has been reported for a plant parasite (*Tylenchus polyphyprus*) which invaded a rye plant collected in Kansas, USA, in 1906 and revived in a herbarium just as the Second World War was ending (Crofton, 1966).

The most dangerous roundworms to man are probably the Old World hookworm (*Ancylostoma duodenale*) and the American hookworm (*Necator americanus*) ('American murderer'), females of which can produce between 25000 and 30000 eggs every day for five years. These parasites gain entrance to the body by burrowing through the skin of the feet and hands. They then travel through the blood vessels to the lungs, thence by the air passages to the alimentary tract. There the adult worms attach themselves to the wall of the upper intestine, where they feed on the blood and secrete a poison that causes anaemia and general debility. According to a United Nations report over 600000000 people suffer from hookworm disease round the world, and in some areas of Africa it is a more serious danger to the health of the native population than the parasite of malaria.

Another strong contender is the tiny (1·5–4 mm *0·05–0·15 in*) Pork-muscle worm (*Trichinella spiralis*). This nematode is usually found in pigs, but it can also be picked up from eating the badly cooked meat of other omnivorous animals – including man. That is why, says Dr Donald Carr (1972), human cannibals who make a habit of eating the raw brains of their enemies inevitably get trichinosis, a disease marked initially by colicky pains, nausea and diarrhoea, and later by muscular pains, fever and laboured respiration. He recommends boiling the war trophies first! It has been calculated that in man five larvae per gram of body-weight can cause death (cf. ten larvae for pigs and 30 larvae for laboratory rats) (Craig & Faust, 1945).

The cosmopolitan *Ascaris lumbricoides*, the largest of the common nematodes of man (up to 350 mm *13·7 in* in length and over 5 mm *0·19 in* in diameter) is also worthy of mention. At the present time about 750000000 people are infected with this parasite which (1) as larvae get into the blood stream via the mouth from contaminated food or drink and are carried to the lungs where they can produce pneumonia and (2) as adult worms in the small bowel cause toxic damage. Females of this species lay a colossal 200000 eggs a day (Brown & Cort, 1927), and it has been computed that the total *Ascaris* population produces 1000000000000000000 eggs each year weighing about 41000 tonne *40268 ton* (Probert, 1972). The number of *Ascaris* worms present in a human victim varies from a single male or female to many hundreds. In may 1974 a man living in the village of Magombo, Malawi, had 936 worms removed from his bowel at Matope Anglican Hospital and made a full recovery. In another case Fulleborn (1932) found 1488 specimens in a patient at autopsy.

Bancroft's filaria (*Wuchereria bancrofti*), found in all the warm regions of the world, is another bad character, man being 'inoculated' with larvae carried by infected mosquitoes which serve as intermediate hosts. The larvae eventually become lodged in the lymph glands, and one of the commonest manifestations of this infection is elephantiasis, which means a certain part of the body swells to incredible size. Sometimes a foot or leg may become so enlarged that it is impossible to lift, and Chandler (1944) mentions a Japanese victim whose scrotum weighed an unbelievable 102 kg *224 lb*!

One of the most amazing creatures in nature is a nematode called *Aphelenchus avenae* which, in times of stress (e.g. lack of oxygen) manufactures its own alcohol! This seems to improve matters, because once it has stored up enough alcohol for its own needs the tiny organism goes into a state of suspended animation called cryptobiosis which can preserve its life indefinitely. When the emergency is over the worm 'surfaces' again, drinks the alcohol, and carries on again as usual – albeit with a warm glow inside.

Another *fortunate* member of the Nematoda is a German version of the Vinegar eel (*Anguillula aceti*). This tiny (1 m *0·03 in*) worm spends its whole life in beer where it feeds on yeast-germs.

Section XIV

RIBBON WORMS

(phylum Nemertina = Rhynchocoela)

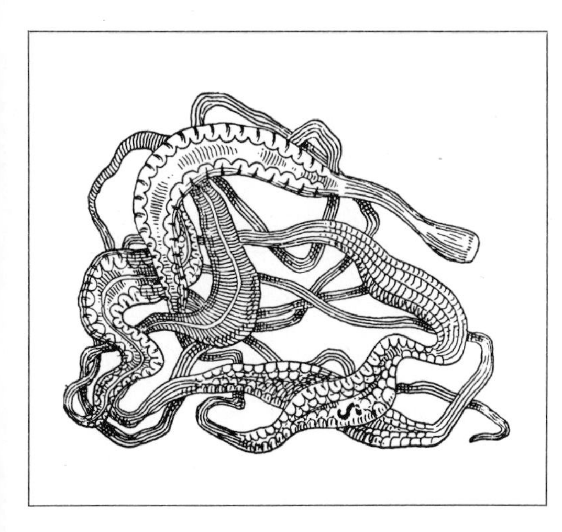

A ribbon worm or nemertine is a carnivorous, soft-bodied marine invertebrate which breathes by means of gills. Its most distinctive feature is the proboscis, a long muscular tube which can be thrown out to grasp prey and drawn back towards the mouth. Young are hatched from eggs or develop within the parent.

Fossil nemertines are unknown, but the earliest forms probably evolved about 350000000 years ago.

There are 750 living species of ribbon worm, and the phylum is divided into two classes. These are: the *Anopla* (proboscis unarmed) and the *Enopla* (proboscis armed).

The longest known species of ribbon worm is the 'Boot-lace worm' (*Lineus longissimus*), which is found in the shallow coastal waters of the North Sea. In 1864 an enormous specimen – it half-filled a jar measuring 203 mm *8 in* in diameter and 127 mm 5 *in* deep – was washed ashore at St Andrews, Fifeshire, Scotland, after a storm and was examined by Prof W C McIntosh of the Gatty Marine Laboratory. He managed to measure 27·32 m *30 yd* of the worm before it ruptured and then gave up, but at that point he said 'the mass was not half uncoiled'.

The Mediterranean species *Nemertes borlasi* is also very long, having been measured up to 12·19 m *40 ft*, and mention should also be made of the Clam worm (*Cerebratulus lactus*), found along the eastern seaboard of the USA, which grows to 3·65 m *12 ft* in length but can contract to 0·6 m *2 ft*. (Most ribbon worms can shrink to less than a third of their normal length.)

The smallest known ribbon worms are some of the pelagic species which measure only a fraction of an inch in length.

Ribbon worms are very voracious animals and can devour creatures much larger than themselves. When food is scarce, however, they get round the problem by 'absorbing' themselves, and Henry (1958) quotes a record of a ribbon worm digesting 95 per cent of its own body in a few months without apparently suffering any ill-effect. As soon as food became available again the lost tissues were restored.

Section XV

TAPEWORMS

(class Cestoda)

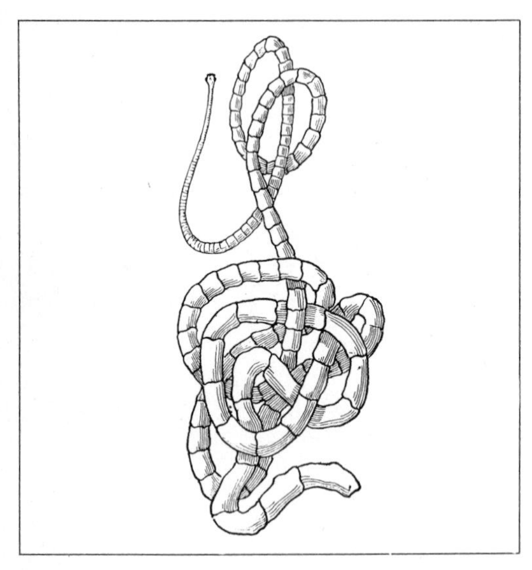

A tapeworm is a ribbon-like invertebrate which lives as an adult in the intestine or bile duct of a vertebrate. Its body is divided into three parts: a minute head with suckers, sucking grooves or hooks for maintaining its place in the gut; a short neck or growing region; and a long series of body sections, each containing both male and female sex organs, which are produced by a budding process of the neck region. Young are hatched from eggs after leaving the parent.

Fossil tapeworms are unknown, but they probably evolved later than the nematode and nemertine worms.

There are more than 3000 known species of tapeworm and the class is divided into ten orders. These are: *Pseudophyllidea*; *Haplobothriodea*; *Tetrarhynchoidea*; *Diphyllidea*; *Tetraphyllidea*; *Lecanicephaloidea*; *Tetrabothrioidea*; *Proteocephaloidea*; *Nippotaenoidea* and *Cyclophyllidea*.

The longest known tapeworm is the cestode *Polygonoporus giganteus*. Individual pieces measuring up to 5 m *16 ft 5 in* in length and 45 mm *1·77 in* in diameter found in the intestine of a sperm whale indicate a total length of *c.* 30 m *98 ft 5 in*. In 1957 another tapeworm (species not named) with an alleged length of 30·48 m *100 ft* was found in the gut of a whale captured off Santa Catalina Island, California, USA.

During the 1957–8 whaling season a new giant Diphyllobothriid cestode (*Multidictus physeteris* sp.) was discovered in the bile ducts of three sperm whales caught by the British ship S.S. *Southern Harvester* in the Bellingshausen Sea, Antarctica. The specimens were all incomplete, but the longest one measured 21 m *68 ft 11 in* and had a maximum diameter of 30 mm *1·18 in* (Clarke, 1962).

The longest species of human tapeworm is the Beef tapeworm (*Taenia saginata*). The usual length is 10–12 m *32 ft 9½ in–39 ft 4½ in*, but females have been measured up to 25 m *82 ft* (Craig & Faust, 1945). It is interesting to note that the heads of this species are (or were) sometimes sold as active ingredients of 'reducing pills' but, although this remedy may be slightly effective inasmuch as it causes general debility, the thought of having a colony of tapeworms clinging to the walls of one's stomach for an indefinite period hardly justifies the means!

The Fish or Broad tapeworm (*Diphyllobothrium latum*) is another veritable monster, reaching a length of 18·28 m *60 ft*. Tarassow (1934) mentions a Russian woman who served as intermediate host to six examples measuring 88·93 m *290 ft* in total length, and he also recorded a case of a 23-year-old man who had 143 worms inside him measuring 117 m *384 ft* in total length. 'Fortunately', writes Chandler (1946), 'in tapeworm infections the size of the worms usually is in inverse proportion to their number.' This tapeworm, as its name suggests, is picked up by eating uncooked or pickled fish, and in some parts of Finland where raw fish livers are considered a delicacy, 80 per cent of the people are infected.

The shortest known tapeworm is *Echinococcus oligarthrus* from the puma and jaguar of Central and South America, which measures only 1·7–2·5 mm *0·06–0·09 in* in length. *E. minimus* from the wolf of Macedonia is nearly as microscopic (Wardle & McLeod, 1952).

Although very little accurate information has been published on the life-spans of tapeworms generally, it is known that human tapeworms can live for a considerable length of time. According to Penfold *et al.* (1937) 83 persons in Australia harboured a beef tapeworm for an average period of 13 years, and one specimen infected a man for 35 years. Riley (1919) gives a record of a fish tapeworm which lived inside its human host for 29 years, and Lawson (1939) cites a figure of 56 years for an enormous *Echinococcus* cyst.

The shortest-lived tapeworm is probably *Ligula intestinalis*, a large, fleshy worm found in the alimentary tract of diving and wading birds, which lives only a few days (Joyeux & Baer, 1929).

Some tapeworms can produce incredibly large numbers of eggs, which is not really surprising since they have little to do except cling on, grow and reproduce. The beef tapeworm, for example, can produce a total of 2500000000 eggs, and a fish tapeworm about 2000000000 eggs over a ten-year period.

The most dangerous tapeworm to man is probably the 8 mm *0·31 in* long hydatid worm *Echinococcus granulosus*. The microscopic eggs of this parasite adhere to the hair of dogs and make their way into human tissue via the mouths of dog patters. Once established in an organ of the body a single egg can multiply asexually into millions of tapeworm heads which form larval cysts. In Barnett's case (1944) a complex of hydatid cysts varying in size from a cherry to a coconut were removed from a 39-year-old shepherd in New Zealand. Altogether they held 55 litres *96¼ pints* of fluid and the total weight was 29 kg *63 lb 13 oz*. Sixteen years after the operation two more cysts the size of an infant's head were removed. In 1938 there were 151 deaths from this infection in Australia and New Zealand (Chandler, 1946).

The only time this class of animal is 'beneficial' to man is when the larvae of marine fish tapeworms get into oysters and act as an irritant, thereby producing a pearl. Most natural pearls in the Orient are formed in this way.

Section XVI

COELENTERATES

(phylum Coelenterata)

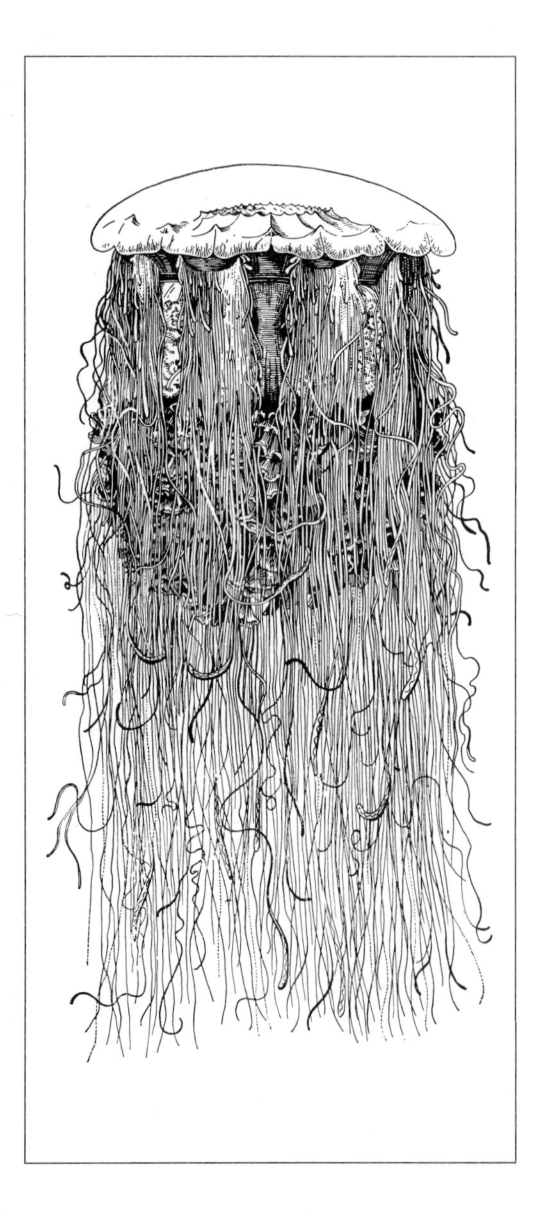

A coelenterate is a soft-bodied marine invertebrate with a radially symmetrical body which is little more than a stomach. It occurs in two forms: the cylindrical polyp which attaches itself to rocks, and the gelatinous umbrella-shaped medusa which is free-swimming. Both types, however, have a mouth fringed by delicate tentacles which can sting and paralyse prey or act as defensive weapons. Young are reproduced from eggs discharged into the sea where they are fertilised (medusa), or from larva buds which detach from the parent and grow directly into adults (polyp).

The earliest known coelenterates lived about 570000000 years ago.

There are about 9400 living species of coelenterates, and the phylum is divided into three classes. These are: the *Hydrozoa* (hydroids); the *Scyphozoa* (jellyfish); and the *Anthozoa* (sea-anemones and corals). The largest class is *Anthozoa* which contains about 6500 species or nearly 70 per cent of the total number.

The largest known true jellyfish is *Cyanea arctica*, which is found in the north-western Atlantic Ocean from Greenland to North Carolina, USA. One examined by Agassiz (1865) in Massachusetts Bay, Massachusetts, USA, had a bell measuring 2·28 m *7 ft 6 in* in diameter and tentacles stretching 36·57 m *120 ft* thus giving a *theoretical* tentacular spread of some 74·67 m *245 ft*.

The largest coelenterate found in British waters is the rare 'Lion's mane' jellyfish (*Cyanea capillata*), which is also known as the Common sea blubber. One individual measured by McIntosh (1885) at the Gatty Marine Laboratory, Fifeshire, Scotland, had a bell diameter of 910 mm *35·82 in* and tentacles stretching over

13·7 m *45 ft*, and Russell (1970) saw the impression of another one in the sand at Newton-by-the-Sea, Northumberland, in September 1967 which must have had a bell diameter of at least 910 mm *35·43 in*. In September 1959 a 139 mm *5½ in* long silver spoon was found inside the stomach of a *Cyanea* washed up on a Suffolk beach.

Some true jellyfishes have a bell diameter of less than 20 mm *0·78 in*.

The most dangerous of all coelenterates are the sea-wasps or box jellies of the genera *Chiropsalmus* and *Chironex* of the Indo-Pacific region, which carry a neurotoxic venom similar in strength to that found in the Asiatic cobra. Cleland & Southcott (1965) have documented 55 fatalities caused by jellyfish stings in the Austro-Asia region (up to March 1963), and during the next three summers four more fatalities occurred in Australian waters. The species responsible for most of these deaths was *Chironex fleckeri*.

The symptoms of box jelly stings are profuse sweating, convulsions and blindness, followed by respiratory paralysis, and Halstead (1956) says death usually occurs in three to ten minutes. Sometimes, however, the period can be much shorter. Clarke (1969) quotes the case of a man stung by a *Chironex* off a Queensland beach who died in agonising pain *within 30 seconds* of being stung, and animals injected in a laboratory with *Chironex* venom diluted 10000 times have died before the hypodermic needle could be removed.

According to the autopsy findings of two fatal cases in Australia (an eleven-year-old girl and a 38-year-old man) the lungs and air passages of both victims were blocked by enormous quantities of frothy mucus. The child had been standing in 76 cm *2 ft 6 in* of water at Tully, North Queensland, and the man in approximately the same depth near Townsville, Queensland. The girl died within ten minutes, and the man in less than 35 minutes (Kingston & Southcott, 1960).

It has been calculated by Southcott (1971) that 'a large *Chironex fleckeri* swimming along may present 200 ft (about 60 m) of trailing tentacle in a volume of sea-water, say a cylinder less than a foot across and 3–6 ft long. Observations on victims of fatal stings indicate that only a small fraction of this (about 20 ft, or 6–7 m) in vigorous contact with the skin is sufficient to cause death.'

Although some progress has been made in the development of antitoxin, no specific treatment is yet available against *Chironex* envenomation, which has been described by Dr Robert Endean, a Queensland biologist, as 'one of the deadliest toxins known to man'. Even if an antidote was available, its practical use would be very restricted and the same authority thinks the eventual answer will be some sort of vaccine for swimmers.

Meanwhile, the only known defence against box jellies is apparently women's pantyhose! Tests have shown that the sting cannot penetrate this material, and it is now regulation wear for Queensland life-savers at surf carnivals.

The most dangerous coelenterate found in British waters is the Portuguese man-o'-war (*Physalia physalis*) which, strictly speaking, is not a jellyfish but a closely related siphonophore. This species has been measured up to 304 mm *12 in* in length and 152 mm *6 in* across, and its tentacles can extend nearly 12·19 m *40 ft*.

Most people stung by this highly coloured float experience a burning pain followed by a large weal which may last for perhaps a week before fading. Others, however, who are particularly sensitive to the venom, which in its crude form is about 75 per cent as toxic as that of the Asiatic cobra, can become seriously ill and may have to spend weeks in hospital. It is also dangerous to small children or adults with weak hearts, and a severe shock could lead to drowning.

One animal not bothered by the sting is the Loggerhead turtle (*Caretta caretta*) which has been observed happily eating its way through a shoal of Portuguese man-o'-war with no worse effect than swollen eyes!

Discoma *sp., the largest known sea anemone (Queensland Government Office)*

On a single day in November 1967 more than 1000 people lazing on Sydney's beaches, New South Wales, Australia, were treated for stings after winds swept hundreds of these creatures ashore.

The 'lion's mane' jellyfish is also a dangerous species. In August 1959 twelve bathers at Aldeburgh Suffolk, were treated for serious stings from this species, but so far no fatalities have been reported.

Not a lot is known about the life-spans of jellyfish, but most small species probably live less than one year.

The largest known sea-anemone is *Discoma* sp. of the Great Barrier Reef, Queensland, Australia, which has an expanded oral disc measuring up to 609 mm *2 ft* in diameter.

The largest sea-anemone found in British waters is *Bolcera tuediae* which has an expanded oral disc measuring up to 300 mm *11·81 in* and inner tentacles measuring 100 mm *3·93 in* or more in length (Stephenson, 1935).

The smallest British sea-anemone is probably *Gonactina prolifera* which does not exceed a total length in extension of 4 mm *0·15 in*.

The most toxic sea-anemone is *Rhodactis howesii*, found on the reefs of American Samoa, SW Pacific which, when cooked, forms part of the native diet. This coelenterate, known locally as 'Matamulu', has been responsible for a number of human fatalities after accidental or deliberate ingestion of the raw material, death being due to respiratory failure (Farner & Lerke, 1963).

Section XVII

EXTINCT ANIMALS

An extinct animal is a creature which no longer exists on the face of the earth in a living state, although at one time it flourished. It is only known from fossil remains, and these consist either of organic material or the impression of organic objects like feet. The word 'fossil' is derived from the Latin verb '*fodere*' meaning 'to dig' and originally applied to almost anything of interest that was dug up out of the ground. For a living thing to become a fossil it must be buried soon after death so that it cannot be eaten by scavengers or destroyed by oxygen-breathing bacteria. The exact age at which an animal qualifies for this category is not known, but it is probably at least 25000 years, and the science dedicated to their study is called palaeontology.

The first dinosaur to be scientifically described was *Megalosaurus* ('large lizard') in 1824. A lower jaw and other bones of this animal had been found before, in 1818, in a slate quarry at Stonesfield, near Woodstock, Oxfordshire, and placed in the University Museum, Oxford, where they were examined by Dean William Buckland. This 6·09 m *20 ft* long bipedal theropod (carnivore) stalked across what is now southern England about 130000000 years ago. In March 1822 Dr Gideon Mantell (or rather his wife Mary), discovered some fossilised teeth of another bipedal dinosaur of the same period (Lower Cretaceous) in the nearby district of Cuckfield, and described them in 1825 under the name *Iguanodon* ('iguana-tooth'). Unlike *Megalosaurus*, however, this 9·14 m *30 ft* long dinosaur was herbivorous. Seven years later Mantell found fragmentary remains of a 9·14 m *30 ft* long armoured dinosaur in the Tilside Forest, Sussex, and called it *Hylaeosaurus* ('toad lizard'). It was not until 1842, however, that the name *Dinosauria* ('fearfully great lizards') was given to these reptiles by Prof (later Sir) Richard Owen, the great English anatomist and vertebrate palaeontologist.

The longest dinosaur so far recorded is *Diplodocus* ('double-beam'), an attenuated sauropod which ranged over what is now western North America about 150000000 years ago. A composite skeleton of three individuals excavated near Split Mountain, Utah, USA, between 1909 and 1922 for the Carnegie Museum of the Natural Sciences in Pittsburgh, Pennsylvania, USA, and subsequently named *Diplodocus carnegii*, measures 26·67 m *87 ft 6 in* in total length (neck 6·7 m *22 ft*, body 4·57 m *15 ft*, tail 15·39 m *50 ft 6 in*) – nearly the length of three London double-decker buses – and has a mounted height of 3·58 m 11 ft 9 in at the pelvis, the highest point on the body. This animal weighed an estimated 10·56 tonne/*ton* when alive (Colbert, 1962).

Britain's longest dinosaur was the sauropod *Cetiosaurus* ('whale lizard'), which roamed over what is now southern England about 165000000 years ago. It measured up to 18·28 m *60 ft* in length and weighed over 15 tonne/*ton*. This

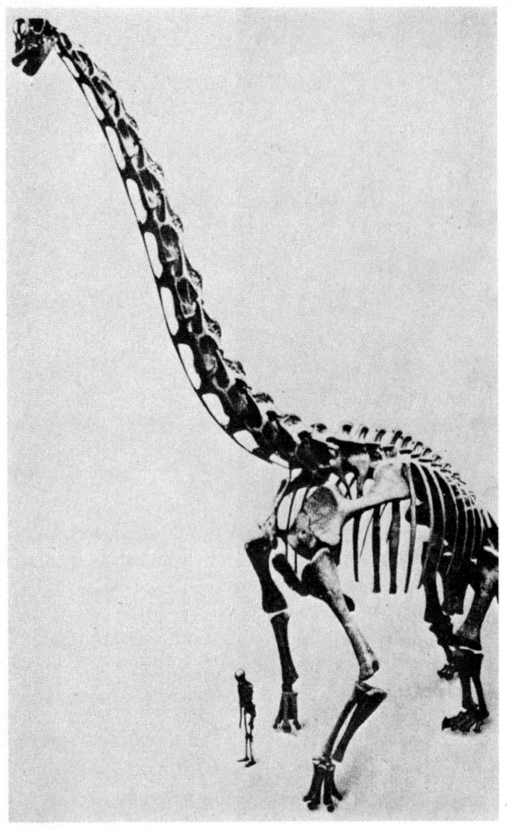

Brachiosaurus, the heaviest land vertebrate of all time (Museum für Naturkunde, Berlin)

reptile was originally described by Owen in 1841 who – on the evidence of a few fragmentary remains – thought it was a crocodile. In May 1898, however, a partial skeleton was found in the No. 1 Brickyard of the New Peterborough Brick Co., Peterborough, Northamptonshire, by Alfred N Leeds, and other remains were later discovered in the quarries of Oxfordshire.

The heaviest of all prehistoric animals, and the heaviest land vertebrate of all times was probably *Brachiosaurus* ('shoulder lizard'), which lived in what is now East Africa (Rhodesia and Tanzania), and Colorado, Utah and Oklahoma, USA, between 135 000 000 and 165 000 000 years ago. A complete skeleton excavated by the German expedition at Tendaguru, in what was then southern Tanganyika, German East Africa, in 1909 and subsequently mounted in the Museum für Naturkunde in East Berlin, Germany, measures 22·7 m *74 ft 6 in* in total length (height at shoulder 6·4 m *21 ft*) and the raised head is 11·88 m *42 ft* above the level of the toes. It weighed a computed 78·26 tonne *77 ton* when alive.

According to Dr Bjorn Kurten (1968), the Finnish palaeontologist, isolated bones have since been discovered at the Tendaguru site, Tanzania, which suggest that some individuals may have weighed as much as 102 tonne *100 ton* and measured over 27·43 m *90 ft* in length – which would also have made *Brachiosaurus* the *longest* of all dinosaurs.

This latter statement is interesting because, in the summer of 1972, the remains of another enormous sauropod new to science but looking something like *Brachiosaurus* in general structure, were discovered in a flood-plain bone jam on the Uncompahgre Plateau, western Colorado, USA, by a small expedition led by Mr James Jensen, Director of the Earth Science Museum at Brigham Young University, Provo, Utah. According to one report the matching scapulae or shoulder blades of this huge quadruped measured 2·43 m *8 ft* in length, the pelvis 1·98 m *6 ft 6 in* in width (cf. 1·21 m *4 ft* in a large sauropod skeleton) and the largest of the vertebrae 1·52 m *5 ft* across, and these extreme measurements were confirmed by Mr Jensen (pers. comm. 7 November 1972).

He also went on to say that because the cervical (neck) series of vertebrae were about three-quarters complete in number he was able to make at least one *substantial* calculation as to its total length: 'Whereas you note *Brachiosaurus* as being mounted with its head 42 ft in the air', he says, 'a conservative estimate of the neck length from the scapulae to skull, based upon the *B.* cervial formulae, adds up to ±39 ft! This is due to the fact that the shortest neck vertebrae collected – one near the anterior end – is 3 ft in length and the longest one is nearly 5 ft long. When you calculate these dimensions in a series having 13 vertebrae it adds up to a long, long neck.'

A *c.* 11·88 m *39 ft* neck presupposes an over-all length of approximately 30·48 m *100 ft*, a shoulder height of *c.* 7·92 m *26 ft* and a raised head measurement of *c.* 16·45 m *54 ft*.

At one time it was generally thought that large sauropods like *Brachiosaurus* were aquatic creatures who spent most of their lives in freshwater lakes or swamps, where their enormous weight would be buoyed up by the water, and the fact that the nostrils and eyes of these dinosaurs were placed high up on the head like those of a hippopotamus certainly seemed to indicate a largely amphibious existence. Dr

Robert T Bakker (1971), however, of the Dept of Vertebrate Palaeontology at Harvard University, Cambridge, Massachusetts, USA, knocks down this theory because he claims *Brachiosaurus et al*, having compact feet like those of the elephant, would have become hopelessly bogged down in a swamp and starved to death. He believes they were dry land animals which roamed around in herds.

Until quite recently it was widely believed that the largest theropod or carnosaur ('meat reptile') was *Tyrannosaurus rex* ('tyrant lizard'), which lived in what are now the states of Montana and Wyoming in the USA about 75 000 000 years ago. This dinosaur measured up to 14·32 m *47 ft* in over-all length, had a bipedal height of up to 5·63 m *18 ft 6 in*, a stride of 3·96 m *13 ft* and weighed a calculated 6·78 tonne/*ton* (Colbert, 1960). It also had a 1·21 m *4 ft* long skull containing serrated teeth measuring up to 184 mm *7·25 in* in length (Kerr, pers. comm. 11 January 1972). It is now known, however, that some other carnosaurs were just as large or even larger than *Tyrannosaurus*.

In 1934 labourers working on Highway 64 near Kenton in Cimarron County, Oklahoma, USA, accidentally unearthed a huge rib 2·13 m *7 ft* long and 0·3 m *1 ft* in circumference. Dr J W Stovall, palaeontologist at Norman University, was hurriedly called to the site of the discovery and over the next few months he and a special crew of men excavated the bones of an *Apatosaurus* (= *Brontosaurus*), a *Ceratosaurus*, a *Camptosaurus*, a *Stegosaurus* and a huge carnosaur new to science. This specimen, which measured 12·8 m *42 ft* in over-all length and had a bipedal height of 4·87 m *16 ft*, was much more massive in proportion to its height than *Tyrannosaurus* and had arms more than twice as long (Ray, 1941).

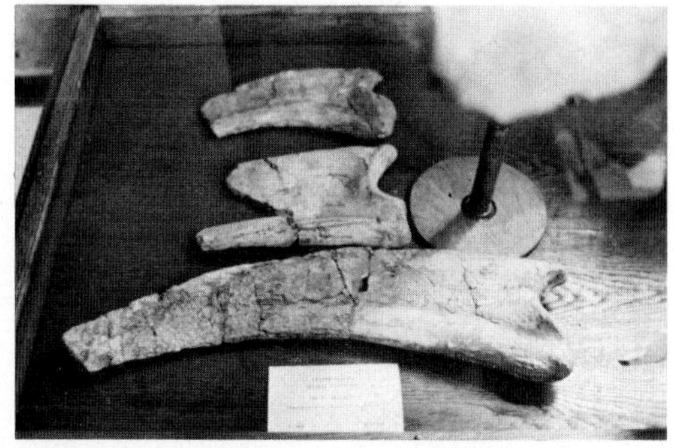

The 700 mm 27·55 in *long dinosaur claw preserved in the Museum of Palaeontology at the Academy of Sciences, Moscow (Victor Louis)*

Stovall named the carnosaur *Saurophagus maximus* ('lizard eater'), but after his death the bones were re-examined and found to be those of a very large *Allosaurus*. Since then the remains of other allosaurs of comparable size have been found in Utah (De Camp & De Camp, 1968).

Mention should also be made of *Tarbosaurus* from the Nemegetu Basin, Mongolia, which was also very similar in shape and size to *Tyrannosaurus*.

In 1930 the British Museum Expedition to East Africa dug up the pelvic bones and part of the vertebrae of another huge carnosaur at the Tendaguru site which must have measured about 16·45 m *54 ft* in over-all length when alive.

During the summers of 1963–5 a Polish-Mongolian expedition led by Dr Zofia Kielan-Jaworowska, the Polish woman palaeontologist, discovered a pair of forelimbs in the Nemegetu Basin, south Gobi Desert, which were 2·59 m *8 ft 6 in* long! Unfortunately nothing else is known about this carnosaur which has been given the generic name *Deinocheirus*, but if the rest of its body was built on the same gigantic scale it must have been a truly colossal animal.

In the Museum of Palaeontology at the Academy of Sciences of the USSR in Moscow there is preserved an *incomplete* dinosaur claw which measures a staggering 700 mm *27·55 in* along the outside curve! (Louis, pers. comm. 1 June 1971). According to Rozhdestvensky (1971) the claw was found in the Nemegetu Basin by a Russian expedition led by Dr I Efremov in 1948 and came from the phalanga of the front foot of a member of the *Therizinosaurus* family – a group of dinosaurs similar to present-day anteaters. Such claws, he said, were presumably used for stripping bark or digging in the ground to find the nests of termites, ants, et cetera. The therizinosaurs, which lived in forest and savannah regions some distance from lakes or inland seas about 70 000 000 years ago, had a feeble skull partly or entirely lacking teeth, and Rozhdestvensky thinks this particular specimen fell prey to a stronger dinosaur when it came to a pool to drink. Similar claw phalangi were discovered in the same area in 1956 and 1960, but these were smaller, i.e. 22–25 cm *8½–10 in*.

Stegosaurus ('plated reptile'), a fairly large, heavily built herbivore which roamed across the Northern Hemisphere about 150 000 000 years ago, has been dubbed the most brainless of all the dinosaurs. This is because it had a poorly organised walnut-sized brain weighing only 70·7 g

$2\frac{1}{2}$ *oz* which represents 0·004 per cent of its body-weight, compared with 0·074 per cent for an elephant and 1·88 per cent for an adult human being. This reputation, however, is largely undeserved because skull cavities show that all dinosaur brains were small, and that even the 81 tonne *80 ton Brachiosaurus* probably had a brain weighing less than 453 g *1 lb*. What *Stegosaurus* did have though was a greatly enlarged ganglia above the shoulders and hips which was 20 times as large as the brain itself, and this development was common among many of the other dinosaurs although not nearly so pronounced. These great nerve centres unconsciously controlled the functioning of the huge legs and tail.

The largest known dinosaur eggs are those of *Hypselosaurus priscus*, a 9·14 m *30 ft* long sauropod which lived about 80000000 years ago. Some specimens found in the Valley of the Durance, near Aix-en-Provence, southern France, in October 1961 would have had – uncrushed – a long axis of 304 mm *12 in* and a shorter axis of 254 mm *10 in*, giving a capacity of 3·3 litre *5·77 pint*. Although these eggs are not in the same size class as those of the extinct Elephant bird (*Aepyornis*) of Madagascar which had a capacity of up to 8 litre *14·0 pint*, they probably represent the maximum size to which a reptilian egg can grow. This is because the shell is much more fragile than that of a bird's egg. In a larger specimen the pressure of the internal fluid would have been too great and the egg would have burst; also, if the shell had been thicker the embryo would not have been able to break out of its limey prison. This suggests that *Brachiosaurus* and *Apatosaurus*, the giants among the sauropods, were viviparous.

The largest known flying reptile is a yet unnamed pterosaur which glided over what is now Texas, USA, about 70000000 years ago. Three partial skeletons – four wings, a neck, hind legs and mandibles – excavated at Big Bend National Park, SW Texas, during the three-year period 1972–4, indicate the animal had a wing-expanse of *c.* 15·54 m *51 ft* (Lawson, 1975), which means it was twice as large as *Pteranodon ingens* (up to 8 m *26 ft 3 in*) the previous record-holder.

It is interesting to note that the bones of this winged dragon were dug out of rocks which are essentially non-marine, so how did a creature this size, living some 400 km *250 miles* from the nearest coastline, ever manage to get airborne? Lawson thinks its mode of life may have been that of a carrion-eater like the vulture, the long neck probing the insides of rotting dinosaur carcasses, but even if this is true, what happened when the scavenger was stalked by a hungry carnosaur? It couldn't flap its wings, says Desmond (1975), because it had neither the muscles nor the mechanical strength for such action – hence the absence of a bird-like keel on the breastbone – and it couldn't run because its legs trailed. One can only assume the Big Bend pterosaur waited with its wings extended and elevators raised until a strong wind lifted it into the air . . . but what if it was a calm day?

As Desmond concludes: 'Lawson's ultimate pterosaur raises more questions than it answers.'

Britain's largest known flying reptile was *Ornithodesmus latidens* which flew over what is now Hampshire and the Isle of Wight about 90000000 years ago. It had a wing-expanse of about 5 m *16 ft 4 in* allowing for the natural curve (Hooley, 1913).

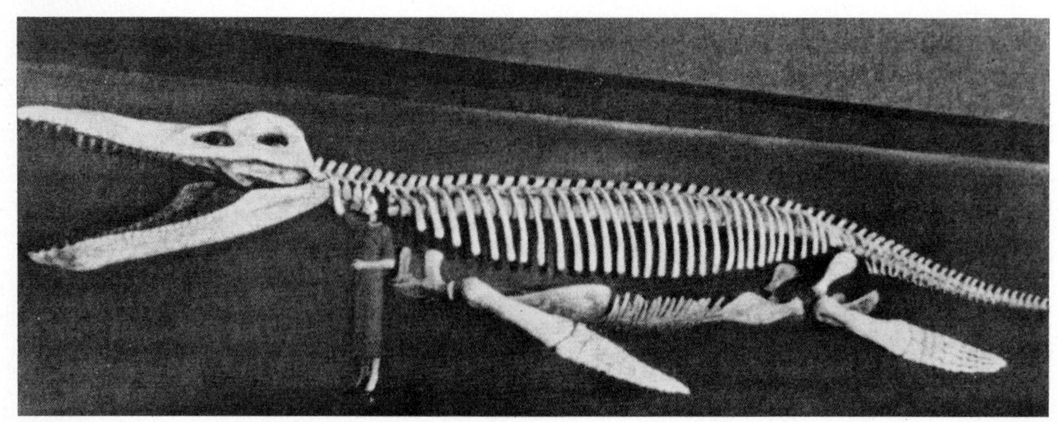

Kronosaurus queenslandicus, *the largest marine reptile ever recorded (Museum of Comparative Zoology, Harvard University)*

The largest marine reptile ever recorded was *Kronosaurus queenslandicus*, a short-necked pliosaur which swam in the seas around what is now Australia about 100000000 years ago. It had a 3·04 m *10 ft* long skull containing 80 spiked teeth and measured up to 15·25 m *50 ft* in over-all length (Fletcher, 1959). This size was closely matched by another short-necked pliosaur *Stretosaurus macromerus*. A headless skeleton from the Kimmeridge Clay of Stretham, Cambridgeshire and Oxfordshire, is not as large as *Kronosaurus*, but a mandible found at Cumnor, Oxfordshire, and now in the University Museum, Oxford, has a restored length of over 3 m *9 ft 10 in* (Tarlo, 1957–9).

The long-necked plesiosaur *Elasmosaurus*, which swam in the seas over what is now the state of Kansas, USA, about 130000000 years ago, measured up to 14·32 m *47 ft* in total length, of which the flexible neck accounted for 7·62 m *25 ft*. Most of the ichthyosaurs measured less than 9·14 m *30 ft*, but *Leptophtergius acutirostris*, which swam in the seas round what is now Europe about 140000000 years ago, had a 2·13 m *7 ft* long skull and measured *c.* 12·19 m *40 ft* in total length.

The largest known crocodile was *Phobosuchus hatcheri* (=*Deinosuchus hatcheri*=*Deinosuchus riograndensis*), which lived in the lakes and swamps of what are now the states of Montana and Texas, USA, about 75000000 years ago. The first remains of this monstrous saurian consisting of two large vertebrae, ribs and a few other fragments, had been discovered in the Judith Riverbeds, Montana. In 1940, however, the American Museum-Sinclair Expedition found large broken sections of the skull and jaw of another specimen completely embedded in hard stone in the 'Big Bend' region of southern Texas just north of the Mexican border, and it took eight months of difficult work before the skull was prepared for study and exhibition at the American Museum of Natural History, New York. According to Barnum Brown (1942) the total length of the crocodile was calculated on the ratio of 304 mm *1 ft* of body to each 25 mm *1 in* of mandibular tooth space, a ratio which is fairly constant among living crocodiles, and the measurement came out at 15·24 m *50 ft*, but Neill (1971) says 'there is no particular reason to think that it was of record length for its species'.

The huge Gharial *Rhamphosuchus*, whose remains have been found in the Siwalik Hills, north India, was also of comparable size, reaching a length of 15·24 m *50 ft*, and

The 1·83 m 6 ft *long skull of* Phobosuchus hatcheri, *the largest known crocodile (American Museum of Natural History)*

Gavialosuchus of Florida, USA, measured about 13·71 m *45 ft* (Neill, 1971). Both saurians lived about 7000000 years ago.

The largest prehistoric marine turtle was probably *Cratochelone berneyi*, which swam in the shallow seas over what is now Queensland, Australia, about 75000000 years ago. In 1914 the fossil remains of a specimen which must have measured at least 3·65 m *12 ft* in over-all length when alive were discovered at Sylvania Station, 32 km *20 miles* west of Hughenden. *Archelon ischyros*, which lived in the shallow seas over what are now the states of South Dakota and Kansas, USA, about 80000000 years ago was also in the same size class. An almost complete skeleton with a carapace (shell) measuring 1·98 m *6 ft 6 in* in length was discovered in August 1895 near the south fork of the Cheyenne River in Custer County, South Dakota. The skeleton, which has an over-all length of 3·45 m *11 ft 4 in* across the outstretched flippers, is now preserved in the Peabody Museum of Natural History at Yale University, New Haven, Connecticut, USA. This turtle is estimated to have weighed 2727 kg *6000 lb* when it was alive, but 1818 kg *4000 lb* is probably a more realistic figure.

The largest prehistoric tortoise was *Colossechelys atlas* which lived in what is now northern India between 7000000 and 12000000 years ago. The first fragmentary remains of this chelonian were discovered in the Upper Siwalik beds near Chandigarh in 1837 by Dr Hugh Falconer, a British scientist, who calculated that the animal must have had a carapace measuring 3·65 m *12 ft* over the curve, but *Colossochelys* was somewhat smaller than this in reality. In 1923 Dr Barnum Brown, Curator of Fossil Reptiles at the

American Museum of Natural History in New York, found the first complete (although fragmented) shell in the same locality and shipped back the pieces to the museum. When reconstructed, this tortoise proved to be an old male whose shell measured 2·23 m *7 ft 4 in* in length over the curve, 1·52 m *5 ft* in width and 89 cm *2 ft 11 in* in height. Its weight was computed to be 955 kg *2100 lb*.

The longest prehistoric snake was the python-like *Gigantophis garstini*, which inhabited what is now Egypt about 50000000 years ago. Parts of a spinal column and a small piece of jaw discovered at El Faiyum indicate a length of about 11·28 m *37 ft*. Another fossil snake, *Madtsoia*, from Patagonia, South America, had an estimated length of 10·05 m *33 ft*.

The largest prehistoric fish was the Great shark (*Carcharodon megalodon*), an ancester of the present-day Great white shark (*C. carcharias*), which abounded in Miocene seas some 15000000 years ago. In 1909 the American Museum of Natural History undertook a restoration of the jaws of this giant shark, basing the size on 101 mm *4 in* long fossil teeth, and found that the jaws measured 2·74 m *9 ft* across and had a gape of 1·83 m *6 ft*. The shark was estimated to have measured 24·38 m *80 ft* in length when alive, but Randall (1973) says the reconstruction was made at least one-third too large in the mistaken belief that *all* the teeth were nearly the same size as the large ones medially in the jaws. 'Actually', he points out, 'the most lateral teeth are very small compared to those at the symphysis.'

According to Randall's own calculations, based on a projection of a curve of tooth size of *C. carcharias* and the enamel height of the largest fossil tooth (115 mm *4·52 in*) in the American Museum of Natural History and the largest (117·5 mm *4·62 in*) in the US National Museum, Washington, DC, the *maximum* length attained by *C. megalodon* was more probably in the region of 13·10 m *43 ft*. This figure, however, is probably a bit on the low side because larger fossil teeth are known. In fact the same writer states that the South Australian Museum has a number of *Carcharodon* teeth collected from Lake Bonney, South Australia with an enamel height of 127 mm *5 in*, and some of the fossil teeth collected at Sharktooth Hill, near Bakersfield, California, USA, are nearly 152 mm *6 in* long and weigh 339 g *12 oz*. The owner of the latter dentition must have measured some 16·16–16·76 m *53–55 ft* in length and weighed at least 20 tonne/*ton*.

The largest prehistoric insect was the Dragon-fly *Meganeura monyi* which lived between 280000000 and 325000000 years ago. Fossil remains, i.e. impressions of wings discovered at Commentry, central France, indicate that it had a wing-expanse of up to 70 cm *27½ in*.

The largest bird on record was the Elephant bird (*Aepyornis maximus*), also known as the 'Roc bird', which lived in southern Madagascar. It was a flightless bird standing 2·74–3·04 m *9–10 ft* in height, and its weight has been computed at 438 kg *965 lb*.

This bird also produced the largest eggs of any known animal and Dr Bjorn Kurten (1968), the Finnish palaeontologist, reckons the size was probably close to the maximum possible. 'In a still larger egg', he writes, 'the pressure of the internal fluid would be so great that the shell would have to be excessively thick, and as a result the young would find it difficult to get out.'

One huge example preserved in the British Museum (Natural History) measures 856 mm *33¾ in* round the long axis with a circumference of 723 mm *28½ in*, giving a capacity of 8·88 litre *2·35 gal*, or seven times that of an ostrich egg. A more cylindrical egg collected by a merchant captain on the SW coast of Madagascar and now in the Academie des Sciences, Paris, France, measures 326 × 390 mm *12⅞ × 15⅜ in* (Heuvelmans, 1958), and probably weighed about 12·2 kg *27 lb* with its contents.

Aepyornis may have survived until *c.* 1660 which, strictly speaking, means it is not really a prehistoric animal.

In August 1974 the fossilised bones of another gigantic flightless bird were discovered near

The huge fossilised egg of an Elephant bird

Alice Springs, central Australia, by an expedition from Queensland Museum, Brisbane. The bird reportedly stood more than 3·04 m *10 ft* tall and was of such massive proportions that it *may* have been the largest bird ever to walk the face of the earth, exceeding even *A. maximus* in size.

The flightless Moa *Dinornis giganteus* of North Island, New Zealand, which was exterminated by the Maoris quite recently in human history, was even taller, attaining a maximum height of 3·96 m *13 ft*, but it was of more slender build and its weight has been computed at 245 kg *520 lb*.

Another ratite, the carnivorous *Phororhacos longissimus*, which stalked across what is now Patagonia, South America, about 10000000 years ago, was also probably heavier than *Dinornis giganteus*. It stood about 3·04 m *10 ft* tall and had a skull as large as that of a horse.

The largest prehistoric bird actually to fly (in terms of wing-span) was probably *Gigantornis eaglesomei*, which soared over what is now Nigeria about 45000000 years ago. It is only known from a breastbone, but the enormous size of this fossil – and its close similarity to the breastbone of the albatross – suggest that the bird had long, narrow wings spanning as much as 6·09 m *20 ft* (Fisher & Peterson, 1964).

The condor-like *Teratornis incredibilis*, which flew over what is now the American state of Nevada between 50000 and 1750000 years ago, was another gigantic flying bird. Remains discovered in Smith Creek Cave, Nevada, in 1952 indicate a wing-expanse of 5 m *16 ft 4¼ in* and a weight of nearly 22·2 kg *50 lb* (Howard, 1952).

In *c.* 1971 a fragment of the beak of a large flying bird was collected in the Fish Creek area of the Anza-Borrego Desert State Park, San Diego County, California, USA. According to Hildegarde Howard (1972) the anterior portion of the upper mandible bore a marked resemblance to that of *T. merriami*, but was about 41 per cent larger; indicating a wing-expanse of *c.* 5·26 m *17 ft 3 in* if the rest of the body was in proportion.

The heaviest flying bird so far recorded is *Osteodontornis orri*, related to the pelicans and storks which flew over what is now the State of California about 20000000 years ago. It had a wing-expanse of *c.* 4·87 m *16 ft* and *may* have weighed as much as 27·2 kg *60 lb*, although a bird of this poundage must have flown only reluctantly – and then for short distances.

The largest prehistoric mammal, and the largest land mammal ever recorded, was

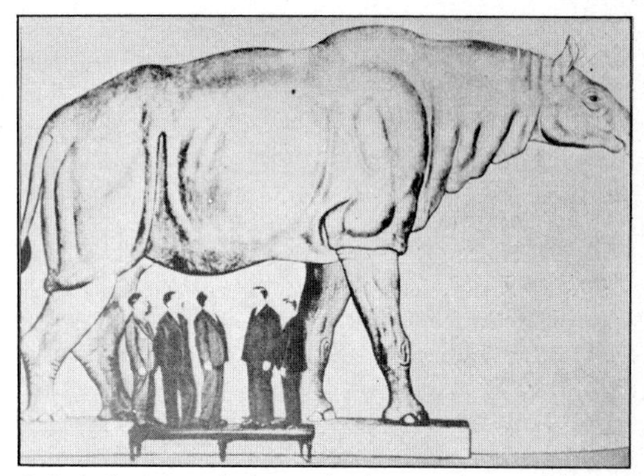

Baluchitherium – the largest land mammal on record (American Museum of Natural History)

Baluchitherium, a long-necked hornless rhinoceros which roamed over what is now central and western Asia and also Europe (Yugoslavia) between 20000000 and 40000000 years ago. The bones of this gigantic browser were first discovered in the Bugti Hills, East Baluchistan, in 1907–8 and described by G Pilgrim who, after examining some teeth, thought they belonged to a giant pig (*sic*). *Baluchitherium* is also represented in *Indricotherium* and *Pristinotherium* of Kazakhstan and *Benaratherium* of South Georgia, USSR. In 1928 the Central Asiatic Expedition of the American Museum of Natural History discovered an incomplete skull – together with part of the lower jaw – measuring over 1·37 m *4 ft 6 in* in length and other bones at Irden Manha in the Gobi Desert, Mongolia, which according to Prof Henry Fairfield Osborn, the great authority on fossil vertebrates, must have come from a *Baluchitherium* measuring 5·41 m *17 ft 9 in* to the top of the shoulder hump (8·23 m *27 ft* to the crown of the head) and 10·66–11·27 m *35–37 ft* in overall length. The animal, named *Baluchitherium grangeri* after its discoverer, Walter Granger, assistant leader of the expedition, was so huge that a procession marching six abreast could have walked under its belly with plenty of room to spare. In life it must have weighed about 16–20 tonne/*ton*.

The only other extinct land mammals that approached *Baluchitherium* in size were some of the giant proboscideans.

The tallest of them all was *Parelephas trogontherii*, which lived about 1000000 years ago in central Europe and North America. A fragmentary skeleton found at Mosbach, Germany, indicates a shoulder height of at least 4·5 m *14 ft 9 in* and de Camp (1965) says the elephant must have weighed about 40000 lb or 17·8 tonne/*ton* when alive. Further on, however, the same writer contradicts himself by pointing out that although this species was exceptionally tall it was short in the body, so a weight of 25000 lb 11·05 tonne/*ton* is probably more feasible. Another mammoth of the same period, *Mammuthus primigenius fraasi*, measured 4·3 m *14 ft 4 in* at the shoulder (Silverberg, 1972). Its remains were first discovered near Steinheim on the Murr, Germany. Both of these mammoths must have measured at least 4·87 m *16 ft* to the top of the hump.

In August 1845 a beautifully preserved mastodon skeleton was discovered in Pleistocene formations on the River Hudson, near Newburgh, New York, USA. The animal was described by Dr John C Warren in a famous memoir in 1852 and it was given the specific name *Mastodon giganteus*. This was changed in 1868 and the new name of *M. americanus* generally adopted. The skeleton was first put on exhibit in the Museum of Comparative Zoology at Harvard University, Cambridge, Massachusetts and, according to one report, the specimen measured nearly 5·48 m *18 ft* at the shoulder. This figure, however, was incorrect, because the head had been set much too high on the body. Shortly afterwards the skeleton was remounted at the Warren Museum in Boston in a more recognisable form but it was not until 1907, when the American Museum of Natural History, New York, purchased the Warren mastodon, as it was known, and Prof Henry Osborne had the bones reassembled again under his own supervision, that justice was finally done and a correct shape achieved. According to Silverberg (1972) the skeleton now has a mounted shoulder height of 4·54 m *14 ft 11 in* but this takes some believing, particularly in view of the fact that Carrington (1962) says the Warren mastodon was only of average size and stood just over 2·74 m *9 ft*. Compare this with the largest known mastodon skeleton in the Geological Museum of Ohio State University, Columbus, which had an estimated shoulder height in life of 3·10 m *10 ft 2 in*.

The longest tusks of any prehistoric animal were those of the Straight-tusked elephant *Hesperoloxodon antiquus germanicus*, which lived in what is now northern Germany about 2000000 years ago. The average length in adult bulls was 5 m *16 ft 4¼ in*. *Parelephas trogontherii*, the steppe mammoth, also had very long tusks and Kurten (1968) says one example from Sussenborn probably measured 5 m *16 ft 4¼ in* when intact. A single tusk of a Woolly mammoth (*Mammuthus primigenius*), preserved in the Franzens Museum at Brno, Czechoslovakia, measures 5·02 m *16 ft 5½ in* along the outside curve (Osborne, 1936–42). In *c.* August 1933 a single tusk of an Imperial mammoth (*Archidiskodon imperator*) measuring 4·87 m *16+ ft* (the anterior end is missing) was discovered by a Mr George B Doughty in Gorza County near Post, Texas, USA. The following year Doughty presented this tusk, which measured 0·6 m *2 ft* in maximum circumference, together with two molars, to the American Museum of Natural History, New York (Holsinger, pers. comm. 28 April 1966).

The heaviest fossil tusk on record is one weighing 150 kg *330 lb* with a maximum circumference of 89 cm *35 in* now preserved in the Museo Civico di Storia Naturale, Milan, Italy. The specimen (in two pieces) measures approximately 3·58 m *11 ft 9 in* in length along the outside curve (Giovanni Pinna, pers. comm. 15 September 1972).

The heaviest recorded mammoth tusks are a pair in the State Museum, the University of Nebraska, Lincoln, Nebraska, USA, which have a combined weight of 226 kg *498 lb* and measure 4·21 m *13 ft 9 in* and 4·16 m *13 ft 7 in* respectively. They were collected in April 1915 from Campbell, Franklin County, Nebraska, during an excavation for a new schoolhouse (Barbour, 1925; Mary Cutler, pers. comm. 22 December 1972).

The Giant deer (*Megaloceros giganteus*), which lived in northern Europe and northern Asia as recently as 50000 BC had the largest antlers of any

The heaviest recorded mammoth tusks on record (State Museum, University of Nebraska)

known animal, the span regularly exceeding 2·43 m *8 ft*. The weight varied between 27·2 and 45·4 kg *60* and *100 lb*. The largest pair listed by Rowland Ward (1928) were found in a marl deposit in Ireland and have a spread of 4·26 m *14 ft* (owner Lord Talbot de Malehide of County Dublin).

The largest extinct whale was *Basilosaurus* (=*Zeuglodon*), which swam in the seas over what are now the American states of Arkansas and Alabama between 38 000 000 and 54 000 000 years ago. In July 1961 a farmer ploughing his field near Millry in Washington County, Alabama, uncovered virtually a complete skeleton which was later presented to the University of Alabama at Tuscaloose. The creature measured 18·28 m *60 ft* in length, had a 1·83 m *6 ft* long skull and weighed an estimated 180 000 lb 81·4 tonne *80·3 ton* when alive (Murray, 1967), but this poundage is much too high for a whale that was very serpentine in appearance. A weight of *c.* 17 tonne/*ton* would be more realistic. Other bones have been found in Arkansas which must have come from an individual measuring 21·3 m *70 ft* in length.

The Giant deer had greatly palmated antlers measuring up to 4.26 m 14 ft *across* (Chorley Guardian)

SELECTED BIBLIOGRAPHY

Agassiz, L (1857). *Contributions to the natural history of the United States of America*, vol. 1–2. Boston.

Ahuja, M L & Singh, Gurkirpal (1954). Snake bite in India. *India J. Med. Res.*, vol. xlii, pp. 661–680.

Amadon, Dean (1943). Bird weights and egg weights. *Auk*, vol. 60, pp. 221–234.

Amaral, Afranio do (1948). Serpentes Gigantes. *Boleti Museu Paraense E. Goeldi*, vol. 10, pp. 211–237.

Anon (1935). Record proboscidean tusk. *Nat. Hist. (New York)*, vol. 35 (4), p. 357.

Anon (1954). 152-year-old sturgeon caught in Ontario. *Comm. Fish. Rev.*, vol. 16, No. 9, p. 28.

Appleby, Leonard G (1971). *British snakes*. London.

Ashley, C W (1938). *The Yankee whaler*. New York.

Augusta, J & Burien, Z (1962). *A book of mammoths*. London.

Babcock, H L (1919). The turtles of New England. *Boston Soc. Nat. Hist.*, vol. 8, pp. 323–431.

Baerg, William J (1958). *The tarantula*. Lawrence, Kansas.

Bainbridge, R (1958). The speed of swimming of fish as related to size and to the frequency and amplitude of tail beat. *J. Exp. Biol.*, vol. 35, pp. 109–133.

Balozet, L (1971). Scorpionism in the Old World, in *Venomous animals and their venoms. 3. Venomous invertebrates* (ed. W Bucherl & E E Buckley). New York–London, pp. 349–371.

Banks, E (1931). Some measurements of the Estuarine Crocodile (*C. porosus*) from Sumatra. *J. Bomb. Nat. Hist. Soc.*, vol. 34, pp. 1086–1088.

Barrett, Charles (1931). Megascolides, the world's biggest earthworm. *Aus. Mus. Mag.* (Sydney), vol. 4, No. 7, pp. 227–230.

Bates, Henry W (1863). *The naturalist on the River Amazon, etc.* London.

Baughman, J L (1955). The oviparity of the Whale shark, *Rhineodon typus*, etc. *Copeia*, No. 1, pp. 54–55.

Bauman, J E (Feb. 1926). Observations on the strength of the chimpanzee and its implementations. *J. of Mamm.*

Behler, J L (1975). The great American snake hunt. *Animal Kingdom*, vol. 78, No. 2, pp. 21–26.

Belkin, D A (1961). The running speeds of the lizards *Dipsosaurus dorsalis* and *Callisaurus draconoides*. *Copeia*, No. 2, pp. 223–224.

Bergman, Sten (1936). Observations on the Kamchatkan bear. *J. of Mamm.*, vol. 17, No. 2.

Berzin, A A (1972). *The Sperm whale* (Israel Program for Scientific Translations). Jerusalem.

Best, Anthony (ed) (1971). *Rowland Ward's Records of Big Game*, 14th Edition (Africa), London.

Blair, A W (1934). Spider Poisoning. Experimental study of the effects of the female *Latrodectus mactans* in Man. *Arch. Int. Med.*, vol. liv, pp. 831–843.

Bogert, C M & del Camp, R (1956). The Gila monster and its allies. *Bull. Amer. Mus. Nat. His.*, vol. cix, pp. 151–154.

Bonner, W N (1971). An aged Grey seal (*Halichoerus grypus*). *J. of Zool.*, vol. 164, No. 22, pp. 261–262.

Bourlière, François (1955). *The natural history of mammals*. London.

Bowers, C A & Henderson, R S (1972). *Project Deep Ops: Deep object recovery with Pilot and Killer whales*. Undersea Surveillance and Ocean Sciences Department, Naval Undersea Center, San Diego, California.

Brander, A Dunbar (1931). An enormous estuarine crocodile (*C. porosus*). *J. Bomb. Nat. Hist. Soc.*, vol. 34, p. 584.

Bristowe, William S (1958). *The world of spiders*. London.

Bristowe, William S (1971). *The world of spiders* (revised edition). London.

Brongersma, L D (1972). European Atlantic turtles. *Zoologische Verhandelingen*, Nr. 121. Leiden.

Brown, Barnum (1931). The largest known land tortoise. *Nat. Hist (N.Y.)*, vol 31, No. 2, pp. 186–188.

Brown, J (1962). The monster fish of American rivers. *Sports Illus.*, vol. 17, No. 12, pp 64–78.

Brown, Leslie & Amadon, Dean (1968). *Eagles, hawks and falcons of the world*. 2 vols. London.

Brown, Leslie (1970). *African birds of prey*. London.

Brown, Robert (1973). Has the Thylacine really vanished? *Wildlife.*, Sept., pp. 416–419.

Brunn, Anton F et al. (1956). *The Galathea deep sea expedition, 1950–52*. London.

Bruyns, W F (1971). *Field guide of whales and dolphins*. Amsterdam.

Bucherl, Wolfgang (1971). Spiders, in *Venomous animals and their venoms. 3. Venomous invertebrates* (ed. W Bucherl and E E Buckley). New York-London.

Bucherl, Wolfgang (1971). Classification, biology and venom extraction of scorpions, in *Venomous animals and their venoms. 3. Venomous invertebrates* (ed. W. Bucherl & E E Buckley). New York-London, pp. 317–347.

Budker, Paul (1971). *The life of sharks*. London.

Burne, E C (1943). A record of gestation periods and growth of trained elephant calves in the southern Shan States, Burma. *Proc. Zool. Soc. London*, vol. 133, p. 27.

Burton, R G (1929). The record Indian crocodile. *J. Bomb. Nat. Hist. Soc.*, Vol. 34, pp. 1086–1088.

Bussmann, J (1946). Beitrag zur Kenntnis der Brutbiologie und des Wachstums des Grosses Buntspechts, *Dryobates major* (L.). *Ornith. Beob.*, vol. 43, pp. 137–156.

Butler, Amos W (1898). The birds of Indiana. *22nd Ann. Rept. Dept. Geol. Nat. Res. Indiana*. Indianapolis, p. 626.

Butler, L S (1932). Studies in Australian spiders. *Proc. Roy. Soc. Vic.*, vol. 44, No. 2, pp. 103–117.

Caras, Roger (1964). *Dangerous to man*. Philadelphia.

Carr, Carlyle (1927). The speed of pronghorn antelope. *J. Mamm.*, vol. 8, No. 33, pp. 249–250.

Carr, Donald (1972). *The deadly feast of life*. London.

Chamberlin, R V & Ivie, W (1935). The Black Widow spider and its varieties in the United States. *Bull. Univ. of Utah*, vol. xxv. Salt Lake City.

Chandler, Asa C (1946). *Introduction to parasitology*. New York–London.

Chaudhuri, D K et al. (1971). Pharmacology and toxicology of the venoms of Asiatic snakes, in *Venomous animals and their venoms. 2. Venomous vertebrates* (ed. W Bucherl & E E Buckley). New York-London.

Chisholm, Alec H (1948). *Bird wonders of Australia*. Sydney.

Clark, Aisla M (1962). *Starfishes and their relations*. Pub. British Museum (Natural History). London.

Clark, Leonard (1954). *The rivers ran east*. New York.

Clarke, James (1969). *Man is the prey*. London.

Clarke, Malcolm R (1966). A review of the systematics and ecology of oceanic squids. *Adv. Mar. Biol.*, vol. 4, pp. 93–300 (ed. Frederick S. Russell). London and New York.

Clarke, Robert (1955). A giant squid swallowed by a sperm whale. *Norsk. Hvalfangst-Tidende*, vol. 44, No. 10, pp. 589–593.

Clay, C L (1911). Some diving notes on cormorants. *Condor*, vol. 13, p. 138.

Cloudsley-Thompson, J L (1958). *Spiders, scorpions, centipedes and mites*. London.

Coates, C W et al. (1937). The electric discharge of the electric eel *Electrophorus electricus* (Linnaeus). *Zoologica (New York)*, vol. 22, No. 1, pp. 1–31.

Colbert, Edwin A (1962). The weights of dinosaurs. *Amer. Mus. Novitates*, No. 2076, pp. 1–16.

Comfort, Alex (1964). *Ageing: the biology of senescence*. London.

Conant, Roger & Hudson, Robert (1949). Longevity records for reptiles and amphibians in the Philadelphia Zoological Garden. *Herpetologica*, vol. 5, pp. 1–8.

Constant, P & Cannonge, B (1957). Evaluation de la vitesse de vol des Miniopteres. *Mammalia*, vol. 21, pp. 310–302.

Cooke, Mary (1933). Speed of bird flight. *The Auk*, vol. 1, pp. 309–316.

Cott, Hugh B (1961). Scientific results of an inquiry into the ecology and economic status of the Nile crocodile (*Crocodilus niloticus*) in Uganda and Northern Rhodesia. *Trans. & Zool. Soc. Lond.*, vol. 29, pt 4, pp. 211–337.

Cottam, Clarence et al. (1942). Flight and running speeds of birds. *Wilson Bull.*, vol. 54, No. 2, pp. 121–131.

Cousins, Don (1972a). Body measurements and weights of wild and captive gorillas, *Gorilla gorilla. Zool. Garten N.F., Leipzig*, vol. 41, pt 6, pp. 261–277.

Cousins, Don (1972b). Gorillas in captivity past and present. *Zool. garten N. F., Leipzig*, vol. 42, pt 5/6 S., pp. 251–281.

Craig, C F & Faust, E C (1945). *Clinical parasitology*, 4th ed. London.

Crandall, Lee S (1964). *Management of wild mammals in captivity*. Chicago–London.

Cross, H & Fischer, P (1862). Nouveaux documents sur les céphalopodes. *J. Conch (Paris)*, vol. 10, pp. 124–140.

Curry-Lindahl, Kai (1972). *Let them live*. New York.

D'Abrera, Bernard (1975). The largest butterfly in the world. *Wildlife*, vol. 17, No. 12.

D'Amour et al. (1936). The Black Widow spider. *Quar. Rev. of Biol.*, vol. xi.

Davis, David H (1964). *About sharks and shark attack*. Pietermartizburg.

Day, Francis (1880–84). *The fishes of Great Britain and Ireland*. 2 vols. London.

Dean, S (1954). Length of python. *North Queensland Nat.*, vol. 22, No. 1, pp. 13–14.

Dell, R K (1970). A specimen of the giant squid *Architeuthis* from New Zealand. *Records Dominion Mus.*, vol. 7, No. 4, pp. 25–36.

Deoras, P J (1963). Studies on Bombay snakes: snake farm venom records and their probable significance, in *Venomous and poisonous animals and noxious plants of the Pacific region*. Oxford-London-New York-Paris, pp. 337–349.

Desmond, Adrian J (1975). *The hot-blooded dinosaurs*. London.

Dewar, John M (1924). *The bird as a diver*. London.

Dice, Lee R (1945). Minimum intensities of illumination under which owls find dead prey by sight. *Amer. Nat.*, vol. 70, pp. 385–416.

Dorst, Jean & Dandelot, Pierre (1970). *A field guide to the larger mammals of Africa*. London.

Dorst, Jean (1971). *The life of birds*, vol. 1. London.

Dowling, Herndon G (1961). How old are they and how big do they grow? *Animal Kingdom*, vol. 64, No. 6, pp. 171–175.

Downey, Maureen E (1972). *Midgardia xandaros*, new genus, new species, a large brisingid starfish from the Gulf of Mexico. *Proc. Biological Soc. Washingt.*, vol. 84, No. 48, pp. 421–426.

Dresner, Simon (1973). King cobra's longevity record. *Intl Zoo News*, No. 13, 21 May.

Droscher, Vitus B (1969). *The magic of the senses*. London.

Dunn, Emmett, R (1944). *Los Generos de Anfibios y Reptiles de Colombia*. III. *Caldasia*, pp. 155–224.

Dymond, J R (1947) Longevity of captive snails. *Can. Field Nat.*, vol. 61, p. 69.

Eaton, Howard E (1910). Birds of New York. *N. Y. State Mus. Mem.*, vol. 12, pt 1, p. 214.

Einarsen, Arthur S (1948). *The pronghorn antelope and its management*. Wildl. Management Instit. (Washington, D. C.).

Ernest, C H & Barbour, R W (1972). *Turtles of the United States*. Univ. Press of Kentucky.

Frost, Nancy (1934). Notes on a giant squid (*Architeuthis* sp.) captured at Dildo, Newfoundland, in December 1955. *Rept. Newfoundland Fish. Comm.*, vol. 22, pp. 100–114.

Evans, Glyn (1972). The prodigious jump of the click beetle. *New Scientist*, 21 Sept., pp. 490–493.

Evans, Glyn (1975). *The life of beetles*. London.

Faber, F (1826). *Ueber das leben hochnordischen Vogel*. Leipzig.

Fairley, N Hamilton (1929). The present position of snake bite and the snake bitten in Australia. *Med. J. Australia*, pp. 296–311.

Feder, Howard M & Christensen, Aage Moller (1966). Aspects of asteroid biology, in *Physiology of Echinodermata* (ed. Richard A. Boolootian). New York-London-Sydney, pp. 87–128.

Fell, H Barraclough (1966). The ecology of Ophiuroids, in *Physiology of Echinodermata* (ed. Richard A. Boolootian). New York-London-Sydney, pp. 129–144.

Firth, Frank E (1940). Giant lobsters. *New Eng. Nat.*, No. 9, pp. 84–87.

Fisher, James & Petersen, Roger (1964). *The world of birds: a comprehensive guide to general ornithology*. London.

FitzSimons, V F M (1970). *A field guide to the snakes of Southern Africa*. London.

Fleay, David (1952). With a Wedge-tailed eagle at the nest. *The Emu*, Vol. 52, pt 1, pp. 1–16.

Follett, W I (1967). Man-eater of the California coast. *Pacific Discovery*, vol. xix, No. 1, pp. 18–22.

Forbush, E H (1922). Some underwater activities of certain waterfowl. *Mass. Dept. Agric. Econ. Biol. Bull.*, No. 8, pp. 32–33.

Forster, R R (1958). The spiders of the Family Symphytognathidae. *Trans. Royal Soc. N.Z.*, vol 86 (3–4), pp. 269–329.

François, D (1960). Freshwater crayfishes. *Aust. Mus. Mag.*, vol. 13, No. 7, pp. 217–221.

Fraser, F C (1974). *Report on Cetacea stranded on the British coasts from 1948 to 1966. No. 14*, Trustees Brit. Mus (Nat. Hist.). London.

Frith, H J & Calaby, J H (1969). *Kangaroos*. London.

Frith, H J (1962). *The Mallee fowl*. Sydney

Fulleborn, F (1932). Ueber Klinik und Bekampfung der Spulwurm-Infektion. *Klin. Wchnschr.*, vol. 11, pp. 1679–1684, 1716–1720.

Gambell, Ray (July 1970). Weight of a sperm whale, whole and in parts. *S. African J. of Sci.*, pp. 225–227.

Gambell, Ray (1972). Why all the fuss about whales? *New Scientist*, 22 June, pp. 674–676.

Gawn, R W L (1948). Aspects of the locomotion of whales. *Nature*, London, vol. 161, No. 4080, pp. 44–46.

Gaymer, R (1968). The Indian Ocean giant tortoise *Testudo gigantea* on Aldabra. *J. Zool., Lond.*, vol. 154, pp. 341–363.

George, J C (1970). *Animals can do anything*. London.

Gertsch, W J (1949). *American spiders*. New York-Toronto-London.

Gibson-Hill, C A (23 Oct. 1948). *Giant King cobra*. Field mag.

Gibson, J D & Sefton, A R (May 1960). Second report of the New South Wales albratross study group. *The Emu*, vol. 60, pp. 125–130.

Grant, E M (1972). *Guide to fishes*. Brisbane.

Grasset, E (1946). La Vipère du Gabon. *Acta Tropica*, vol. III, p. 101.

Gray, James (1968). *Animal locomotion*. London.

Griffin, D J (March 1970). Australian crabs. *Aust. Nat. Hist.*, vol. 16, No. 9, pp. 304–308.

Griffin, Donald R & Hitchcock, Harold B (1965). Probable 24-year longevity records for *Myotis lucifugus*. *J of Mamm.*, vol. 46, No. 2.

Gudger, E W (1944). The giant freshwater perch of Africa. *Sci. Monthly*, Vol. lviii, pp. 269–272.

Gudger, E W (1945). Is the giant catfish, *Silurus glanis*, a predator on man? *Sci. Monthly*, vol. lxi, No. 6, pp. 451–454.

Guggisberg, C A W (1975). *Wild cats of the world*. London.

Gunter, Gordon (1941). Occurrence of the manatee in the United States with records from Texas. *J. of Mamm.*, vol. 22, No. 1.

Guppy, Nicholas (1963a). The dreaded Piranha. *Animals*, vol. 2, No. 1, pp. 16–19.

Guppy, Nicholas (1963b). The largest snake in the world. *Animals*, 30 July, pp. 165–167.

Hall, F G (1937). Adaptations of mammals to high altitudes. *J. of Mamm.*, vol. 18, No. 4.

Halstead, B W & Ryckman, R (1949). Injurious effects from contacts with millipedes. *Med. Arts and Sciences*, vol. 3, pt. 1, p. 16.

Hamlet, Sybil E (1968). Oldest zoo resident. *Intl Zoo News*, vol. 15, No. 3, p. 83.

Harting, J E (1906). *Recreations of a naturalist*. London.

Harvey, M (1874). Gigantic cuttlefishes in Newfoundland. *Ann. Mag. Nat. Hist.*, Ser. 4, vol. 13, pp. 67–70.

Hayward, B & Davis, R (1964). Flight speeds in Western bats, *J. of Mamm.*, vol. 45, No. 2, pp. 236–241.

Headley, H S (1927). The record Australian python. *The Field*, vol. 150, No. 3896, p. 319.

Heerden, J van (1937). Erdwurms die allerbeste tuiniers. Knewels van 22 voet aangetref in Transvaal. *Die Costerlig*, after 1937. Port Elizabeth. (Earthworms the very best gardeners – Giants of 22 feet found in the Transvaal.)

Heezen, Bruce C (1957). Whales entangled in deep-sea cables. *Deep-Sea Res. (London)*, vol. 4, pp. 105–115.

Henshaw, H W (1920). Autobiographical notes. *Condor*, vol. 22, pp. 3–10.

Herrick, F H (1911). Natural History of the American lobster. *Bull. Bur. Fish (Washington, DC)*, vol. 29, pp. 149–408.

Hervy, G F & Hems, J (1968). *The Goldfish.* London.

Heuvelmans, Bernard (1968). *In the wake of the sea-serpents.* New York.

Hill, J E (1957). Record ivory in the collection of the British Museum (Natural History). *Tanganyika Notes and Records*, No. 46, pp. 29–31.

Hill, John E (1974). A new family, genus and species of bat (Mammalia: Chiroptera) from Thailand. *Bull. Brit. Mus. (Nat. Hist.)*, vol. 27, pp. 301–336.

Hingston, R W G (1936). Earth's highest animals. *Zoo Mag.*, vol. 1, No. 1, pp. 90–91.

Hooley, R W (1913). *Qrtly J. Geog Soc.*, vol. lxix.

Horring, R (1919). Fugle. 1. Andefugle og honsefugle. *Danmarks Fauna (Kobenhavn)*, Nr. 23.

Hosking, Brian (1953). The intrinsic range and speed of flight of insects. *Trans. Roy. Entomological Soc. Lond.*, vol. 104, pp. 223–520.

Howard, Hildegarde (1952). The prehistoric avifauna of Smith Creek Cave, Nevada, with a description of a new gigantic raptor. *Bull. Southern Calif. Acad. Sci.*, vol. 51, pp. 50–54.

Howard, Hildegarde (1972). The incredible Teratorn again. *Condor*, vol. 74, pp. 341–344.

Howarth, T G (1967). Expensive butterflies. *Animal Mag.*, vol. 10, No. 3, p. 141.

Hubbard, Clarence (1947). *Fleas of western North America.* Ames, Iowa.

Hurrell, H G (1968). Pine Martens. *Forestry Commission: Forest Rec.*, No. 64.

Ichihara, T (1961). Blue whales in the waters around Kerguelen Island. *Norwegian Whal. Gaz. (Oslo)*, vol. 50, No. 1, pp. 1–22.

Isemonger, R M (1962). *Snakes of Africa.* Johannesburg.

Jones, Marvin L (1972a). Longevity of mammals in captivity. *Intl. Zoo News*, No. 104.

Jones, Marvin L (1972b). Longevity of mammals in captivity. *Intl. Zoo News*, No. 107.

Jourdain, F C F (1913), in *The British bird book* (ed. F. B. Kirkman *et al*), vol. 4, pp. 444–446. London.

Joyeux, C & Baer, J (1936). Recherches biologiques sur la ligule intestinale: re-infestation parasitaire. *Compt. rend soc. biol.*, vol. 121, pp. 67–68.

Kaestner, Alfred (1968). *Invertebrate zoology.* New York-London-Sydney.

Kimura, Seiji & Nemoto, Takahisa (1956). Notes on a minke whale kept alive in aquarium. *Sci. Rep. Whales Res. Inst.*, vol. 11, pp. 181–189.

Kinghorn, J R (1930). What is the life span of a bird? *Aus. Mus. Mag. (Sydney)*, vol. 4, No. 2, pp. 43–46.

Kirk, T W (1888). Brief description of a new species of large decapod (*Architeuthis longimanus*), *Trans. N.Z. Inst.*, vol. 20, pp. 34–39.

Kistchinski, A A (1972). Life history of the Brown bear in north-east Siberia, in Bears – their biology and management. *Int. Union for Cons. Nature and Natural Resources*, No. 23.

Klauber, Laurence M (1956). *Rattlesnakes, their habits, life histories and influence on mankind.* 2 vols. Berkeley and Los Angeles.

Koford, Carl B (1953). *The California Condor.* New York-Toronto-London.

Kolb, A (1955). Wie schnell fliegt eine Fledermaus? *Saugetierk.* Mitteil. 3, pp. 176–177.

Kooyman, G L et al. (Oct. 1971). Diving behaviour of the Emperor penguin, *Aptenodytes forsteri. The Auk*, pp. 775–795.

Kopstein, F (1932). *Bunarus javanicus*, een nieuwe Javaansche Giftslang. Mededelling over een doodelijke Bungarusbeet. *Geneeskunde Tijdschrift Nederland-Indies*, vol. lxxii, pp. 136–140.

Kurten, Bjorn (1968). *Pleistocene mammals of Europe.* London.

Kurten, Bjorn (1968). *The age of dinosaurs.* London.

Lane, Frank W (1957). *Kingdom of the Octopus.* London.

Laurie, A H (1933). Some aspects of respiration in blue and fin whales. *Disc. Repts.*, No. 7, pp. 363–406.

Laws, R M (1953). Elephant sea (*Mirounga leonina* Linn.) 1. Growth and age. *Sci. Repts. Falkland Is. Dep. Surv. (London)*, No. 8.

Laws, R M et al. (1967). Estimating live weights of elephants from hind-leg weights. *E. Afr. Wildl. J.*, vol. 5, pp. 105–106.

Lawson, Douglas (1976). Pterosaur from the Latest Cretaceous of West Texas: Discovery of the largest flying creature. *Science*, vol. 187, No. 4180.

Lederer, Gustav (1944). Nahrungserwerb, Entwicklung, Paarung and Brutfursorge von *Python reticulatus* (Schneider). *Zool. Jahrbucher (Anatomie) (Jena)*, vol. 68, pp. 363–398.

Lehane, Brendan (1969). *The compleat flea.* London.

Leighton, G (1901). *The life history of British serpents and their local distribution in the British Isles.* London.

Lewis, George (1955). *Elephant tramp.* New York.

Locket, G H & Millidge, A F (1951–53). *British spiders.* 2 vols. London.

Ljungstrom, P O & Reinecke, A J (1969). Ecology and natural history of the microchaetid earthworms of South Africa. *Pedobiologia*, Bd. 9, S. 152–157.

Lockley, R M (1970). The most aerial bird in the world. *Animals*, vol. 13, No. 1, pp. 4–7.

Loomis, H F (1936). The millipedes of Hispaniola with descriptions of a new family, new genera, and new species. *Bull. Mus Comp. Zool. Harv.*, vol. 80, pp. 1–191.

Loveridge, Arthur (1945). *Reptiles of the Pacific world.* New York.

MacGinitie, G E & MacGinitie, N (1968). *Natural history of marine animals* (2nd ed.). New York-London-Toronto.

Marki, F & Witkop, B (1963). The venom of the Colombian arrow poison frog *Phyllobates bicolor. Separatum Experientia (Basel)*, vol. 19, p. 329.

Mazzotti, L & Bravo-Becherelle, M (1963). Scorpionism in the Mexican Republic, in *Venomous and poisonous animals and noxious plants of the Pacific region* (ed. H L Keegan & W V Macfarlane). London, pp. 119–131.

McIlhenny, Edward A (1935). *The alligator's life history.* Boston.

McKeown, K C (1936). *Spider wonders of Australia.* Sydney.

McLaren, I A (1958). The biology of the Ringed seal (*Phoca hispida* Schreber) in the Eastern Canadian Arctic. *Bull. Fish. Res. Bd. Can.*, No. 118, vii+pp. 1–97.

McMichael, D F (1964). The identity of the venomous octopus responsible for a fatal bite at Darwin, Northern Territory. *J. Malac. Soc. Aus.*, vol. 1, No. 8, pp. 23–24.

Meinertzhagen, Richard (1955). The speed and altitude of bird flight (with notes on other animals). *Ibis*, vol. 97, No 1, pp. 81–117.

Merfield, Fred G (1956). *Gorillas were my neighbours.* London.

Micks, Don. W (1963). The current status of necrotic arachnidism in Texas, in *Venomous and poisonous animals and noxious plants of the Pacific region* (ed. H L Keegan & W V Macfarlane). Oxford-London-New York-Paris, pp. 153–159.

Minton, S A & Minto, M (1971). *Venomous reptiles.* London.

Mitsukuri, K & Ikeda, S (1895). Notes on a giant cephalopoda. *Zoo. Mag. Tokyo*, No. 7, pp. 39–50.

Mole, R R (1924). The Trinidad snakes. *Proc. Zool. Soc. Lond.*, pp. 235–278.

Mosauer, Walter (1935). How fast can snakes travel? *Copeia*, No. 1, pp. 6–9.

Murie, Olaus J (1951). *The Elk of North America.* Harrisburg & Washington.

Murray, Marian (1967). *Hunting for fossils.* New York.

Murphy, Robert Cushman (1914). Notes on the sea elephant *Mirounga leonina* (Linné). *Bull. Amer. Mus. Nat. Hist.*, vol. 33, pp. 63–79.

Napier, J R & Napier, P H (1967). *A handbook of living primates.* London.

Neill, Wilfred T (1971). *The last of the ruling reptiles.* New York and London.

Nice, M. (1953). The question of ten-day incubation periods. *Wilson Bull.*, vol. 65, pp. 81–93.

Nishiwaki, M (1972). General biology, in *Mammals of the sea (Biology and Medicine)*, ed. Sam H. Ridgway, Springfield, Illinois.

Norwood, V G (1964). *Drums along the Amazon.* London.

Nott, J Fortune (1886). *Wild animals.* London.

Oliver, James A (1958). *Snakes in fact and fiction.* London.

Osgood, W H (1943). The mammals of Chile. *Publ. Field Mus. Zoo.*, Ser. 30, pp. 1–268.

Penfold, W J et al. (1937). A survey of the incidence of *Taenia saginata* infestation in the population of the State of Victoria from January 1934 to July 1935. *Med. J. Australia 23rd year*, vol. 1, pp. 283–285.

Peterson, Randolph L (1955). *North American moose.* Toronto.

Piers, Harry (1934). Accidental occurrence of the man-eater or Great white shark, *Carcharodon carcharias* (Linn.), in Nova Scotian waters. *Proc. Nova Scotian Inst. Sci.*, vol. 18, No. 3, pp. 192–203.

Pillai, N G (1941). On the height and age of an elephant. *J. Bomb. Nat. Hist. Soc.*, vol. 42, No. 4, pp. 927–928.

Pinney, Peter (1976). *To catch a crocodile.* London-Sydney-Melbourne-Singapore-Manila.

Pitman, C R S (1938). *A guide to the snakes of Uganda.* Kampala.

Pocock, R I (1927). A record Australian python. *The Field*, vol. 149, No. 3879, p. 707.

Pope, Elizabeth C (1958). Giant earthworms. *Aust. Mus. Mag. (Sydney)*, vol. 12, No. 10, pp. 309–311.

Poynton, J C (1964). Amphibia of Southern Africa. *Ann. Natal Mus.*, vol. 17, p. 67.

Powell, A N W (1958). Call of the Tiger. New York.

Probert, A J (1972). *Parasites.* Harmondsworth, Middlesex.

Probst, R T & Cooper, E L (1954). Age, growth and productions of the Lake sturgeon (*Acipenser fulvescens*) in the Lake Winnebago region, Wisconsin. *Trans. Amer. Fish Soc.*, vol. 84, pp. 207–227.

Rand, A Stanley (1952). Jumping ability of certain anurans, with notes on endurance. *Copeia*, No. 1, pp. 15–20.

Randall, John E (1973). Size of the Great white shark. *Science*, vol. 181, pp. 169–170. 13 July.

Ray, Carlton (1966). Stalking seals under Antartic ice. *Nat. Geogr. Mag. (Washington D.C.)*, vol. 129, No. 1, pp. 54–55.

Ray, G E (1941). Big for his day. *Nat. Hist (N.Y.)*, vol. 48, pp. 36–39.

Rees, W J (1949). Note on the hooked squid, *Onychoteuthis banksi. Proc. malac. Soc. Lond.*, vol. 28, pp. 43–45.

Rees, W J & Maul, G E (1956). The Cephalopoda of Madeira. *Bull. Brit. Mus. (Nat. History) (Zool.)*, vol. 3, pp. 259–281.

Reid, H Alistair (1956). Sea-snake bite research. *Trans. Roy. Soc. Trop. Med. & Hyg.*, p. 525.

Reid, H A (1956). Three fatal cases of sea snakebite. *Venoms A.A.A.S.*, Pub. No. 44.

Reid, H A (1959). Sea-snake bite and poisoning. *The Practitioner*, vol. 183, p. 530.

Remington, C L (1950). The bite and habits of a giant centipede (*Scolopendra subspinipes*) in the Philippine Is., *Amer. J. Trop. Med.*, vol. 30, pp. 453–455.

Riess, B H et al. (1949). The behaviour of two captive specimens of the lowland gorilla, *Gorilla gorilla gorilla* (Savage and Wyman). *Zoologica (New York)*, vol. 34, No. 13, pp. 111–117.

Riper, Walker van (1954). Measuring the speed of a Rattlesnake's strike. *Animal Kingdom*, vol. 57, pp. 50–53.

Risting, Sigurd (1922). Av. Hvalfangstens Historie. *Publikatio Nr. 2. Fra Kommander Chr. Christensens Hvalfangstmuseum. I. Sandefjord*, p. 1.

Risting, S (1928). Whales and whale foetuses. Statistics of catch and measurements collected from the Norwegian Whalers' Association 1922–1925. *Rapp. Cons. Explor. Mer.*, vol. 50, pp. 1–122.

Robson, G C (1933). On *Architeuthis clarkei*, a new species of giant squid, with observations on the genus. *Proc. zool. Soc. Lond.*, pp. 681–697.

Rose, Walter (1950). *The reptiles and amphibians of Southern Africa.* Cape Town.

Rothschild, M and Clay, T (1952). *Fleas, flukes and cuckoos.* London.

Rozhdestvensky, A K (1970). On the gigantic claws of some enigmatic Mesozoic reptiles. *Paleontological Mag.*, No. 1, pp. 131–141.

Rydzewski, W (1962). Longevity list for birds. *The Ring*, vol. 3 (33), pp. 147–152.

Salou, G (1951). Envenimation par *Echis carinatus* dans la Région de Sokode (Nord-Togo). *Med. Trop.*, vol. xi, pp. 655–660.

Sanderson, Ivan (1937). *Animal treasure.* London.

Sars, G O (1879). Reports made to the Department of the Interior of investigations of the salt-water fisheries of Norway during the years 1874–1877. *Trans. Herman Jacobson. Rep. U.S. Fish Comm. for 1877*, pp. 663–705.

Sasaki, M (1929). A monograph of the dibranchiata cephalopods of the Japanese and adjacent waters. *J. Fac. Agric. Hokkaido Imp. Univ.*, 20 Suppl. pp. 1–357.

Savory, Theodore H (1928). *The biology of spiders.* London.

Savory, T H (1945). *The spiders and allied orders of the British Isles.* London.

Savory, T H (1966). Britain's most elusive spider. *Animals*, vol. 13, No. 1, pp. 500–501.

Scheffer, V B & Rice, D W (1963). A list of the marine mammals of the world. *U.S. Fish and Wildl. Ser. Spec. Sci. Rept. – Fish*, No. 431.

Scheffer, Victor B (May 1969). Super flea. *Nat. Hist. New York.*

Scheffer, Victor (October 1970). Clicke of the killer. *Nat. Hist. New York.*

Scheithauer, Walter (1967). *Hummingbirds, flying jewels.* London.

Schmidt, Karl & Inger (1957). *Reptiles of the world.* London.

Schneider, Dietrich (1974). The sex-attractant receptor of moths. *Sci. Amer.*, vol. 231, No. 1, pp. 28–35.

Schneirla, T C (1971). *Army ants.* San Francisco.

Schomburgk, R (1922). *Travels in British Guiana.* London.

Schorger, A W (1947). The deep diving of the Loons and Old-Squaw and its mechanism. *Wilson Bull.*, vol. 59, No. 3, pp. 151–159.

Schottler, Werner H (1952). Problems of antivenin standardization. *Bull. World Health Organisation*, pp. 293–320.

Schultz, Harold (1963). Voracious fish of the Amazon. *Animals*, vol. 2, No. 19, pp. 526–529.

Schultz, Leonard & Stern, Edith (1948). *The ways of fishes.* New York-Toronto-London.

Sikes, Sylvia K (1972). *Lake Chad.* London.

Silverberg, Robert (1970). *Mammoths, mastodons and man.* London.

Simon, E S (1943). Life span of some wild animals in captivity. *J. Bomb. Nat. Hist. Soc.*, vol. 44, pp. 117–118.

Singh, Gyan (1961). The Eastern Steppe Eagle (*Aquila n.* Hodgson) on the South Col of Everest. *J. Bomb. Nat. Hist. Soc.*, vol. 58 (No. 1), p. 270.

Small, George L (1971). *The Blue Whale.* New York-London.

Smith, Hugh M (1925). The Whale shark *Rhineodon* in the Gulf of Siam. *Science*, N.S., vol. 62, p. 438.

Smith, K G V (ed) (1973). *Insects and other arthropods of medical importance.* Trustees Brit. Mus (Nat. Hist). London.

Smythies, E (1942). *Big game shooting in Nepal.* Calcutta.

Sotavalta, O (1947). The essential factor regulating the wing-stroke frequency of insects in wing mutilation and loading experiments and in experiments at subatmospheric pressure (Contributions to the problem of insect flight. II). *Ann. Zool. Soc. Zool-bot, fenn Vanamo*, vol. 15, pp. 1–67.

Soule, Gardner (1974). *Wide ocean – discoveries at sea.* Folkestone.

Stahnke, H (1963). Some pharmacological and biochemical characteristics of *Centruroides sculpturatus* Ewing scorpion venom. *Proc. Int. pharmac. Meet., Prague*, pp. 63–70.

Steenstrup, J (1962). The cephalopod papers of Japettus Steenstrup (ed. Agnete Volsoe *et al.*). Copenhagen.

Steinbach, Martin J & Money, K E (1973). Eye movements of the owl. *Vision Res.*, vol. 13, pp. 889–891.

Stephens, F (1895). Notes on the California vulture. *Auk*, vol. 12, pp. 81–82.

Stephenson, J (1930). *The Oligochaeta.* Oxford.

Stewart, Margaret (1966). Herpetofauna of the Nyika Plateau (Malawi and Zambia). *Ann. Natal Mus.*, vol. 18, p. 301.

Summers-Smith, J D (1963). *The House sparrow.* London.

Swaroop, S & Grab, B (1954). Snakebite mortality in the world. *Bull. WHO, League of Nations*, vol. 10, No. 1, pp. 35–76.

Tamiya, N & Puffer, H (1974). *Toxicon*, vol. 12, pp. 85–87.

Tao, L (1930). Notes on the ecology and the physiology of *Caudina chilensis* (Muller) in Mutsu Bay. *Proc. 4th Pacific. Sci. Cong.*, vol. 3, pp. 7–11.

Tarassow, W (1934). Beitrage zum Problem des Kampfes gegen *Diphyllobothrium latum* (L.) im Nord-Westgebiet. *Arch Schiffs-Tropenhyg.*, vol. 38, pp. 477–486.

Tarlo, L B (1957–59) *Palaeontology*, vol. 1. Trustees Brit. Mus. (Nat. Hist).

Tarrant, W P (1883). Fishing and catching ducks. *Ornith. and Ool.*, vol. 8, p. 3.

Thiele, J (1921). Die Cephalopoden der Deutsch Sudpolar-Expedition 1901–1903. *Dt. Sudpol Exped. 16 (Zoology Bd 8)*, pp. 433–465.

Thomas, Richard (1965a). A new gecko from the Virgin Islands. *Quart. J. Florida Acad. Sci.*, vol. 28, No. 1, pp. 117–122.

Thomas, Richard (1965b). The genus *Leptotyphlops* in the West Indies, etc., *Breviora, No. 222, Mus. Comp. Zool. Harv. Univ.*, Cambridge, Mass.

Thompson, M C (1961). The flight speed of a Red-breasted merganser. *Condor*, vol. 63, p. 265.

Thorp, R W & Woodson, W D (1945). *Black Widow – America's most poisonous spider*. Univ. N. Carolina Press.

Tickell, W L N (1968). The biology of the great albatrosses, *Diomedea exulans* and *Diomedea epomophora*, in *Antarctic bird studies* (ed. Oliver Austin), pp. 1–54. Baltimore.

Tomilin, A G (1967). *Mammals of the U.S.S.R. and adjacent countries*. Vol. 9. *Cetacea*. Jerusalem (Israel Program for Scientific Translations).

Troughton, E (1965). *The furred animals of Australia*. Sydney-London.

Tweedie, Michael (1967). The largest fish of all time. *Animals*, vol. 10, No. 6, pp. 258–259.

Tyne, Josselyn van & Berger, A J (1971). *Fundamentals of ornithology*. New York.

Verany, J B (1851). *Mollusques Méditerranéens. 1. Céphalopodes*. Geneva.

Verrill, A E (1879–81). The cephalopods of the north-eastern coast of America. *Trans. Conn. Acad. Arts. Sci.*, vol. 5, pp. 177–446.

Verrill, A E (1897a). A gigantic cephalopod on the Florida coast. *Am. J. Sci.*, Ser. 4, pp. 3, 79, 162–163 and 355–356.

Verrill, A E (1897b). The Florida monster. *Science (N.Y.)*, vol. 5, pp. 393 and 476.

Verrill, A H (1929). *Thirty years in the jungle*. New York.

Vladykov, V D & McKenzie R A (1935). The marine fishes of Nova Scotia. *Proc. Nova Scotian Inst. Sci.*, vol. 19, No. 1, pp. 17–113.

Waldon, C W (1950). The largest whale ever weighed. *Nat. Hist. (New York)*, vol. 59, No. 9, pp. 393–399.

Wardle, R & McLeod J (1952). *The zoology of Tapeworms*. Minneapolis.

Waterman, J A (1938). Some notes on scorpion poisoning in Trinidad. *Trans. R. Soc. trop. Med. Hyg.*, vol. 31, pp. 607–624.

Wheeler, Alwyne (1969). *The Fishes of the British Isles and North-West Europe*. London.

Whitehead, G Kenneth (1972). *Deer of the world*. London.

Willoughby, David P (January 1950). The gorilla – largest living primate. *Sci. Monthly*, pp. 48–57.

Williams, C B (1965). *Insect migration*. London.

Wolff, N O & Githens, T S (1939). Record venom extraction from Water moccasin. *Copeia*, No. 1, p. 52.

Wood, F G & Gennero, J F (March 1971). An octopus trilogy. *Nat. Hist. New York*, pp. 14–23, 24, 84–87.

Wood, Gerald L (1972). *The Guinness Book of Animal Facts and Feats* (1st ed.). Enfield.

Worrell, Eric (1963). *Reptiles of Australia*. Sydney.

Yablokov, A V (1958). O stroenii zubnoi sistemy i tipakh zubov u kitoobraznykh (Dentition and Types of Teeth in Whales). *Byulleten MOIP. Otdel biologicheskii*, No. 2.

Yalden, B W & Morris, P A (1975). *The lives of bats*. London-Vancouver.

Yonge, C M (April 1975). Giant clams. *Sci. Amer.*, pp. 96–103.

Young, J (1944). Giant nerve-fibres. *Endeavour*, vol. 3, pp. 108–113.

Zahl, Paul A (July 1967). The Giant Goliath Frog. *Nat. Geog. Mag.*

Zenkevich, L A et al. (1955). Issledovaniya donnoi fauny Kurilo-Kamchatskoi vpadiny (Investigations on the Bottom Fauna of the Kuril-Kamchatka Trench). *Trudy Instituta Okeanografii Akademii Nauk.*, vol. 12.

Zolotow, M (1974). *John Wayne – shooting star*. London–New York.

INDEX

D

N

O

P

Q

R

S

T

U

V

W

X

Y

Z

Compiled by John Hood-Williams